T0210990

Lecture Notes in Artificial Intelligence   13788

Subseries of Lecture Notes in Computer Science

Series Editors

Randy Goebel
University of Alberta, Edmonton, Canada
Wolfgang Wahlster
DFKI, Berlin, Germany
Zhi-Hua Zhou
Nanjing University, Nanjing, China

Founding Editor

Jörg Siekmann
DFKI and Saarland University, Saarbrücken, Germany

More information about this subseries at https://link.springer.com/bookseries/1244

Ana Cristina Bicharra Garcia ·
Mariza Ferro · Julio Cesar Rodríguez Ribón (Eds.)

Advances in Artificial Intelligence – IBERAMIA 2022

17th Ibero-American Conference on AI
Cartagena de Indias, Colombia, November 23–25, 2022
Proceedings

 Springer

Editors
Ana Cristina Bicharra Garcia ⓘ
Federal University of Rio de Janeiro
Rio de Janeiro, Brazil

Mariza Ferro ⓘ
Fluminense Federal University
Niterói, Brazil

Julio Cesar Rodríguez Ribón ⓘ
University of Cartagena
Cartagena de Indias, Colombia

ISSN 0302-9743   ISSN 1611-3349   (electronic)
Lecture Notes in Artificial Intelligence
ISBN 978-3-031-22418-8   ISBN 978-3-031-22419-5   (eBook)
https://doi.org/10.1007/978-3-031-22419-5

LNCS Sublibrary: SL7 – Artificial Intelligence

This Springer imprint is published by the registered company Springer Nature Switzerland AG
The registered company address is: Gewerbestrasse 11, 6330 Cham, Switzerland

Preface

We welcome you to the proceedings of the 17th Ibero-American Conference on Artificial Intelligence (IBERAMIA 2022). After a four-year break imposed by the COVID-19 pandemic, IBERAMIA returned to its in-person mode. This has been a period of great advances in the field of AI in response to the pandemic challenges. Besides the advances on AI technologies and applications, researchers have looked at the new challenges concerning the social impact. Papers submitted to IBERAMIA 2022 reflected these advances, as can be seen in the proceedings of the conference presented in this volume.

IBERAMIA 2022 was held in Cartagena de Indias (Colombia) during November 23–25, 2022, organized by the Universidad de Cartagena and the Sociedad Colombiana de Computación IA chapter. IBERAMIA is the biennial Ibero-American Conference on Artificial Intelligence. The conference is sponsored by the main Ibero-American Societies of Artificial Intelligence (AI) and gives researchers from Portugal, Spain, and the Latin American countries the opportunity to meet with AI researchers from all over the world.

Since its first edition in Barcelona in 1988, IBERAMIA has continuously expanded its scope to become a well-recognized international conference where the AI community shares the results of their research. Since 1998, the works accepted for the conference have been published in the Springer Lecture Notes in Computer Science series.

The organizational structure of IBERAMIA 2022 followed the standard of the most prestigious international scientific conferences. The scientific program led to fruitful debates among the researchers on the main topics of AI. As usual, the program of the conference was organized in several track areas, each coordinated by area chairs who were in control of the reviewing process. The full list of the area Chairs, Program Committee (PC) members, and additional reviewers can be found in the Organization pages.

IBERAMIA 2022 received 67 papers with widespread contributions from Latin America and from other countries all over the world. From that initial set of 67 submissions, 33 papers were accepted as full papers and four were accepted as short papers. Acceptance decisions involved the collaboration of the three reviewers per paper. When necessary, additional reviews were requested to obtain a clear decision on a particular work.

This IBERAMIA edition had a large number of submissions in machine learning and robotics. The tracks organized for this edition were the following: Applications of AI, Ethics and Smart Cities, Green AI, Machine Learning, Natural Language Processing, Simulation and Forecasting, and Robotics and Computer Vision. The tracks were designed to reflect the areas of the accepted papers.

We would like to express our sincere gratitude to all the people who helped make IBERAMIA 2022 happen. First of all, we want to thank the authors who contributed their high-quality work to the conference and for their cooperation in the preparation of this volume. We also want to give special thanks to the area chairs, the members of the Program Committee and the additional reviewers for the quality of their work which undoubtedly helped with the difficult task of evaluating and selecting the papers for the conference. We also thank Francisco Garijo and Federico Barber (IBERAMIA's Executive Board) for their continuous support in administrative matters, as well as for supporting the website of the conference. We also want to acknowledge Springer's EquinOCS for the facilities provided to support the submission and review of the papers, as well as for the preparation of the proceedings. Finally, it is important to mention that nothing would have been possible without the initiative and dedication of the Organizing Committee from the Universidad de Cartagena. We are very grateful to all the people who helped in the large variety of organizing tasks.

November 2022

Ana Cristina Bicharra Garcia
Mariza Ferro
Julio Cesar Rodríguez Ribón

Organization

Program Committee Chairs

Bicharra Garcia, Ana Cristina	Universidade Federal do Estado do Rio de Janeiro, Brazil
Ferro, Mariza	Universidade Federal Fluminense, Brazil
Rodríguez Ribón, Julio Cesar	University of Cartagena, Colombia

Area Chairs

Correia, Luís	Universidade de Lisboa, Portugal
Costaguta, Rosanna	Universidad Nacional de Santiago del Estero, Argentina
Hernandez-Orallo, Jose	Universitat Politecnica de Valencia, Spain
Marcacini, Ricardo	Universidade de São Paulo, Brazil
Martinez Carranza, Jose	Instituto Nacional de Astrofisica Optica y Electro, Mexico
Martí, Luis	Inria, Chile
Molina, Jose M.	Universidad Carlos III de Madrid, Spain
Novais, Paulo	Universidade do Minho, Portugal
Pavón, Juan D.	Universidad Complutense de Madrid, Spain
Reyes Ballesteros, Alberto	Instituto Nacional de Electricidad y Energías Limpias, Cuernavaca, Mexico
Rezende, Solange	Universidade de Sao Paulo, Brazil
Sanchez-Pi, Nayat	Inria, Chile
Schiaffino, Silvia	UNCPBA and CONICET, Argentina
Viterbo, José	Universidade Federal Fluminense, Brazil

Program Committee

Barber, Federico	Universitat Politécnica de València, Spain
Becerra Duran, Israel	CIMAT, Mexico
Bernardini, Flavia	Universidade Federal Fluminense, Brazil
Braud, Agnès	University of Strasbourg, France
Cardoso Durier da Silva, Fernando	IBM, Brazil
Correia, Luís	Universidade de Lisboa, Portugal
Cozman, Fabio	Universidade de São Paulo, Brazil
D'Antonio Maceiras, Sergio	Universidad Politécnica de Madrid, Spain
Daudt, Fabio	Universidade Federal do Estado do Rio de Janeiro, Brazil

Furtado Silva, Diego	Universidade de São Paulo, Brazil
Garijo, Francisco	Sociedad Iberoamericana de Inteligencia Artificial, Spain
González, Enrique	Pontificia Universidad Javeriana, Colombia
Gravano, Agustin	Universidad Torcuato Di Tella, Argentina
Gutiérrez-Giles, Alejandro	Universidad Nacional Autónoma de México, Mexico
Hernandez-Orallo, Jose	Universitat Politecnica de Valencia, Spain
Jorge, Alípio	University of Porto, Portugal
Lescano, Germán	Universidad Nacional de Santiago del Estero, Argentina
Lopez-Luna, Aaron	Instituto Nacional de Astrofísica, Óptica y Electrónica, Mexico
Machado, José	Universidade do Minho, Portugal
Marcacini, Ricardo	Universidade de São Paulo, Brazil
Martinez Carranza, Jose	Instituto Nacional de Astrofísica, Óptica y Electrónica, Mexico
Martinez-Plumed, Fernando	Technical University of Valencia, Spain
Montes, Manuel	National Institute of Astrophysics, Optics and Electronics, Mexico
Monteserin, Ariel	Universidad Nacional del Centro de la Provincia de Buenos Aires, Brazil
Moreno, Mailyn	Universidad Tecnológica de La Habana "José Antonio Echeverría", Cuba
Moura, Raimundo	Universidade Federal do Piauí, Brazil
Negrete Villanueva, Marco Antonio	Universidad Nacional Autónoma de México, Mexico
Oliveira Junior, Carlos Roberto	Instituto Federal de Educação, Ciência e Tecnologia do Rio de Janeiro, Brazil
Omicini, Andrea	University of Bologna, Italy
Paes, Aline	Universidade Federal Fluminense, Brazil
Pardo, Thiago	Universidade de São Paulo, Brazil
Pavón, Juan	Complutense University of Madrid, Spain
Pintas, Julliano	Petrobras, Brazil
Pinto, Fernando	Universidade Federal do Rio de Janeiro, Brazil
Pinto, Gabriel	Universidade Federal do Estado do Rio de Janeiro, Brazil
Reali Costa, Anna Helena	Universidade de São Paulo, Brazil
Recio-Garcia, Juan A.	Complutense University of Madrid, Spain
Reyes Ballesteros, Alberto	Instituto Nacional de Electricidad y Energías, Mexico
Rezende, Solange	Universidade de Sao Paulo, Brazil
Ribeiro, Luiz	Universidade Federal do Estado do Rio de Janeiro, Brazil
Rodríguez Ribón, Julio Cesar	Universidad de Cartagena, Colombia
Santos, Paulo	Universidade Federal do Estado do Rio de Janeiro, Brazil

Sinoara, Roberta	Instituto Federal de São Paulo, Brazil
Siqueira, Sean	Universidade Federal do Estado do Rio de Janeiro, Brazil
Souza, Rodrigo Clemente	Universidade Federal do Paraná, Brazil
Tommasel, Antonela	Universidad Nacional del Centro de la Provincia de Buenos Aires, Argentina
Traver, Javier	Universitat Jaume I, Spain
Zorrilla, Rocio	Laboratório Nacional de Computação Científica, Brazil
Zubiaga, Arkaitz	Queen Mary University of London, UK

Reviewers

Calçada, Dario	Universidade Estadual do Piaui, Brazil
Fdez-Riverola, Florentino	Universidade de Vigo, Spain
Garcia, Nuno	Universidade de Lisboa, Portugal
Lobato, Fábio	Universidade Federal do Oeste do Pará, Brazil
Pesquita, Cátia	Universidade de Lisboa, Portugal
Rossi, Rafael	Universidade Federal do Mato Grosso do Sul, Brazil
Yokoyama, André M.	National Laboratory for Scientific Computing, Brazil

Contents

Ethics and Smart City

Green and Sustainable AI

Machine Learning

Natural Language Processing

Robotics and Computer Vision

Simulation and Forecasting

Short Papers

Applications of AI

Gait Patterns Coded as Riemannian Mean Covariances to Support Parkinson's Disease Diagnosis

Juan Olmos[1], Juan Galvis[2], and Fabio Martínez[1]([✉])

[1] Biomedical Imaging, Vision and Learning Laboratory (BIVL2ab),
Universidad Industrial de Santander (UIS), Cra 27 Calle 9 Ciudad Universitaria,
Bucaramanga, Colombia
jaolmosr@correo.uis.edu.co, famarcar@saber.uis.edu.co
[2] Departamento de Matemáticas, Universidad Nacional de Colombia,
Carrera 45 No. 26–85, Edificio Uriel Gutiérrez, Bogota D.C., Colombia
jcgalvisa@unal.edu.co

Abstract. Gait is one of the main Parkinson disease (PD) biomarkers that support diagnosis and allows to measure neuromotor progression. The gait analysis is nonetheless limited to scarse observational scales or quantified from kinematic features, obtained from marker-based setups. These classical methodologies alter natural locomotion gesture and may limit the quantification of important PD signs. This work explores a new markerless gait representation that code spatio-temporal video patterns in mean covariances. To start, a per-frame covariance matrix is computed from a set of deep convolutional features to describe each recorded locomotion. These matrices are in the Riemannian manifold, from which, a geometric mean is computed to compactly describe gait and postural patterns. Afterwards, a projection into a log Euclidean space allows to train a supervised learning algorithm to automatically discriminate between Parkinson and control population. An study of 22 participants (11 parkinson patients and 11 control adults) was herein used to evaluate the classification achieving remarkable results. Interestingly enough, a geometric low dimensional projection enhance the discrimination performance and allow a potential use to support PD diagnosis.

Keywords: Covariance mean · Deep features · Riemannian manifold · Parkinson disease · Diagnosis support

1 Introduction

Parkinson's disease (PD) is the second most common neurodegenerative disorder, affecting between 2–3 % of the population, over 65 years of age worldwide [18]. The pathological hallmark of the disease from early stages is characterized by neuronal loss in the substantia nigra, which causes dopamine deficiency [18]. This deficiency explains the major symptoms of Parkinson's disease, including motor

A. C. Bicharra Garcia et al. (Eds.): IBERAMIA 2022, LNAI 13788, pp. 3–14, 2022.
https://doi.org/10.1007/978-3-031-22419-5_1

disabilities such as tremor, rigidity, postural instability, and slowness of movement (bradykinesia) [10]. Besides, the movement disorders of the upper and lower limbs are also strongly related to the disease, affecting locomotor gait patterns [8]. During gait, it is possible to observe alterations such as arm swing, reduced footprint, decreased ground clearance, slowness, and stiffness in displacement [5]. Additionally, other related neuromotor alterations can be amplified such as the head stability and the stiffness of body segments. Hence, the analysis and measure of such patterns is fundamental to quantify disease progression, to properly characterize the disease and even to approach the pattern exaggeration to early approximate the PD diagnosis. Nonetheless, the current diagnosis is commonly subject to observational analysis, reporting errors up to 24% [18]. In more sophisticated scenarios, the gait analysis is supported by marker based strategies that associate joint dynamics to a set of devices attached to the body [3]. These methods however highly depend on the correct placement of the markers, may be invasive and affects the naturalness of the patient's movements. Likewise, these approaches rarely take into account some tremor and the stiffness of body segment patterns [6,22].

From 2D video recordings, markerless strategies have become a powerful alternative to quantify kinematic descriptors and to find new hidden relationships correlated with locomotor diseases [2,6]. Specifically to characterize PD, some strategies have formulated video-based descriptors to quantify local velocity field patterns [21] or using long motion trajectories to approximate gait kinematics [7]. Besides, the use of video recordings have allowed the analysis of different biomechanical features of gait, such as stride length, gait speed and cadence [24]. In the same way, the deep learning methods have been introduced to extract features and support the parkinsonian gait analysis and diagnosis. For example, convolutional neural networks (CNN) have been proposed to classify gait video sequences by learning spatio-temporal regions that better describe locomotor abnormalities [6]. In addition, deep learning-based pose estimator have been proposed to quantify the cadence of gait steps through sequential gait features [19]. In general, these markerless strategies were designed to analyze and quantify locomotor alterations in specific body segments. Additionally, the use of kinematic feature descriptors can be redundant and computationally costly to learn and to run in real scenarios of daily practice. From this, the use of compact matrices for the feature description has been proposed as a statistical method of excellent reliability to select representative spatio-temporal features for the analysis of parkinsonian gait [1]. These representations are however dedicated only to carried out feature analysis and as filter methodologies of recorded data, loosing their ability to compactly describe significant features that quantify disease-related locomotor abnormalities.

This work introduces a compact video descriptor to support the diagnosis of Parkinson's disease based on a geometrical mean covariance description. The proposed approach take advantage of deep features primitives to robustly represent postural gait patterns at each frame of recorded video sequence. These features were captured from the first layer of a pre-trained convolutional neural network (CNN). At each time, these deep features are summarized as frame-covariance matrices. Then, a video results in a sequence of covariance matrices that together form a sequence over a Riemannian manifold that code postural dynamic changes during locomotion. This way, we introduce a compact video descriptor by computing the geometric mean, that is, a matrix with minimal distance w.r.t the video on the manifold. Once the mean covariance descriptor is obtained, a machine learning algorithms were trained under a supervised assumption to classify between Parkinson and control population. Additionally, we project whole resultant descriptors into a low dimensional space to analyze the distribution geometry of evaluated population. This projection result interesting as observational alternative to support diagnosis regarding closest points in the space.

2   Proposed Approach: The Covariance Mean as Parkinsonian Descriptor

2.1   Temporal Gait Representation and Deep Features

We can consider a video D as a sequence of k frames $\{I_t\}_{t=1}^{k} \subset \mathbb{R}^{W \times H}$, where each frame can be described with a set of features $F_t = \{F_t^{(1)} \ldots F_t^{(N)}\}$. For a compact description, each feature can be represented as column vectors to calculate the covariance matrix of these features [16]. The covariance between two features $F^{(i)}$ and $F^{(j)}$ is $C_{ij} = \frac{1}{M-1} \sum_{\ell=1}^{M} \left(F_\ell^{(i)} - m_i\right) \left(F_\ell^{(j)} - m_j\right)^T$, where $m_i = \frac{1}{M} \sum_{\ell=1}^{M} F_\ell^{(i)}$. Therefore, for each frame I_t we can calculate an associated covariance matrix $C_{I_t} = [C_j]_{i=1,j=1}^{N,N}$. This matrix is symmetric semi-positive definite and belongs to the space of symmetric positive definite (SPD) matrices (denote S_{++}^n), which is a Riemannian manifold [13,15].

Particularly, in this work, the N features F_t corresponds to deep activations (deep features), computed from pre-trained networks. Specifically, each frame of a video is mapped onto the first layers of known and pre-trained deep convolutional neural networks (CNNs), which have been implemented for a general natural image classification problem. In Fig. 1 are illustrated sample deep features from a DenseNet CNN [9]. The representations based on partial and general learned convolutional schemes allow to capture variations underlying the data samples [4,11,20]. In the Parkinson context that can be relevant to capture postural representation related to the gait alterations. Particularly, in this work we compute deep features from the first ReLu layer block for the MobileNet (32 features of size $W = 112 \times H = 112$) and the Densenet (64 features of size $W = 112 \times H = 112$).

2.2   Video Sequences and Riemannian Manifold

Hence, any video is expressed as a sequence $\{C_{I_1}, \cdots, C_{I_k}\}$ of SPD covariance matrices, that can be seen as a sequence of points on the Riemannian manifold S_{++}^n (see Fig. 2). With respect to the geometry of the space, it is known that

Fig. 1. Sample deep features extracted from the DenseNet [9].

any two matrices in S_{++}^n can be joined with a unique geodesic, the shortest path on this space [13]. With these geodesics, given any $P \in S_{++}^n$ there are two important Riemannian maps to measure distances on this space. The exponential $\exp_P$ and the logarithm map $\log_P$, which projects vectors from the tangent space $T_P S_{++}^n = S^n$ (Symmetric matrices) to S_{++}^n and elements from S_{++}^n to $T_P S_{++}^n$ respectively (see Fig. 2). Pennec in [17], Moakher in [14] and Fletcher in [23] introduce different ways to obtain the geodesics in S_{++}^n from a *affine-invariant* Riemannian metric. Based on this metric, exponential and logarithm map can be defined. Given $P \in S_{++}^n$ and a tangent vector $V \in T_P S_{++}^n$, the exponential map of V on P is given by:

$$\exp_P(V) = P^{\frac{1}{2}} \exp\left(P^{-\frac{1}{2}} V P^{-\frac{1}{2}}\right) P^{\frac{1}{2}}, \tag{1}$$

and the logarithm map of $Q \in S_{++}^n$ on P is given by:

$$\log_P(Q) = P^{\frac{1}{2}} \log\left(P^{-\frac{1}{2}} Q P^{-\frac{1}{2}}\right) P^{\frac{1}{2}}. \tag{2}$$

See Fig. 2 for a geometric illustration of these maps.

2.3 Geometric Mean Descriptor on the Riemannian Manifold

To calculate a covariance mean we can use geometric and statistical behaviours to describe locomotion patterns in a simple descriptor. For instance, a variational formulation of the geometric mean $\arg\min_{X \in S_{++}^n} \frac{1}{2k} \sum_{i=1}^{k} d_R(X, C_i)^2$, where $d_R(X, C_i)$ is a Riemannian distance. For this problem, Pennec in [17] introduces the Fréchet mean as the unique global minimum in S_{++}^n using the Riemannian distance $d_R(P, Q) = \|\log_P(Q)\|_P$ that represents the length of the geodesic that joins P and Q. Therefore, taking $\{C_i\}_{i=1}^{k} \subset S_{++}^n$, Pennec propose a Newton gradient descent algorithm to solve the minimization problem above, defined as

$$\mu_{t+1} = \exp_{\mu_t}\left(\frac{1}{k} \sum_{i=1}^{k} \log_{\mu_t}(C_i)\right), \tag{3}$$

where μ_t is the t-th approximation of the geometric mean. In this regard, using (3) a video D can be compactly described with a covariance mean matrix

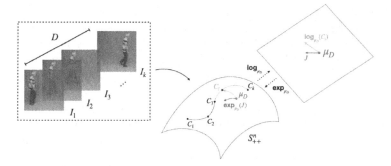

Fig. 2. Covariance mean video descriptor. A video can be described as a sequence of covariance matrices. The covariance mean μ_D describes all D. The exponential $\exp_P : T_P S^n_{++} \rightarrow S^n_{++}$ maps vectors from the tangent space $T_P S^n_{++} = S^n$ (Symmetric matrices) to S^n_{++} and the logarithm $\log_P : S^n_{++} \rightarrow T_P S^n_{++}$ maps matrices from the manifold S^n_{++} to the tangent space S^n.

$\mu_D \in S^n_{++} \subset \mathbb{R}^{N \times N}$ (see Fig. 2). For instance, a video with K frames and F features of dimension $N \times N$ per frame results in a descriptor of dimension $K \times F \times N \times N$, using the covariance mean instead, we have a compact descriptor of dimension $F \times F$. From this, taking into account the geometry, a video was reduced to a single point on the manifold. Therefore, a set of videos represented in this way on the manifold can allows a better discrimination performance among Parkinson and control populations. Once the geometric mean is computed, because it is a symmetric matrix, we take the upper triangular part and resize it into a vector. To operate with common machine learning classifiers, it is necessary that the descriptors lie in a Euclidean space. Therefore, before resizing, other most common approaches map the covariance mean matrix to the tangent space $T_{id} S^n_{++}$ with the logarithm map.

2.4 Classification and Low-Dimensional Visualization

In order to discriminate between Parkinson and control populations we design a binary classification task. To this end, gait videos $V_1, V_2, \ldots, V_m$ were described with m covariance mean matrices, which in turn are resized in m vectors $\{X_1, X_2, \ldots, X_m\}$ respectively. Finally, different algorithms can be proposed for learning to predict the correspondence of a given vector X_i to the Parkinson or Control class.

The proposed approach has also the capability to explain disease associations by projecting descriptors in a low dimensional space. This representation is a potential diagnosis tool that may be used to associated motion patterns that are close in the space. For a low-dimensional visualization, we implement a uniform manifold approximation and projection (UMAP) [12], that take advantage of a local manifold data approximations to associate a low-dimensional representation. The UMAP could be able to get down important geometric differences between the location of the two classes of the Parkinson Population.

Additionally, a distance based metric is proposed to measure the intra-class sparsity and the inter-class separation. With the class-labeled data, we consider two disjoint sets, the Parkinson descriptors P and control subjects descriptors C. Therefore we define the inter-class metric as the Riemannian distance between P and C, as $d_R(P,C) = \frac{1}{|P||C|} \sum_{x \in P} \sum_{y \in C} d_R(x,y)$. Similarly, the intra-class metric $d_R(C,C)$ or $d_R(P,P)$ measure the distance between points on the same set. Both mentioned metrics, with a well low-dimensional representation allows understanding where is located certain patient with respect to the known Parkinson population.

2.5   Experimental Setup

Data. In this study was recorded a total population of 11 Parkinson patients (average age 72.3 ± 7.4) and 11 control patients (average age 72.2 ± 6.1), walking in a sagital perspective. Each patient was recorded eight times, 4 times to the right and 4 times to the left, for a total of 176 sequences with an average duration of $4\,\mathrm{s}$ (~ 100 frames, approximately), an spatial resolution of 520×520 pixels and a temporal resolution of 25 frames per second. This dataset is age matched and subjects with different disease degree progression were selected to include inter-subject variability. With the help of a physical therapist, PD patients were categorized into the Hoehn-Yahr rating scale. A total of five patients were categorized in stage two, six patients in stage three, and two patients in stage four. This study was approved by Ethic committee. Participants were recruited from a local foundation and a local elderly institution. A written informed consent was obtained for every participant. Data available once the paper has been accepted.

Deep features and covariance mean configuration. The sixteen most representative deep features from MobileNetV2 and twenty from DenseNet architecture were selected from first layer of each net respectively. In such case, the representative deep features correspond to non zero feature maps. A set of sample deep features are showed in Fig. 1. These features are able to decompose relevant locomotion information, such as postural configurations, gradient of silhouettes, and other feature maps focusing specific body parts like the low body, the trunk and the upper body silhouette of the patients. From these deep features we construct four experiments:

- A covariance mean computation in the manifold S_{++}^n, using MobileNetV2 features, with a resultant spatial dimension for covariances of 16×16, obtaining a covariance mean video descriptor with size 120.
- A covariance mean computation in the manifold S_{++}^n, using DenseNet features, with a resultant spatial dimension for covariances of 20×20, obtaining a covariance mean video descriptor with size 190.
- A covariance mean computation in the tangent space $T_{Id}S_{++}^n$, using MobileNetV2 features, with a resultant spatial dimension for covariances of 16×16, obtaining a covariance mean video descriptor with size 120.

- A covariance mean computation in the tangent space $T_{Id}S_{++}^n$, using DenseNet features, with a resultant spatial dimension for covariances of 20×20, obtaining a covariance mean video descriptor with size 190.

Classification. The resultant mean covariances, computed on the manifold and the tangent space, were used as videos descriptors to discriminate the Parkinson gait from a control population. In such case, we implement four machine learning classifiers: the Gaussian Naive Bayes (GaussianNB), the Logistic Regression (Logistic), the RandomForest (RF) and the Support Vector Machine (SVM). A cross-validation *leave one out patient out* scheme was implemented to validate the proposed approach. In each iteration, the 8 video descriptors of each patient were taken for testing and the rest ($168 = 21 \times 8$) of the descriptors for training. The predictions performance was evaluated with the accuracy, precision, and recall metrics. This validations process was run 100 times to avoid random bias taking the average and standard deviation of the classification metrics.

3   Evaluation and Results

The proposed approach was firstly validated with respect to the capability to support Parkinson classification, from observed gait recordings. Table 1 summarizes the achieved classification metrics scores using the video descriptors on the manifold. In general, all classifiers have a remarkable performance on discrimination task, which suggest a proper clustering of both classes, where a simple linear separation achieve accuracy scores up to 96%. The Random Forest classifier achieves the best classification metrics using the MobilNet deep features, with an average accuracy of 99.9%, a precision of 99.61% and a recall of 99.56%. In such case, the approach only fail in the classification of only one video video sequence of a control patient. Particularly, this subject during recording video has old-age advanced motor symptoms, which may be associated to the disease, such as bradykinesia and trunk stiffness, and therefore the classifier associated such patterns to the Parkinson subject.

Table 1. Average accuracy, precision and recall using different methods to classify the covariance mean video descriptors on the Riemannian manifold.

Method	Deep features	Accuracy	Precision	Recall
GaussianNB	DenseNet	96.59 ± 13.15	95.45 ± 14.37	93.75 ± 20.12
	Mobilnet	94.32 ± 15.41	93.18 ± 17.16	90.34 ± 24.48
Logistic	DenseNet	93.18 ± 22.84	93.18 ± 22.84	92.05 ± 25.435
	Mobilnet	96.02 ± 18.23	97.73 ± 10.41	95.74 ± 19.53
RF	DenseNet	99.10 ± 4.99	98.14 ± 9.47	97.69 ± 11.80
	Mobilnet	$\mathbf{99.90 \pm 1.19}$	99.61 ± 4.38	99.56 ± 4.96
SVM	DenseNet	82.39 ± 31.00	79.55 ± 32.54	75.28 ± 37.19
	Mobilnet	92.61 ± 21.54	88.64 ± 25.81	87.22 ± 28.11

In a second validation, the video-descriptors were firstly mapped to the tangent space and then projected to the machine learning algorithms. This mapping may lost geometrical manifold properties but can be a more natural representation for optimization methods in classification tasks. Table 2 summarizes the achieved classification metrics scores, where Gaussian Naive Bayes achieve a perfect score classification, suggesting a linear partition of data class clusters. The SVM and the logistics regression report slightly lower scores than in the manifold geometry, but with an overall remarkable performance. Contrary, the Random Forest classifier is robust to space representation, achieving equivalent results in the tangent space, using the configurations with both deep features. In this configuration, a total of five video sequences were missclassified as a Parkinson. As in the previous experiment, the control subject exhibit motor patterns that may be associated to the disease.

Table 2. Average accuracy, precision and recall using different methods to classify the covariance mean descriptors on the tangent space.

Method	Deep features	Accuracy	Precision	Recall
GaussianNB	DenseNet	**100 ± 0**	100 ± 0	100 ± 0
	Mobilnet	84.09 ± 30.89	81.82 ± 32.14	78.409 ± 36.23
Logistic	DenseNet	97.73 ± 10.41	97.72 ± 10.41	96.59 ± 15.62
	Mobilnet	97.16 ± 13.02	97.73 ± 10.41	96.31 ± 16.94
RF	DenseNet	98.41 ± 8.43	98.11 ± 9.53	97.32 ± 13.60
	Mobilnet	97.16 ± 13.02	97.73 ± 10.41	96.31 ± 16.92
SVM	DenseNet	94.89 ± 18.33	93.18 ± 17.16	90.62 ± 24.48
	Mobilnet	75.57 ± 38.52	72.73 ± 39.10	69.60 ± 41.56

Under same experimental setup, the proposed approach (in manifold and in tangent plane) outperform the baseline approach that follow a Spatio-temporal 3D ConvNet [6], to classify Parkinson patients. The baseline uses a end-to-end video learning to discriminate such patients, achieving an average accuracy of 90%, while the proposed approach achieve scores up to 99%. For this remarkable performance, crude and adjusted odds ratios were computed to evaluate possible confounding variables. Specifically, Gender and age were considered. Before consider these variables, the crude ratio was 9.35 and taking the variables into account the adjusted ratio was 8.13. Indicating that age and gender variables do not influence the results of the evaluation and therefore is not confounded association.

A constant feature in clinical routine is the limited availability of patient data to train computational models and to update technological architectures. Hence, the proposed computational alternatives in this scenarios should be sufficiently robust to tackle such limited setup conditions. In consequence, a second evaluation of the proposed approach consist on train the machine learning approach but with limited samples, varying from 21 to 7 patients in training. To implement this experiment, the best configuration in each descriptor representation was implemented. Hence, the Mobilnet deep features and RF classifier was implemented for the manifold space, while the DenseNet features and a Gaussian classifier was used as the configuration for tangent space. Figure 3 illustrates the achieved performance for both configurations (manifold and tangent space), showing a notable performance, even for very strong train configurations with only seven patient for train. In extreme conditions (seven patients to train), the best score is achieved for descriptors taken from manifold geometry (around 95% in recall), while in the tangent plane is achieved only (70% in recall), which exhibit the richness of the manifold to capture motor geometry, that achieve a better representation of the disease. Contrary, in tangent plane, the logarithm map could be affecting the dispersion of descriptors, making the classification task more difficult when there is a lack of data. The proposed approach, according to the experiment, shows a robust capability to represent Parkinson disease even in scarce scenarios, with a relative independence to the classifiers and the boundary hypothesis of each of these strategies.

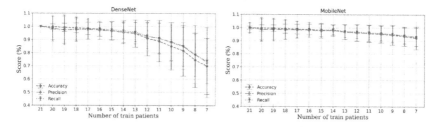

Fig. 3. Classification scores varying the number of training patients. Left, *tangent space* DenseNet deep features and Gaussian classifier. Right, *manifold* Mobilnet deep features and Random Forest classifier.

Table 3. Intra/inter class metric of the descriptors using a euclidean distance in the manifold (d_E in S_{++}^n), in the tangent space (d_E in $T_{Id}S_{++}^n$) and using the Riemannian distance in the manifold (d_R in S_{++}^n).

	d_E in S_{++}^n	d_E in $T_{Id}S_{++}^n$	d_R in S_{++}^n
Intra class Parkinson	$23,07$	$44,647$	$73,421$
Intra class No Parkinson	$23,98$	$6,22$	$126,9$
Inter class	$31,48$	$40,63$	$209,342$

Finally, a low dimensional space was built from resultant covariance mean compact descriptors $U = \{\mu_1, \ldots, \mu_{176}\}$, for respective video sequences herein considered. Using the manifold UMAP projection (see in Fig. 4) we can visualize the projection constructed from the original covariances and using the respective log map $\log(U)$ in the tangent space $T_{Id}S_{++}^2$. As expected, the resultant

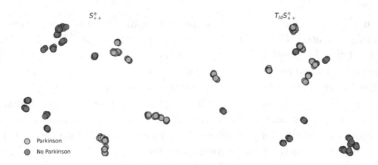

Fig. 4. UMAP embedding of the covariance mean descriptors in the manifold (left) and in the tangent space (right). Red dots represent the descriptors of parkinson patients and blue the no parkinson. (Color figure online)

low dimensional projections reveals a sparse and discriminative data separation, according to Parkinson and control labels. It should be remarked the clear separation, specially using the original descriptors, i.e., on the Riemmanian manifold. This properly separation can be explained by the intrinsic geometry that result from minimization among samples to find the correspondence in low-dimensional space. To quantify the separation among classes in projected spaces an euclidean distance (d_E) was measured among samples of the same class (intra-class) and regarding the other class (inter-class). The Table 3 evidences the proper separation among samples of different classes for whole built spaces. Interestingly enough, computing the Riemannian distance (d_R) mentioned in Sect. 2.3, the intra/inter class metrics is more reliable results with better contrast between samples of same class with respect to the other class.

4 Discussion and Conclusions

In this work was designed and proposed a markerless method to model kinematic motion alterations involved in the Parkinson gait, as an alternative to support diagnosis and characterization of the disease. The proposed approach uses compact covariance means to represent each gait video sequence, capturing principal pose geometry during time, and allowing to project such information in a Riemmanian manifold space. The sparsity of the studied population on this spaces allowed low dimensional projection and more accurate measures of these kind of descriptors to finally support the diagnosis of Parkinson's disease. The achieved results evidence that codified descriptor captures the key kinematic concepts of the Parkinsonian gait, allowing a robust representation, where performance in classification task has classification scores up to 90%. Interestingly enough, the proposed approach show a remarkable representation of the disease, being sufficient few patient samples to train a classifier. This fact is very important to transfer into a clinical practice, where the trade-off between approach sensibility and resources for training and updating, are essential to support routine.

Also, the low-dimensional projection of the descriptors allows visualizing the distribution of the population, being an observational alternative to analyze disease patterns. These projections may support diagnosis and following by comparing the patient points with other patients in the manifold space. The projection from Riemannian manifold exhibit a better separation and discrimination of population points. To quantify the distribution of the data, the intra/inter class Riemannian distance, measured the separation between the set of Parkinson's patients and healthy patients, showing a remarkable discrimination. Together, these results represents a potential alternative to project new patients and find closes samples that eventually can share diagnosis and motor symptoms. Moreover could help to a treatment planning.

The covariance representation show remarkable properties regarding the accuracy, the simplicity, the low-dimensional nature, that result ideal to transfer in real scenarios. Future works include the study of this motion descriptors over larger dataset with the information about stage of the disease. Also, from the achieved descriptors it will be demanding to back-propagate information to explain the kinematic and postural features that have major association with the disease, to better explain the capabilities of the proposed approach.

Acknowledgments. The recorded dataset was possible thanks to the support of a Parkinson foundation, an Adults Institution, and an University for supporting the research project. Full names of the institutions will be filled in the case of acceptance.

References

1. Arcolin, I., Corna, S., Giardini, M., et al.: Proposal of a new conceptual gait model for patients with Parkinson's disease based on factor analysis. BioMed. Eng. OnLine **18**(1), 70 (2019)
2. Castelli, A., Paolini, G., Cereatti, A., et al.: A 2D Markerless Gait Analysis Methodology: Validation on Healthy Subjects. Comput. Math. Methods Med. **1** (2015)
3. Davis, R.B., Õunpuu, S., Tyburski, D., et al.: A gait analysis data collection and reduction technique. Human Movement Sci. **10**, 575–587 (1991)
4. Donahue, J., Jia, Y., Vinyals, O., et al.: Decaf: A deep convolutional activation feature for generic visual recognition. In: ICML. pp. 647–655 (2014)
5. Duncan, R.P., Leddy, A.L., Cavanaugh, J.T., et al.: Balance differences in people with parkinson disease with and without freezing of gait. Gait & posture **42**(3), 306–309 (2015)
6. Guayacán, L.C., Rangel, E., Martínez, F.: Towards understanding spatio-temporal parkinsonian patterns from salient regions of a 3D convolutional network. In: EMBC (2020)
7. Guayacán, L.C., Valenzuela, B., Martinez, F.: Parkinsonian gait characterization from regional kinematic trajectories. In: 14th International Symposium on Medical Information Processing and Analysis, vol. 10975 (2018)
8. Halliday, S.E., Winter, D.A., Frank, J.S., et al.: The initiation of gait in young, elderly, and Parkinson's disease subjects. Gait & Posture **8**(1), 8–14 (1998)
9. Huang, G., Liu, Z., Van Der Maaten, L., et al.: Densely connected convolutional networks. In: CVPR (2017)

10. Jw, B.: Motor deficiency in Parkinson's disease. Acta Neurobiologiae Experimentalis **58**(1), 79–93 (1998)
11. Long, M., Cao, Y., Wang, J., et al.: Learning transferable features with deep adaptation networks. In: ICML (2015)
12. McInnes, L., Healy, J., Melville, J.: Umap: Uniform manifold approximation and projection for dimension reduction. arXiv preprint arXiv:1802.03426 (2018)
13. Minh, H.Q., Murino, V.: Covariances in Computer Vision and Machine Learning. Morgan & Claypool Publishers (2017)
14. Moakher, M., Zerai, M.: The riemannian geometry of the space of positive-definite matrices and its application to the regularization of positive-definite matrix-valued data. J. Math. Imag. Vision **40**, 171–187 (2011)
15. Pennec, X.: Probabilities and Statistics on Riemannian Manifolds?: A Geometric approach. Tech. rep, INRIA (2004)
16. Pennec, X.: Intrinsic Statistics on Riemannian Manifolds: Basic Tools for Geometric Measurements. J. Math. Imag. Vision **25**, 127–154 (2006). https://doi.org/10.1007/s10851-006-6228-4
17. Pennec, X., Fillard, P., Ayache, N.: A riemannian framework for tensor computing. International J. Comput. Vision **66**, 41–66 (2006). https://doi.org/10.1007/s11263-005-3222-z
18. Poewe, W., Seppi, K., Tanner, C.M., et al.: Parkinson disease. Nature Rev. Disease Primers **3** 17013 (2017)
19. Sato, K., Nagashima, Y., Mano, T., et al.: Quantifying normal and parkinsonian gait features from home movies: Practical application of a deep learning-based 2d pose estimator. PloS one vol.14 (2019)
20. Sharif Razavian, A., Azizpour, H., Sullivan, J., et al.: Cnn features off-the-shelf: an astounding baseline for recognition. In: CVPR. pp. 806–813 (2014)
21. Soran, B., Hwang, J., Lee, S., Shapiro, L.: Tremor detection using motion filtering and svm. In: ICPR. pp. 178–181 (2012)
22. Taha Khan, P.G., Nyholm, D.: Computer vision methods for parkinsonian gait analysis: A review on patents. Recent Patents on Biomedical Engineering (Discontinued) **6**(2), 97–108 (2013)
23. Thomas Fletcher, P., Joshi, S.: Riemannian geometry for the statistical analysis of diffusion tensor data. Signal Process. **87**, 250–262 (2007)
24. Verlekar, T.T., Soares, L.D., Correia, P.L.: Automatic classification of gait impairments using a markerless 2d video-based system. Sensors **18**(9), 2743 (2018)

Forroset: A Multipurpose Dataset of Brazilian Forró Music

Lucas Ferreira-Paiva[1](✉)(iD), Elizabeth Regina Alfaro-Espinoza[2](iD),
Pablo de Souza Vieira Santana[3](iD), Vinicius Martins Almeida[4](iD),
Amanda Bomfim Moitinho[3](iD), Leonardo Bonato Felix[3](iD),
and Rodolpho Vilela Alves Neves[3](iD)

[1] Programa de Pós-Graduação em Ciência da Computação,
Departamento de Informática, Universidade Federal de Viçosa, Viçosa, Brazil
lucas.paiva@ufv.br
[2] Programa de Pós-Graduação em Bioinformática, Instituto de Ciências Biológicas,
Universidade Federal de Minas Gerais, Belo Horizonte, Brazil
elizaespinoza@ufmg.br
[3] Departamento de Engenharia Elétrica, Universidade Federal de Viçosa, Viçosa,
Minas Gerais, Brazil
{pablo.santana,amanda.moitinho,leobonato,rodolpho.neves}@ufv.br
[4] Centro de Ciências Exatas e Tecnológicas - Engenharia de Computação,
Centro Universitário de Viçosa, Viçosa, Brazil
viniciusmartins@unicosa.com.br

Abstract. Forró is an important genre that has been developing the cultural identity of Brazil and it is one of the most consumed by Brazilians on Spotify. However, the lack of datasets and their specificity leads to less research about this genre. In order to overcome this issue, it is presented a set of data roughly compounded by 3000 songs named Forroset, which provides editorial information, audio features, information of rhythm, and audio files from Spotify. Furthermore, over 1400 lyrics of songs were obtained by the Vagalume platform. When Forroset is compared to other sets of data, it was seen that our dataset is more powerful regarding the diversity of information heading to comprehensive problems resolution.

Keywords: Music information retrieval · Spotify data · Database · Dance teaching · Machine learning · Music industry

1 Introduction

Forró is an important Brazilian musical genre that is popular throughout all socioeconomic layers and has its traditional matrix recognized as the Intangible Cultural Heritage of Brazil [11]. With its early origins in a party, it has contributed to the construction of northeastern and national identity for more

---

Supported by CAPES. Financing code 001 and Fellowship 88887.517813/2020-00.

A. C. Bicharra Garcia et al. (Eds.): IBERAMIA 2022, LNAI 13788, pp. 15–26, 2022.
https://doi.org/10.1007/978-3-031-22419-5_2

than a century. Furthermore, forró is becoming one of the most played genres on Spotify in Brazil [16], the world's largest streaming music platform.

Forró could be classified in three musical genres: "Pé-de-Serra", "Universitário", and "Eletrônico" [18]. Except "Forró Eletrônico", the core instrumental structures are the same: the zabumba, triangle, and accordion compose this basic structure, while the singer's voice completes the musicality and rhythm.

Forró has been explored in several fields, including applications to aid individuals in dancing [6,17,19–21], genre recognition [3,4], and the evaluation of paradigm-breaking in a musical context [2]. The following gaps detected in the forró datasets identified that limit research findings in this genre are: *(i)* An unavailability or discontinuity of data access; *(ii)* failure to follow FAIR Data principles (Findable, Accessible, Interoperable, and Reusable) in dataset construction and sharing; and *(iii)* the dataset information is insufficient to be used in different applications.

To handle these limitations, Forroset is introduced, a dataset that contains extensive information on over 2900 forró music. The present dataset offers information ranging from the song's authors to technical details such as the beat position and length. Forroset also includes over 1400 song lyrics and gives MP3 files for all dataset songs, allowing users to develop new applications with forró.

2   Related Forró Datasets

Four datasets containing forró music were identified. Their descriptions, as well as their FAIR's shortcomings, are presented in Table 1. The LMD - Latin Music Database [23] is the precursor and has prompted several musical works of genre classification that includes forró music. This dataset, despite being very important to investigating different Latin American genres, it is no longer available. The BSL - Brazilian Songs Lyrics [8] on Kaggle and the BLD - "Brazilian Lyrics Dataset" [3], are two datasets that include extensive lyrics content collected from Vagalume. Lastly, the FVD - Forró em Vinil Dataset [5] has a large number of songs that include spectral and other metadata. This dataset lacks audio files, which may be obtained manually from the "Forró em Vinil" website[1].

Table 1. Forró related datasets. FAIR issues identified: F1 (not persistent identifier), A1 (non low-level protocol), I2 (no documentation), I3 (non qualified cross-reference), R1.1(unknown licence), R1.2. (undetailed provenance).

Name	Tracks	Features	Genres	FAIR
LMD	313	artist, title, genre, MP3 file	10	Not available
BSL	1000	artist, title, genre, lyrics	9	F1, I2, I3, R1.1, R1.2
BLD	11862	artist, title, genre, lyrics	14	F1, A1, R1.1
FVD	27352	artist, title, year, similarity network, mfcc	1	A1, I3

---

[1] https://www.forroemvinil.com/.

There is a large volume of songs in the available datasets, but provide few features and application opportunities. BSL and BLD, for example, are aimed at classifying music genres from the lyrics of the songs. Furthermore, no available dataset actually provides audio files of the songs, as the FVD songs need to be searched on the website. It is worth mentioning that the latter does not have songs after the year 2000, therefore, it does not contain songs related to the rebirth of forró with the emergence of Forró Universitário. Finally, all datasets have FAIR's shortcomings.

3   Forroset Creation

Forroset contains six kinds of information: General Information (GI), Audio Features (AF), Audio Analysis (AA), Filters and Organization (FO), Lyrics (L) and MP3 files. File sets as GI, AF, and MP3 were collected via Spotify API[2] and Spotipy package[3], the AA through Spotify API, Spotipy and Librosa package[4], the Lyrics using Vagalume API[5] and package[6]. Furthermore, we obtained FO by Spotify data transformations and manual annotation. Figure 1 depicts a concise representation of the Forroset development.

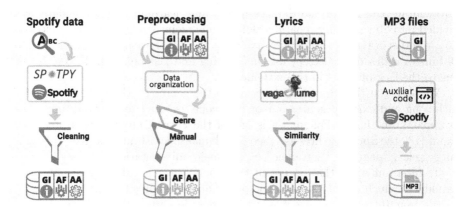

Fig. 1. Main steps for obtaining Spotify and Vagalume data.

3.1   Spotify Data

The Spotify tabular data collection was carried out in Spotify's database. To search for songs, we chose 5 keywords that refer to Forró Pé-de-Serra e o Forró

---

[2] https://developer.spotify.com/documentation/web-api/.

[3] https://spotipy.readthedocs.io/en/2.19.0/.

[4] https://librosa.org/doc/latest/index.html.

[5] https://api.vagalume.com.br/.

[6] https://github.com/diegoteixeir4/python-vagalume.

Universitário: "Forró Pé de Serra", "Forró Universitário", "Forró Tradicional", "Xote", and "Baião"; plus the names of 20 forró artists/groups. In the next step, songs with the same identifier, identical names, explicit content, non-playable, and without popularity and tempo values were completely removed from the dataset.

GI was collected through a keyword search, while AF and part of AA from the Spotify database via song identifiers were got separately. All Spotify tabular data is given for the entire song, although the audio files supplied only contain 30 s of a track which it is unknown if it is the beginning, middle or end of the song. This aspect prevents the application of the dataset for beat tracking since it is not possible to match the beat annotations of the entire track with the 30-s sample.

To overcome this setback, the librosa package for audio file beat recognition was used. In order to enable the algorithm's convergence, it was provided to the algorithm an initial stance of the song's tempo collected by Spotify to estimate the beats. Finally, a discrepancy score consisting of the error between the average distance of librosa and Spotify annotated beats was created.

3.2   Preprocessing Data

To facilitate the use of the dataset in machine learning tasks, the data was divided into 20 randomly separated folds while maintaining a uniform distribution in terms of tempo and popularity. The tempo of the songs were divided into 10 bins of width of 11, except for the starting and last bins, which were larger due to the lack of songs in the extremes.

We observed that Spotify only provides the genres by the artists/bands for each song and existing genres out of the scope ascribed to Forró authors. For example, even though Falamansa is one of the Forró Universitário's precursor bands [18], the Spotify genres ascribed to Falamansa artist include "axe", "brazilian reggae", "pagode", "sertanejo" e "sertanejo universitario".

To filter out songs that do not fit the scope, a metric called "genre filt" was developed, which calculates the ratio of the artist's genres that fit the scope on a scale from 0 to 1. The songs having a score of less than 0.3 on this scale were automatically excluded. Additionally, the songs from the 20 forró bands used for keyword selection stage, weren't filtered and a score of 1.1 was assigned to them.

Another point that was addressed is the presence of tracks with more than one song, which can be a problem, as it is not possible to identify from which part of the track the 30s is. Therefore, we manually evaluated the 100 most popular songs (according to Spotify popularity score), from each tempo bin. A binary score called "manual filt" was established, and the tracks that are in the scope received a score of 1. On the other hand, the tracks that weren't recognized as Forró Pé-de-Serra/Universitário, or with more than one song per track, were given a score of zero.

3.3   Vagalume Lyrics

Vagalume is a Brazilian website for music lyrics. The search for the songs lyrics was performed using the information of the song and artist/group name provided by Spotify. In cases when the search was successful, we picked up the track artist, track title, lyric and its Vagalume's access link.

To ensure that the expected song and the received lyrics are consistent, the similarity was calculated using the python package difflib[7]. When the track title and artist similarities were less than 90%, the lyrics were manually evaluated, removing all confirmed divergent lyrics.

3.4   Getting MP3 Files

Forroset contains a python helper code that uses the preview URL present in the GI to automatically get 30 s samples for all Forroset songs. Samples are downloaded in MP3 format at 22.05 kHz to a specified directory.

3.5   Ethics of Data Collection

Regarding the ambiguity found in the Terms of Services (TOS) of Social Media [7], as required by the Vagalume API TOS, the link to the song's lyrics on the Vagalume website is provided. Furthermore, according to the section *IV.3.a.i* of the Spotify API TOS[8], the GI, AF and AA data, necessary to operate Forroset, are compiled.

Furthermore, it is argued in this paper that Forroset was constructed in an ethical manner [10], considering that the data collected is publicly accessible on both sites and it is worth mentioning that the data does not contain sensitive personal information of any kind. Moreover, with this paper and the Completed Transparency Report, we are providing the detailed description of the aims, construction details, limitations and applications of Forroset.

3.6   FAIR Principles Implementation

An attempt to ensure that Forroset adheres to all FAIR principles[9] was done, making it available on GitHub and Zenodo platforms, satisfying the Findable and Accessible requirements. To join to the Interoperable and Reusable principles, the cross-reference identifiers for data retrieved from Spotify and Vagalume, as well as documentation related to the dataset's usage are provided. Finally, the Reusable criteria was fulfilled by assigning the dataset the Creative Commons Attribution License (CC BY 4.0).

---

[7] https://docs.python.org/3/library/difflib.html.

[8] https://developer.spotify.com/terms/.

[9] https://www.go-fair.org/fair-principles/.

4   Data Details

Forroset comprises a tabular file with 40 columns and a python code that downloads the Spotify previews in MP3 format from the URLs provided in the "preview url" column of the dataset. All information in GI, FO, AF, AA, and L groups are present in Table 2.

Table 2. Forroset tabular groups information.

Infos	Features	Tracks
GI	Track, track id, artist, artist id, popularity, album, album id, uri, track year, duration and preview url	2977
AF	Energy, liveness, tempo, valence, acousticness, instrumentalness, key, time signature, danceability, loudness, speechiness and mode	2977
AA	Beats start, duration and confidence; bars start, duration and confidence; tatums start, duration and confidence	2977
	Librosa beats start and discrepancy	2976
FO	Tempo bins, tempo bins max, genre filt, folds	2977
	Manual filt	1000
L	Lyrics	1415

4.1   General Information

The search performed on 2021/04/11 returned 9043 tracks. Following data cleaning, 5373 tracks from 1032 different artists were found. After removing artist with a high propensity for having songs that were out of scope, Forroset had 2977 tracks from 82 distinct artists. The Forroset artists are shown in Fig 2, where the word cloud indicates all of them and the bar graph depicts the top 10 artists with the most tracks.

Fig. 2. Artists ranked according to their number of tracks on Forroset.

Figure 3 shows the total songs by popularity and year. In addition, the number of songs per album is displayed in ascending order. Histograms are shown for all (in orange), songs with lyrics (in green), and songs that have been manually evaluated (in purple). Forroset contains 502 albums, of which 260 were manually analyzed, and the songs with lyrics are from 284 different albums.

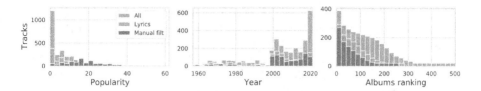

Fig. 3. Histograms with popularity, albums, and year for each subset.

4.2 Spotify Features

For each song, the dataset includes the feature information provided by the Spotify API. The number of songs per feature bin is presented in Fig. 4. Histograms are shown for all AF information group and subsets.

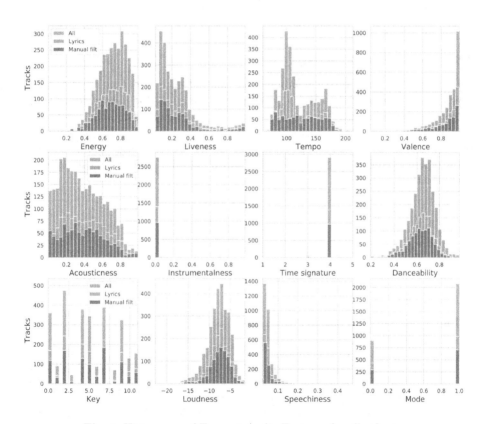

Fig. 4. Histograms of Forroset Audio Features for all subsets.

The audio features collected via the Spotify API are briefly described in the following lines. More information is available in the API documentation.

- **Energy:** A perceptual measure of intensity and activity.
- **Liveness:** Higher liveness numbers indicate a greater likelihood that the track was performed live.
- **Tempo:** The overall estimated tempo of a track in beats per minute (BPM).
- **Valence:** The musical optimism given by a song.
- **Acousticness:** The degree to which the music is acoustic. This feature predicts whether or not a recording has no voices.
- **Time signature:** A notation standard that specifies the number of beats in each bar.
- **Danceability:** It describes a track's suitability for dancing.
- **Key:** The key of the track is given by matching the integers to pitches using standard Pitch Class notation.
- **Loudness:** The overall volume of a track measured in decibels (dB).
- **Speechiness:** Detects the presence of spoken words in a track.
- **Mode:** A track's mode denotes the type of scale from which its melodic content is formed.

4.3   Spotify Audio Analysis

Forroset includes the beginnings and durations of each bar, beat, and tatum, over the entire song, for all songs. Each metric is assigned to a confidence level ranging from 0 to 1, with 1 being the highest level of confidence. In addition, the beat starts for the 30 s audio samples estimated by Librosa and a discrepancy score is provided. Figure 5 depicts the occurrences of the three events during a 15-second extract of Falamansa's song Xote dos Milagres, the most popular track in the dataset.

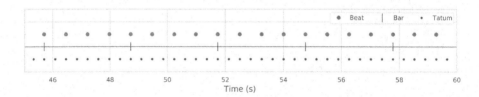

Fig. 5. Bar, beat and tatum of Falamansa's song Xote dos Milagres of Forroset.

The following is a brief description of the described rhythmic structures:

- **Bar:** A time signature is a method of quantitatively organizing the sounds of a musical composition into beats and pauses. A bar in Forró is often made up of four beats.
- **Beat:** A beat is the fundamental time unit of music. Typically, beats are multiples of tatums.
- **Tatum:** A tatum is the lowest regular pulse train that a listener infers instinctively from the time of observed musical events.

4.4 Filters and Organization

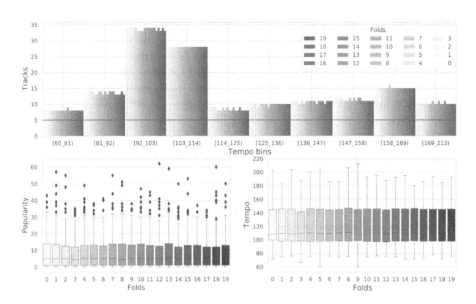

Fig. 6. Separating and balancing folds over time and popularity. The blue line on the top histogram represents the 50 songs for fold that have been manually reviewed. (Color figure online)

Regarding the 1000 tracks manually analyzed, 848 are from the Forró Universitário and Forró Pé-de-Serra scope. The histogram at the top of Fig. 6 shows how songs are divided into 20 balanced folds based on tempo. At the bottom, the box-plots compare the popularity and tempo of each fold, allowing the balance of the folds to be seen concerning these characteristics.

4.5 Lyrics

The search for title and artist in Vagalume's API, performed on 2021/10/12, yielded lyrics for 1435 tracks in which 20 are manually removed. In a unique list format, the title, artist, lyric, and url were created. Figures 3 and 4 highlight the information for songs that have lyrics.

4.6 MP3 Files

Using our auxiliary code, the 2959-audio-files were automatically downloaded. It took roughly six hours to complete the access to all the audios in the dataset. This extra time is due to the use of delays between queries to avoid overloading Spotify's platform.

5   Forroset's Potential Applications

Forroset was created with the goal of supporting Forró research. Information from multiple sources and kinds were organized around a unique identifier per song to allow its usage in several tasks, as well as the use of over one type of information to explore the same task. The variety of information allows the contribution to areas that have already been explored for forró such as Music Information Retrieval and Dance Teaching, in addition, it can motivate the beginning of research in unprecedented areas in the literature such as the Music Industry.

5.1   Forró Industry

Identifying a future hit song is a task of great interest to the music industry. However, the success of a song can be related to several factors, making this a task widely studied [9,12,24]. In Forroset, the Spotify popularity score is provided and can be estimated using the information contained in GI, AF and AA, the lyrics and the audio files. This information could be used together or separately, allowing, besides predicting the popularity, the selection of the best information. In this way, music recommendation is a another complex task that Forroset can be used for. Because most of the data come from Spotify, Forroset can be highly valuable for Spotify-integrated apps such as the one shown in Álvarez et al. (2020) [1].

5.2   Dance Teaching

The use of computer models to help people dance forró has been recently explored. One of the approaches consists of building applications that help teachers to assess how their students are dancing [19–21]. In this case, Forrosset contributes by providing the rhythmic information present in AA and AF. A facilitator is that researchers will be able to play the same versions of songs present in Forroset through Spotify. Another recent initiative is the use of forró rhythm prediction models to pass this rhythm through tactile stimuli for deaf people [6,17]. For this application, Forroset can contribute with the audio files and the respective rhythmic notes of tempo, bar and beat. Furthermore, it allows the application of deep models, previously impossible due to the absence of annotated datasets [6]. For both applications presented, the manual filter can be especially important, since they need songs with a well-defined rhythmic structure, as in the case of Forró Pé-de-Serra and Universitário.

5.3   Music Information Retrieval

Forroset is useful in beat tracking as it provides audio files and annotated beats. In this case, features can be extracted from the audio data to adjust neural network models [15]. This approach can be similarly used for tempo estimation [22]. Forroset has useful information for classifying musical genres such as the audio files [13], the lyrics of the songs [3] and the Spotify audio features [14]. It can be added to other datasets for the same purpose.

6 Conclusions

This paper describes Forroset, a Forró dataset designed to promote studies specially on the musical information recognition, dance teaching and music industry from this valuable Brazilian genre, particularly the subgenres Forró Universitário and Forró Pé-de-Serra. The dataset includes multiple audio information received from Spotify, song lyrics obtained via Vagalume, and information relating to manual and automatic filters. When compared to the other Forró datasets found, Forroset has fewer tracks. However, it contains the most diverse information, allowing for more comprehensive problem-solving.

The main limitations are the small number of tracks, the non-manual rating of all of them, and the approximate measures in the beat annotation. In future works, it will be useful to manually analyze the remaining songs, and expand the dataset to other Forró classes, such as adding identifiers for each subgenre.

7 Availability

Forroset data and code can be accessed at https://github.com/lucas-fpaiva/Forroset.

References

1. Álvarez, P., Zarazaga-Soria, F.J., Baldassarri, S.: Mobile music recommendations for runners based on location and emotions: The dj-running system. Perv. Mob. Comput. **67**, 101242 (2020)
2. Andrade, F.F.N., Figueiredo, F., Silva, D., Morais, F.: Measuring disruption in song similarity networks. In: Proceedings of International Society for Music Information Retrieval (2020)
3. de Araújo Lima, R., de Sousa, R.C.C., Lopes, H., Barbosa, S.D.J.: Brazilian lyrics-based music genre classification using a blstm network. In: International Conference on Artificial Intelligence and Soft Computing, vol. 12415, pp. 525–534 (2020)
4. Esparza, T.M., Bello, J.P., Humphrey, E.J.: From genre classification to rhythm similarity: computational and musicological insights. J. New Music Res. **44**(1), 39–57 (2015)
5. Falcão, F.V.: Dataset forró em vinil (2021). https://doi.org/10.5281/zenodo.5125205
6. Ferreira-Paiva, L., Lopes, H.G., Alfaro-Espinoza, E.R., Félix, L.B., Neves, R.V.A.: Towards a device for helping deaf people to dance: estimation of forro bar length using artificial neural network. IEEE Latin Am. Trans. **20**(6), 970–976 (2022)
7. Fiesler, C., Beard, N., Keegan, B.C.: No robots, spiders, or scrapers: legal and ethical regulation of data collection methods in social media terms of service. In: Proceedings of the International AAAI Conference on Web and Social Media, vol. 14, pp. 187–196 (2020)
8. Figueira, J.L.: Brazilian songs lyrics (2018). https://bit.ly/kaggle-BSL
9. Ge, Y., Wu, J., Sun, Y.: Popularity prediction of music based on factor extraction and model blending. In: 2020 2nd International Conference on Economic Management and Model Engineering, pp. 1062–1065 (2020)

10. Gelinas, L., Pierce, R., Winkler, S., Cohen, I.G., Lynch, H.F., Bierer, B.E.: Using social media as a research recruitment tool: ethical issues and recommendations. Am. J. Bioethics **17**(3), 3–14 (2017)
11. IPHAN: Matrizes Tradicionais do Forró recebem título de Patrimônio Cultural do Brasil (2021). https://bit.ly/iphan-forro
12. Lee, J., Lee, J.S.: Music popularity: metrics, characteristics, and audio-based prediction. IEEE Trans. Multimedia **20**(11), 3173–3182 (2018)
13. Liu, C., Feng, L., Liu, G., Wang, H., Liu, S.: Bottom-up broadcast neural network for music genre classification. Multimedia Tools Appl. **80**(5), 7313–7331 (2021)
14. Luo, K.: Machine Learning Approach for Genre Prediction on Spotify Top Ranking Songs. Master's thesis, University of North Carolina (2018)
15. MatthewDavies, E., Böck, S.: Temporal convolutional networks for musical audio beat tracking. In: 2019 27th European Signal Processing Conference, pp. 1–5 (2019)
16. Mondelli, M.L.B., Jr., L.M.R.G., Ziviani, A.: O que os países escutam: Analisando a rede de gêneros musicais ao redor do mundo. In: Anais do VII Brazilian Workshop on Social Network Analysis and Mining (2018)
17. Paiva, L.F., Lopes, H.G., Felix, L.B., Neves, R.V.: Estimação do compasso musical do forró utilizando rede perceptron multicamadas. In: Congresso Brasileiro de Automática-CBA (2020)
18. Quadros Junior, A.C., Volp, C.M.: Forró universitário: a tradução do forró nordestino no sudeste brasileiro. Motriz. J. Phys. Educ. UNESP, 117–120 (2005)
19. dos Santos, A.D.P., Loke, L., Yacef, K., Martinez-Maldonado, R.: Enriching teachers' assessments of rhythmic forró dance skills by modelling motion sensor data. Int. J. Human-Comput. Stud. **161**, 102776 (2022)
20. Santos, A.D.P.d., Tang, L.M., Loke, L., Martinez-Maldonado, R.: You are off the beat! is accelerometer data enough for measuring dance rhythm? In: Proceedings of the 5th International Conference on Movement and Computing (2018)
21. Santos, A.D.P.d., Yacef, K., Martinez-Maldonado, R.: Let's dance: how to build a user model for dance students using wearable technology. In: Proceedings of the 25th Conference on User Modeling, Adaptation and Personalization, pp. 183–191 (2017)
22. Schreiber, H., Müller, M.: A single-step approach to musical tempo estimation using a convolutional neural network. In: Proceedings of International Society for Music Information Retrieval, pp. 98–105 (2018)
23. Silla Jr, C.N., Koerich, A.L., Kaestner, C.A.: The latin music database. In: Proceedings of International Society for Music Information Retrieval, pp. 451–456 (2008)
24. Zangerle, E., Vötter, M., Huber, R., Yang, Y.H.: Hit song prediction: leveraging low-and high-level audio features. In: Proceedings of International Society for Music Information Retrieval, pp. 319–326 (2019)

Impact of ECG Signal Preprocessing and Filtering on Arrhythmia Classification Using Machine Learning Techniques

Hermes Andrés Ayala-Cucas[1][ID], Edison Alexander Mora-Piscal[1][ID],
Dagoberto Mayorca-Torres[1,3(✉)][ID], Diego Hernán Peluffo-Ordoñez[2][ID],
and Alejandro J. León-Salas[3][ID]

[1] Universidad Mariana, Grupo de investigación de Ingeniería Mecatrónica,
Pasto, Colombia
hayala@umariana.edu.co, dago.mayorca.torres@gmail.com
[2] Modeling, Simulation and Data Analysis (MSDA) Research Program, Mohammed
VI Polytechnic University, Ben Guerir, Morocco
[3] Departamento de Lenguajes y Sistemas Informáticos, Universidad de Granada,
C/Periodista Daniel Saucedo Aranda s/n, 18071 Granada, Spain

Abstract. Cardiac arrhythmias are heartbeat disorders in which the electrical impulses that coordinate the cardiac cycle malfunction. The heart's electrical activity is recorded using electrocardiography (ECG), a non-invasive method that helps diagnose several cardiovascular diseases. However, interpretation of ECG signals can be difficult due to the presence of noise, the irregularity of the heartbeat, and their nonstationary nature. Hence, the use of computational systems is required to support the diagnosis of cardiac arrhythmias. The main challenge in developing AI-assisted ECG systems is achieving accuracies suitable for application in clinical settings. Therefore, this paper introduces a software tool for classifying cardiac arrhythmias in ECG recordings that uses filtering, segmentation, and feature extraction of the QRS interval. We use the MIT-BIH Arrhythmia Database, which has 48 records of five different types of arrhythmias. We evaluate the data using supervised machine learning techniques such as k-Nearest Neighbors (KNN), Random Forest (RF), Multilayer Perceptron (MLP), and the Naive Bayesian classifier. This paper shows the impact of selecting and employing filtering and feature extraction methods on the performance of supervised machine learning algorithms compared with benchmark approaches.

Keywords: Electrocardiogram (ECG) · Cardiac arrhythmia · Feature extraction · Supervised machine learning · Performance measures

1 Introduction

The interior of the human heart is divided into four chambers: right and left atria at the top and right and left ventricles at the bottom. The right heart

A. C. Bicharra Garcia et al. (Eds.): IBERAMIA 2022, LNAI 13788, pp. 27–40, 2022.
https://doi.org/10.1007/978-3-031-22419-5_3

is responsible for receiving and pumping blood into the lungs, where it receives oxygen and gives off carbon dioxide. Then the left heart pumps oxygenated blood into the aorta to be redistributed for systemic circulation. Therefore, detecting pathologies that alter the heart's physiology, such as cardiac arrhythmias, is crucial [1]. Arrhythmias are cardiovascular diseases described as irregularities in the heartbeat. Cardiovascular diseases, which include arrhythmias, are the leading cause of death worldwide; about 17.5 million people die from these cardiac disorders representing 31% of deaths worldwide [2]. The electrocardiogram (ECG) is currently the most straightforward and used test for assessing heart rhythm and diagnosing arrhythmias since it is a non-invasive method. In an ECG, we can observe the constant changes of the different waves, intervals, and segments present in the signal. From these recordings, it is possible to study, analyze and identify irregular patterns in the ECG signal. However, the recorded data are massive, and analysis by cardiologists becomes a time-consuming and costly activity. In addition, ECG recordings are often accompanied by various electrocardiographic noises, further complicating the interpretation of ECG beats. Consequently, classifying or categorizing the different signal beats using advanced computer diagnostic tools is necessary [3]. Thus, the development of computational systems and the adoption of artificial intelligence techniques to support the detection and diagnosis of cardiovascular diseases have increased.

Several studies have developed efficient methods based on supervised machine learning techniques to identify arrhythmias such as Random Forests [4], K-Nearest Neighbor (KNN) classification techniques [5], Multi-Layer Perceptron (MLP) [6,7], and Naive Bayesian classifier (NB) [8]. Sahoo et al. [9] propose a three-stage QRS complex feature detection algorithm. First, they use the multiresolution wavelet transform (DWT) to remove noise from the ECG signal. Then, they use machine learning techniques to identify four types of cardiac abnormalities: normal (N), left bundle branch block (LBBB), right bundle branch block (RBBB) and beats (P). Similarly, Madan et al. [10] develop a hybrid deep learning-based approach for cardiovascular disease detection and classification using several configurations. Initially, their team selects 2D Scalogram images to reduce signal noise and extract features. In the second step, deep learning models, such as convolutional neural networks (CNN) and short-term memory networks (LSTM), identify abnormal heartbeats.

In arrhythmia classification, filtering and feature extraction techniques aim for models to achieve accuracies suitable for application in clinical settings. In this sense, this paper presents a novel system that uses a series of digital filters such as Wavelet filter, FIR filter, and Savitzky Golay filter to remove the many types of noises in the ECG signal. These signal processing techniques reach an optimal balance between robustness, signal variability and accuracy. Moreover, selecting ECG heartbeat interval features is critical for recognizing cardiac arrhythmias, so these features are extracted using the discrete wavelet transform (DWT). This transform produces more accurate time-frequency analysis results and decomposes the signal into several levels to determine which provides the most relevant features. After that, the features are classified using machine learning techniques such as K-Nearest Neighbors, Random Forest, Perceptron Multilayer, and Naive Bayesian, as these are essential for making reliable decisions when analyzing large data sets and events. The model and the different ECG signal processing techniques can assess cardiac rhythm irregularities to identify the type of arrhythmia a patient may manifest and implement prevention strategies. This study expands our knowledge of the incidence, characteristics and intervention methods for each type of arrhythmia. Moreover, this system contributes to increasing the data on the analysis of ECG potentials and compares them with the already extensive literature on cardiac arrhythmias analysis and prevention methodologies. Thus, this work can lead to finding possible variations according to the type of person.

The rest of this paper is structured as follows: Sect. 2 specifies the methodological design and the stages of data processing, segmentation, feature extraction, and classification. Section 3 details the results and discussions, while Sect. 4 describes the conclusions and future work. One of the main contributions of this study is the development of a methodology based on conventional methods.

2   Methodology

Figure 1 shows the methodological scheme of the proposed method for recognising and classifying cardiac arrhythmias. The following subsections specify the database we use and the signal preprocessing, beat segmentation, feature extraction, and classification steps.

2.1   MIT-BIH Arrhythmia Database

We choose the data from the MIT-BIH cardiac arrhythmia database for this study [11]. The database contains 48 half-hourly two-channel ambulatory ECG recordings collected from 47 individuals tested by the BIH Arrhythmia Laboratory between 1975 and 1979. The recordings were digitized using analogue Holter equipment with a sampling rate 360 Hz and 11-bit resolution over a 10 mV range. The database includes about 110.000 recordings with five classes of arrhythmias: Nonectopic Beats (N), Fusion Beats (F), Supraventricular Ectopic Beats (S), Ventricular Ectopic Beats (V), and Unknown Beats (U). Cardiologists perform classification and annotation of the recordings of each beat.

Table 1 shows the specifications of the different beats from the MIT-BIH database and summarizes the five types of ECG beat samples used in this study.

2.2   Preprocessing

ECG recordings obtained from the MIT-BIH arrhythmia database contain various sources of noise that hinder the proper recognition of some irregular beats. Often, the ECG signal data recorded when manipulating the electrodes results in signal contamination caused by different types of noise. These noises are typically classified into mains disturbances (60 Hz), baseline deviation, muscle artifacts,

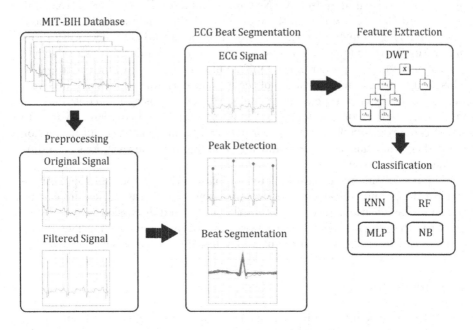

Fig. 1. Methodological scheme for the detection of cardiac arrhythmias.

Table 1. MIT-BIH arrhythmia beats classification per ANSI/AAMI EC57:1998 standard database.

AAMI classes	MIT-BIH annotation	Type
Non-ectopic beat (N)	N	Normal beat
	L	Left bundle branch block
	R	Right bundle branch block
	j	Nodal (junctional) escape
	e	Atrial escape beat
Supra-ventricular ectopic beat (S)	A	Aberrated atrial premature
	a	Atrial premature
	S	Supraventricular premature
	J	Nodal (junctional) premature
Ventricular ectopic beat (V)	V	Ventricular escape
	E	Premature ventricular contraction
Fusion beat (F)	F	Fusion of ventricular and normal
Unknown beat (U)	U	Unclassifiable
	p	Paced
	f	Fusion of paced and normal

and disturbances generated by electrical equipment [12]. Consequently, a preprocessing step that removes baseline noise and filters the ECG signal is necessary.

The techniques used for this stage are the discrete wavelet transform (DWT),
the FIR digital filter, and the Savitzky-Golay (SG) digital filter.

Discrete Wavelet Transform (DWT). The DWT technique is an effective
noise reduction and feature extraction method in signal analysis [13]. The main
wavelet function used is the Daubechies family (db6) to decompose the signal into
nine levels [14]. From this decomposition, the ninth level approximation subband
in the frequency range from 0 to 0.351 Hz is the baseline offset and is discarded
for noise-free signal reconstruction. In addition, we assume ECG signals between
90–180 Hz and 45–90 Hz contain no relevant information because noise is usually
located in these high-frequency bands. Therefore, we discard these signals for
detecting abnormal beats. Finally, we reconstruct the ECG signal with the detail
coefficients of levels 3, 4, 5, 6, 7, 8, and 9 [14].

FIR Digital Filter. The FIR digital filter is a filter that responds to a limited
number of non-zero terms if its input is an impulse, i.e., it is stable, and the signal
is not distorted. In addition, these filters can be designed to be phase linear, so
they do not cause phase lags in the signal. According to Costa *et al.* [15], the
preprocessing is divided into two phases: first, baseline removal is performed
using a trend function to remove the linear trend and some DC components of
the ECG signal; subsequently, a moving average filter with a window size of 5
samples obtains the smoothest signal, removing the low-frequency noise present
in the ECG signal. In the second phase, Costa *et al.* [15] use a cascade filter
composed of a low-pass FIR filter (order 30) with a cutoff frequency of 0.25 Hz,
and a high-pass FIR filter (order 6) with a cutoff frequency 30 Hz.

Savitzky-Golay (SG) Digital Filter. This digital filter is used for its ability
to preserve the attributes of the original signal after its application. A third-
order high-pass Butterworth filter with a cutoff frequency 1 Hz removes low-
frequency interference, effectively removing baseline noise from the ECG signal.
When filtering the ECG signal, low and high-frequency interferences are reduced
by polynomial order Savitzky-Golay (SG) filtering with a window dimension
between 5 and 21, according to [16].

Figure 2 shows the original ECG signal with noise and the signal resulting
from the application of the three digital filtering techniques.

2.3   ECG Beat Segmentation

At this stage, to perform the segmentation of the ECG signal beats, the location of the QRS complex and landmarks is paramount. Specifically, the QRS
complex denotes ventricular depolarization; three waves form it, the Q wave
presents a negative deflection, the R wave reflects a positive deflection, and the

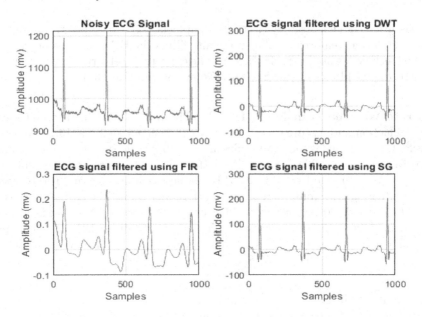

Fig. 2. Removal of noise from ECG signal (Non-ectopic beat).

S wave describes a negative deflection [1]. One aspect to consider is that QRS morphology is highly variable and depends on the point of origin of the beat.

For this reason, the QRS complex is an essential part of the ECG signal that must be analyzed and evaluated. In this regard, we determine the QRS complex using the MIT-BIH dataset's annotations as the developed algorithm detects the positions of each R peak of the ECG signal (midpoint of the QRS complex). For this purpose, we take 99 samples from the left and 100 from the right sides, forming a segment of 200 samples. This process aims to divide the ECG signal into individual beats with a normalized length. The result of the segmentation step is shown in Fig. 3.

2.4   Feature Extraction

This phase aims to correctly select the features with the most helpful information from the individual segments and obtain an acceptable classification performance as a result [17]. In this sense, such features are extracted using the discrete wavelet transform (DWT), which consists of decomposing a signal using approximate versions of the wavelet shape [18]. Feature extraction depends on the type of signal to be studied. Several wavelet families, such as Haar, Daubechies, and Symlet, provide valuable information for recognising different patterns. Therefore, we select the Daubechies wavelet family of order 2 (db2) for feature extraction because it has a morphological distribution similar to the signal of an electrocardiogram [19]. Signal decomposition is performed in up to five levels generating a series of coefficients that set up a field of charac-

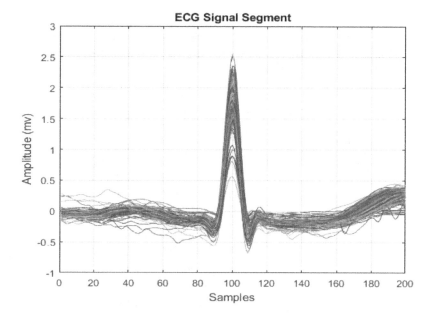

Fig. 3. 200-sample segment of the ECG signal (Non-ectopic beat).

teristics of different types of families. The decomposed signal consists of a series of detail coefficients (CD1 to CD5) and one more approximation coefficient (A5) for a total of 6 levels of decomposition. Four statistical features are estimated for each level: Maximum, Minimum, Mean, and Standard Deviation, collecting a total of 24 features extracted from each beat segment. Finally, these features become the input for the machine learning classifiers.

2.5   Classification

This paper employs and evaluates the following supervised learning algorithms: K-Nearest Neighbor classifier, a nonparametric instance-based algorithm whose classification is based on the training data to predict a new instance's label [20]. The algorithm determines the nearest neighbor in the data set and establishes the degree of similarity. Random Forest (RF) is an algorithm that traces all possible paths of an attribute through a recursive partition to classify the data [21]. One aspect to note is that the RF algorithm generates a C number of decision trees from a set of training samples; each tree in the forest randomly selects the number of samples to create a new training set. Multilayer perceptron (MLP) belongs to the family of supervised neural network algorithms [22], whose main function is to generate a model that correctly maps the input values using past information. The Naive Bayesian classifier is an algorithm with a statistical approach [23]. This algorithm uses the covariance matrix values to produce different probability functions in each class. It thus determines the values with the highest discriminant power for classifying the samples.

In this study, we have considered the most optimal parameters of the algorithms available by Weka software to classify the five classes of arrhythmias considered in this study. Table 2 shows the best parameters of each classifier for the best configuration presented later in the results.

Table 2. List of parameters for each classifier.

Classifier	Parameters
K-Nearest Neighbor	Number of neighbors = 3
	Distance metric = Euclidean
	Distance weighting = No
Random Forest	MaxDepth = 0
	NumDecimalPlaces = 2
	NumFeatures = 0
	NumIterations = 100
Multilayer perceptron	Activation = 'Relu'
	hiddenlayers = 'constant'
	learningRate = 0.3
	momentum = 0.2
	TrainingTime = 500
Naive Bayesian	Default parameters

2.6 Experimental Design

This section describes the design of the experiments and the metrics to evaluate the performance of the developed algorithms. The analyzed dataset contains 109.463 records, including five arrhythmia types (N, S, V, F, U). Three filtering techniques (DWT, FIR, SG), one feature extraction technique (DWT), and four machine learning algorithms (KNN, MLP, RF, NB) with 10-fold cross-validation are applied for a total of 12 experiments, as shown in Fig. 4.

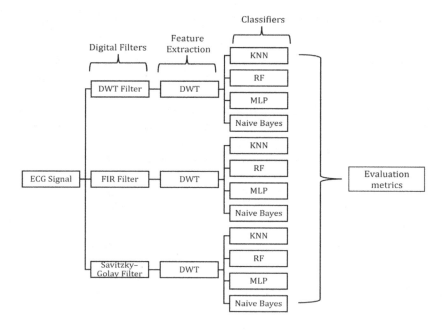

Fig. 4. Experimental diagram

In this paper, we assess the performance of the classification models using accuracy (Acc), sensitivity (Se), and F1-score. These metrics are defined based on four parameters, as shown in equations (1), (2) and (3).

$$Acc = \frac{TP + TN}{TP + TN + FP + FN} \tag{1}$$

$$Se = \frac{TP}{TP + FN} \tag{2}$$

$$F1 - Score = \frac{TP}{TP + 0.5(FP + FN)} \tag{3}$$

where TP corresponds to correctly classified beats, FN to unclassified beats, TN to correctly unclassified beats, and FP to incorrectly classified beats.

3   Results

Table 3 shows the results of the 12 experiments performed, considering the stages of filtering, feature extraction, and classification techniques.

The best configuration for cardiac arrhythmias classification is using the FIR filter for noise reduction, DWT for feature extraction, and applying the k-Nearest Neighbors (KNN) algorithm, as shown in Table 3. This setup achieves an accuracy, sensitivity and F1-score of 98.10%, 98.10%, and 98.10%, respectively.

Table 3. Classification results with four different classifiers.

Classifier	DWT	FIR	SG	Features	Accuracy	Sensitivity	F1-Score
	Yes	No	No	DWT	97.40%	97.50%	97.40%
KNN	No	Yes	No	DWT	98.10%	98.10%	98.10%
	No	No	Yes	DWT	98.00%	98.10%	98.00%
	Yes	No	No	DWT	97.40%	97.50%	97.30%
RF	No	Yes	No	DWT	97.90%	98.00%	97.80%
	No	No	Yes	DWT	98.10%	98.10%	97.90%
	Yes	No	No	DWT	96.20%	96.40%	96.10%
MLP	No	Yes	No	DWT	96.60%	96.70%	96.40%
	No	No	Yes	DWT	97.00%	97.10%	96.90%
	Yes	No	No	DWT	89.40%	56.60%	66.80%
Naive Bayesian	No	Yes	No	DWT	82.40%	76.40%	78.60%
	No	No	Yes	DWT	89.20%	55.60%	66.20%

Meanwhile, using the Savitzky-Golay (SG) filter, DWT feature extraction, and the Random Forest (RF) algorithm is the second best configuration to identify cardiac arrhythmias, achieving an accuracy, sensitivity, and F1-score of 98.10%, 98.10%, and 97.90%, respectively. It should be noted that, unlike other studies, the metrics are computed considering the five classes of arrhythmias (N, S, V, F, U).

3.1 Discussions

We identify studies that use similar approaches and techniques on the MIT-BIH database to compare them with our results and the configurations proposed in this study. Table 4 lists several studies and their performance evaluation with the respective values for each metric.

Ye et al. [24] propose a method for classifying five arrhythmias; using techniques such as DWT along with PCA and ICA, they achieve a 94.00% accuracy. Similarly, Ayar et al. [25] apply a feature extraction approach based on Genetic Algorithms (GA) and the Random Forest (RF) algorithm to reach a 97.98% accuracy. On the other hand, Yang et al. [5] evaluate the parametric features and visual patterns of ECG signal morphology to train the KNN algorithm and achieves an overall accuracy of 97.70%. Next, Saenz-cogollo et al. [26] highlight a detection system that combines several features, including R-R interval duration, amplitude differences, and QRS time characteristics, to classify six classes of arrhythmias using the Random Forest (RF) classifier. As a result, they obtain an overall accuracy of 96.14%. On the other hand, some approaches use deep learning techniques, e.g., Wu et al. [27] uses convolutional neural networks (CNN) and DWT to detect five classes of cardiac arrhythmias achieving an average accuracy of 97.20%. Finally, Mazidi et al. [28] use a feature extraction method employing TQWT and statistical feature calculation to classify the ECG signal into normal and abnormal classes using the KNN classifier, achieving an accuracy of 97.81%.

The experimental results show that the proposed method achieves higher performance metrics values than the benchmark approaches described in Table 4 for the same database. Thus, we can argue that the proposed configuration of signal processing techniques is a better classification method for cardiac arrhythmias. Nonetheless, the amount of data or beats for analysis is a critical factor to consider for a fair comparison. Unlike other studies, we implement three main strategies: first, we adopt all patient records from the MIT-BIH arrhythmia database (109.463 records) to validate our detection system. Secondly, several techniques are implemented in the preprocessing, feature extraction, and classification stages. We perform 12 experiments to observe which signal preprocessing configuration delivers the best cardiac arrhythmia classification and identification results. Finally, we aim to provide a reliable medical diagnostic system to detect all classes of cardiac arrhythmias recommended by the ANSI/AAMI EC57:1998/(R) standard. The validation of this work is based on these three criteria, allowing us to develop a more practical and reliable system for diagnosing an actual event since patients present a broad spectrum of morphologies.

Table 4. Comparison of classification performance methods and the proposed approach.

Literature	Year	Features	Classifier	Classes	Accuracy
Ye et al. [24]	2012	DWT+ICA+PCA	SVM	5	94.00%
Ayar et al. [25]	2018	GA+DT	RF	6	97.98%
Yang et al. [5]	2020	Parameter + Visual pattern of ECG Morphology	KNN	15	97.70%
Saenz-cogollo et al. [26]	2020	R–R-intervals, HBF and time domain morphology	RF	6	96.14%
Wu et al. [27]	2021	DWT	CNN	5	97.20%
Mazidi et al. [28]	2022	TQWT+ statistical features	KNN	2	97.81%
Proposed	2022	DWT	KNN	5	98.10%
			RF		98.10%

4   Conclusions

Electrocardiographic recordings of the heart rhythm's electrical activity provide information about this vital organ's status. In this work, we assess the performance of a novel classification system by analyzing five classes of cardiac arrhythmias. Our method uses a combination of digital filters, feature extraction methods (DWT), and four different classifiers (KNN, RF, MLP, and Naive Bayesian) to provide easy-to-interpret results with acceptable values of accuracy, sensitivity, and F1 score. In addition, each experiment allows us to improve the methods and propose several alternatives at all stages of the modeling process. For future work, we suggest testing new filtering and classification techniques for arrhythmia detection with considerable reliability and accuracy to improve diagnosis in clinical settings.

Acknowledgments. The authors would like to acknowledge the valuable support given by the SDAS Research Group (https://sdas-group.com/).

References

1. Luis, F., Moncayo, G.: Libro de la salud cardiovascular del Hospital clínico san carlos y la Fundación BBVA. 1nd edn., Madrid (2009)
2. Benjamin, E. J., Virani, S. S., Callaway, C. W.: Heart disease and stroke statistics - 2018 update: a report from the American heart association. Circulation **137**(12) (2018). https://doi.org/10.1161/CIR.0000000000000558
3. Pozo-Ruiz, S., Morocho-Cayamcela, M.E., Mayorca-Torres, D., H. Peluffo-Ordóñez, D.: Parkinson's disease diagnosis through electroencephalographic signal processing and sub-optimal feature extraction. In: Rocha, Á., Ferrás, C., Méndez Porras, A., Jimenez Delgado, E. (eds.) ICITS 2022. LNNS, vol. 414, pp. 118–127. Springer, Cham (2022). https://doi.org/10.1007/978-3-030-96293-7_12
4. Alqudah, A.M., Albadarneh, A., Abu-Qasmieh, I., Alquran, H.: Developing of robust and high accurate ECG beat classification by combining Gaussian mixtures and wavelets features. Aust. Phys. Eng. Sci. Med. **42**(1), 149–157 (2019). https://doi.org/10.1007/s13246-019-00722-z
5. Yang, H., Wei, Z.: Arrhythmia recognition and classification using combined parametric and visual pattern features of ECG morphology. IEEE Access **8**, 47103–47117 (2020). https://doi.org/10.1109/ACCESS.2020.2979256
6. Ramkumar, M., Ganesh Babu, C., Vinoth Kumar, K., Hepsiba, D., Manjunathan, A., Sarath Kumar, R.: ECG cardiac arrhythmias classification using DWT, ICA and MLP neural networks. J. Phys. Conf. Ser. **1831**(1), 1–13 (2021). https://doi.org/10.1088/1742-6596/1831/1/012015
7. Cepeda, E., Sánchez-Pozo, N.N., Peluffo-Ordóñez, D.H., González-Vergara, J., Almeida-Galárraga, D.: ECG-based heartbeat classification for arrhythmia detection using artificial neural networks. In: Gervasi, O., Murgante, B., Hendrix, E.M.T., Taniar, D., Apduhan, B.O. (eds) Computational Science and Its Applications - ICCSA 2022, ICCSA 2022, Lecture Notes in Computer Science, vol. 13376, Springer, Cham (2021). https://doi.org/10.1007/978-3-031-10450-3_20
8. Bhoi, A. K., Sherpa, K. S., Khandelwal, B.: Ischemia and arrhythmia classification using time-frequency domain features of QRS complex. Procedia Comput. Sci. **132**(Iccids), 606–613 (2018). https://doi.org/10.1016/j.procs.2018.05.014
9. Sahoo, S., Kanungo, B., Behera, S., Sabut, S.: Multiresolution wavelet transform based feature extraction and ECG classification to detect cardiac abnormalities. J. Int. Meas. Confeder. **108**, 55–66 (2017). https://doi.org/10.1016/j.measurement.2017.05.022

10. Madan, P., Singh, V., Singh, D.P., Diwakar, M., Pant, B., Kishor, A.: A hybrid deep learning approach for ECG-based arrhythmia classification. Bioengineering **9**(4), 1–13 (2022). https://doi.org/10.3390/bioengineering9040152
11. MIT-BIH Arrhythmia Database. https://www.physionet.org/content/mitdb/1.0. 0/. Accessed 27 May 2022
12. Ortega, C. D., Ibarra-piandoy, A., Viveros-villada, E., Mayorca-torres, D.: Prototipo para la adquisición y caracterización de señales electromiográficas superficiales del movimiento de flexión-extensión de los dedos de la mano. Iberian J. Inf. Syst. Technol., 52–65 (2020)
13. Sharma, P., Dinkar, S.K., Gupta, D.V.: A novel hybrid deep learning method with cuckoo search algorithm for classification of arrhythmia disease using ECG signals. Neural Comput. Appl. **33**(19), 13123–13143 (2021). https://doi.org/10. 1007/s00521-021-06005-7
14. Elhaj, F.A., Salim, N., Harris, A.R., Swee, T.T., Ahmed, T.: Arrhythmia recognition and classification using combined linear and nonlinear features of ECG signals. Comput. Methods Prog. Biomed. **127**, 52–63 (2016). https://doi.org/10.1016/j. cmpb.2015.12.024
15. Costa, R., Winkert, T., Manhães, A., Teixeira, J.P.: QRS peaks, P and T waves identification in ECG. Procedia Comput. Sci. **181**(2019), 957–964 (2021). https:// doi.org/10.1016/j.procs.2021.01.252
16. Kumar, C., Kolekar, M.H.: Cardiac arrhythmia classification using tunable Q-wavelet transform based features and support vector machine classifier. Biomed. Signal Process. Control **59**, 101875 (2020). https://doi.org/10.1016/j.bspc.2020. 101875
17. Rodriguez-Sotelo, J.L., Peluffo-Ordoñez, D., Cuesta-Frau, D., Castellanos-Domínguez, G.: Unsupervised feature relevance analysis applied to improve ECG heartbeat clustering. Comput. Methods Programs Biomed. **108**(1), 250–261 (2012). https://doi.org/10.1016/j.bspc.2020.101875
18. Khorrami, H., Moavenian, M.: A comparative study of DWT, CWT and DCT transformations in ECG arrhythmias classification. Expert Syst. Appl. **37**(8), 5751–5757 (2010). https://doi.org/10.1016/j.eswa.2010.02.033
19. Ranaware, P. N., Deshpande, R. A.: Detection of arrhythmia based on discrete wavelet transform using artificial neural network and support vector machine. In: International Conference on Communication and Signal Processing, pp. 1767–1770 (2016)
20. Xiang, Y., Lin, Z., Meng, J.: Automatic QRS complex detection using two-level convolutional neural network. J. BioMed. Eng. Online **17**(1), 1–17 (2018). https:// doi.org/10.1186/s12938-018-0441-4
21. Pandey, S.K., Janghel, R.R., Vani, V.: Patient specific machine learning models for ECG signal classification. Procedia Comput. Sci. **167**(2019), 2181–2190 (2020). https://doi.org/10.1016/j.procs.2020.03.269
22. Ramkumar, M.H., Ganesh Babu, C., Ganesh Babu, K., Hepsiba, D., Manjunathan, A., Sarath Kumar, R.: ECG cardiac arrhythmias classification using DWT, ICA and MLP neural networks. J. Phys. Conf. Ser. **1831**(1), 1–13 (2021). https://doi. org/10.1088/1742-6596/1831/1/012015
23. Nascimento, N.M.M., Marinho, L.B., Peixoto, S.A., do Vale Madeiro, J.P., de Albuquerque, V.H.C., Filho, P.P.R.: Heart arrhythmia classification based on statistical moments and structural co-occurrence. Circ. Syst. Signal Process. **39**(2), 631–650 (2019). https://doi.org/10.1007/s00034-019-01196-w

24. Ye, C., Kumar, B. V. K. V., Coimbra, M. T.: Combining general multi-class and specific two-class classifiers for improved customized ECG heartbeat classification. In: Proceedings - International Conference on Pattern Recognition ICPR, pp. 2428–2431 (2012)

25. Ayar, M., Sabamoniri, S.: An ECG-based feature selection and heartbeat classification model using a hybrid heuristic algorithm. Inf. Med. Unlocked **13**, 167–175 (2018). https://doi.org/10.1016/j.imu.2018.06.002

26. Saenz-cogollo, J. F., Agelli, M.: Investigating feature selection and random forests for inter-patient heartbeat classification. In: Algorithms, pp. 2–13. (2020). https://doi.org/10.3390/a13040075

27. Wu, M., Lu, Y., Yang, W., Wong, S.Y.: A study on arrhythmia via ECG signal classification using the convolutional neural network. Front. Comput. Neurosci. **14**(January), 1–10 (2021). https://doi.org/10.3389/fncom.2020.564015

28. Mazidi, M. H., Eshghi, M., Raoufy, M. R.: Premature ventricular contraction (PVC) detection system based on tunable Q-factor wavelet transform. J. Biomed. Phys. Eng. **12**(1), 61–74 (2022). https://doi.org/10.31661/jbpe.v0i0.1235

Learning Automata Using Dimensional Reduction

David Kuboň$^{(\boxtimes)}$, František Mráz , and Ivan Rychtera

Charles University, Prague, Czech Republic
{dkubon,mraz,rychtera}@ksvi.mff.cuni.cz

Abstract. One-dimensional (string) formal languages and their learning have been studied in considerable depth. However, the knowledge of their two-dimensional (picture) counterpart, which retains similar importance, is lacking. We investigate the problem of learning formal two-dimensional picture languages by applying learning methods for one-dimensional (string) languages. We formalize the transcription process from an input two-dimensional picture into a string and propose a few adaptations to it. These proposals are then tested in a series of experiments, and their outcomes are compared. Finally, these methods are applied to a practical problem and learn an automaton for recognizing a part of the MNIST dataset. The obtained results show improvements in the topic and the potential in using the learning of automata in fitting problems.

Keywords: Learning · Grammatical inference · Automata · Formal languages · Picture languages

1 Introduction

A considerable amount of research has been done on the field of one-dimensional formal languages, which now have a substantial position in the foundations of Computer Science. However, much less is known about formal languages in two dimensions, even though they have both theoretical and practical importance comparable to one dimension. As an example, we could list automatic detection of different shapes (e.g., road signs), or more generally, any problem on two-dimensional data which has some pattern regularity [11].

In some literature and also throughout this paper, the terms two-dimensional and *picture* languages will be used interchangeably. To distinguish them from pictures in the wider, common sense, formal picture languages have a formally exact mathematical description and are not defined as sets of pictures containing, for example, cars, which cannot be defined rigorously. Therefore, the usage of deep neural networks, typically very efficient with recognizing objects in images

Supported by Charles University Grant Agency (GAUK) project no. 1198519 and SVV-260588.

[5], mostly fails to learn picture languages in the formal sense. However, powerful models of automata exist that work on picture languages but lack the efficiency and determinism needed for more practical applications.

Several papers have already been published [7] focusing on finding methods to learn picture languages from positive and negative examples. This process of learning a model (a grammar) for a target language based on some information about the words of the language is called grammatical inference [4]. There are multiple known algorithms of grammatical inference for a number of classes of languages in the one-dimensional domain of the problem, but almost no knowledge in two (or more) dimensions.

The manner in which pictures can be formally represented differs in the literature. One option is generative, which describes the way a picture can be generated from a string. Freeman [2] introduced an 8-letter alphabet with moves representing all the eight directions (north, south, east, west, northeast, southeast, northwest, and southwest). Later Maurer et al. [10] simplified the alphabet into a 4-letter alphabet (up, down, left, right). Both alphabets can represent a way in which a picture is drawn. In order to generate colored pictures, the latter approach was extended with labels by Costagliola [1]. In either of these representations, a picture language is a set of strings describing all pictures in the language.

A second way to represent a picture is closer to the common form – a rectangular array of symbols that could be interpreted as colors of pixels in the image. In this representation, a picture language is the set of pictures accepted by an automaton working on two-dimensional inputs. Examples of such automata are a non-deterministic online tessellation automaton [3], an even more powerful sgraffito automaton [12], and a two-dimensional limited context restarting automaton [6]. As outlined earlier, the problem with these automata is their high complexity, as the problem of deciding whether an input image is accepted by any of them is NP-complete.

This paper follows up on an earlier study [7] that proposed a new representation for picture languages. It follows the second way and represents the pictures using a function R that rewrites any two-dimensional picture p into a string $R(p)$, and a one-dimensional language L. The set of all pictures p for which $R(p)$ is in L defines the picture language. Here, we propose a more formal transcription-evaluation framework and conduct multiple experiments with different languages, learning algorithms, and transcription mechanisms.

The paper is structured as follows: Sect. 2 introduces basic definitions for pictures, and picture languages, in Sect. 3, we present definitions related to the transcription-evaluation framework, Sect. 4 describes the experiments and the obtained results, and Sect. 5 concludes the paper.

2   Definitions

We define picture languages in a fashion that corresponds to pictures in common sense. A *picture* p over a finite alphabet Σ is a two-dimensional rectangular array of elements from Σ – see [3]. We say that p has dimensions (m, n), if it has m rows and n columns. Then $p_{i,j}$ from Σ, $0 \leqslant i < m, 0 \leqslant j < n$ denotes the symbol at position j in row i. The set of all rectangular pictures over Σ of dimensions (m, n) will be denoted as $\Sigma^{m,n}$ and the set of all rectangular pictures over Σ of any dimension will be denoted as $\Sigma^{*,*}$. A *picture language* is any subset of $\Sigma^{*,*}$.

Any automaton working on a picture p of dimensions (m, n) needs to know where is the border of the picture, therefore the picture is typically surrounded by sentinels #, where $\# \notin \Sigma$. Delimited picture p is called a *boundary picture* $\widehat{p}$ over $\Sigma \cup \{\#\}$ of dimensions $(m + 2, n + 2)$:

#	#	#	$\cdots$	#	#	#
#						#
$\vdots$			P			$\vdots$
#						#
#	#	#	$\cdots$	#	#	#

A particular class of formal languages is the class of *locally testable languages*, which will be used later in the paper.

Definition 1. *For a positive integer k, a language L is k-locally testable if there exist sets A, B, and C of words of length k such that a word w of length at least k belongs to L if its prefix of length k belongs to A, its suffix of length k belongs to B and all its infixes of length k belong to C. A language is locally testable if it is k-locally testable for some k.*

3 Transcription-Evaluation Framework

We dedicate this section to methods for recognizing picture languages that transform pictures into strings and then use a string automaton to recognize them. The idea is to leverage our understanding in tackling problems in the domain of one-dimensional languages to help us in the more complex domain of picture languages.

The approach presented in [7] has shown to be promising; thus, we will explore it more in this paper. In the article, various methods of transcribing the pictures into strings are used and followed by an algorithm to construct a deterministic finite automaton (DFA) to classify the resulting strings.

Rather than row-by-row or column-by-column, the proposed transcription methods have used a particular order that concatenated together the contents of 3-by-3 windows of the picture separated by a special symbol. This, the authors theorized, should have led to a better generalization of other simpler approaches.

Some preliminary experiments have shown that rather than simply rewriting the symbols in the original picture into one dimension, we can achieve better performance by transcribing the picture into a string over another larger alphabet. A letter of this alphabet should reflect the whole contents of a 3-by-3 window. As many DFA learning algorithms use a prefix tree built from obtained sample words, the large alphabet reduces the depth of the prefix trees and speeds up their processing.

Definition 2. *Let Σ and Γ be alphabets, and Σ does not contain the symbol $\#$. Then Transducer-Evaluator machine for picture languages (TEMPL) is a pair $M = (T, E)$, where T is a map from Σ^{**} to Γ^{*} and E is a string automaton accepting a language $L(E) \subset \Gamma^{*}$. We say that M accepts the picture language $L(M) = L(T, E) = \{p \in \Sigma^{**} | T(p) \in L(E)\}$.*

In general, a transducer can be any mapping of pictures into strings. In this paper, we will limit the transducer to a two-part process. First, a scanning strategy is designed using a simple automaton. Then a constant dictionary is used to map fixed-size fragments of the picture into substrings according to the order determined by the scanning strategy. We call these parts *scanner* and *sequence dictionary*.

A scanning sequence for a picture p of dimension (m, n) is a finite sequence, where each element is a pair of integers $(i, j); 0 \leqslant i < m, 0 \leqslant j < n$. A scanning sequence can be obtained, e.g., by recording each position in a picture row-by-row and left-to-right. However, any fixed strategy can limit the power of TEMPL. Hence, we introduce a more general way of producing a scanning sequence.

Definition 3. *A four-way scanner automaton (4SA) $M_r = (Q, \Sigma, \Delta, q_0, q_h, \delta)$, where Q is a set of states, Σ is an input alphabet, $\Delta = \{l, r, u, d\}$, q_0 is the starting state, q_h is the halting state, and $\delta \subset Q \times (\Sigma \cup \{\#\}) \to Q \times \Delta \times ((\mathbb{N}^0 \times \mathbb{N}^0) \cup \{\varepsilon\})$ is the transfer function.*

In this paper, we add an additional constraint on δ: $\forall q \in Q; \forall a, b \in \Sigma; \delta(q, a) = \delta(q, b)$. The scanner can only differentiate if it reads a symbol from the picture or a border symbol. Hence if two pictures have the same dimensions, the scanner produces for them identical scanning sequences.

Configuration and transitions of the 4SA are similar to those of the four-way finite automaton working on two-dimensional inputs [3]. Additionally, the 4SA in each transition determines whether to output its current position on the boundary picture $\hat{p}$ as the next element of the scanning sequence. The output is part of the configuration as well.

A computation of a 4SA starts in the initial state at the top-left corner of the input picture and ends by entering the halting state q_h. The output of the computation is a scanning sequence. We call the elements of this sequence *anchors*.

We place an additional constraint on the scanner: the output scanning sequence must contain each position of the picture p exactly once.

A sequence dictionary is a tuple $D = (\Sigma, \Gamma, w, t, k)$ where Σ is the input alphabet, Γ is the output alphabet, w is a sequence of relative positions of length l and $t : (\Sigma \cup \{\#\})^l \to \Gamma^k$ is a map. k is a constant.

On an input picture p and a pair (i, j) from a scanning sequence, we apply the sequence dictionary in the following way. We concatenate symbols from the positions relative to (i, j) according to the sequence of relative positions w into a word $r \in (\Sigma \cup \{\#\})^l$. Then we append $t(r)$ to the output. The output is a word over Γ.

Perhaps the most apparent relative position sequence to try is one that encompasses a 3-by-3 window around the anchor. But other relative position sequences are possible.

Suppose a sequence dictionary $D = (\{a\}, \{0, 1\}, \{(-1, -1), (-1, 0), (0, -1),$ $(0, 0)\}, t)$, where for every word $v \in \{a, \#\}^4$ it holds $t(v) = 1$ if $v = \#\#\#a$ and $t(v) = 0$ otherwise. The sequence of the dictionary corresponds to an upper-left 2-by-2 square relative to the anchor.

Let p be a square picture over $\{a\}$ of dimensions $(2, 2)$ and $\{(1, 1), (2, 2)\}$ be a scanning sequence. The strings obtained from the relative positions to both anchors are $\#\#\#a$ and $aaaa$, resulting in the output word 10 (see Fig. 1).

$$\begin{array}{cccc} \#_1 & \#_2 & \# & \# \\ \#_3 & a_4 & a & \# \\ \# & a & a & \# \\ \# & \# & \# & \# \end{array} \qquad \begin{array}{cccc} \# & \# & \# & \# \\ \# & a_1 & a_2 & \# \\ \# & a_3 & a_4 & \# \\ \# & \# & \# & \# \end{array}$$

Fig. 1. The boundary picture $\hat{p}$ with symbols marked at relative positions $\{(-1, -1), (-1, 0), (0, -1), (0, 0)\}$ with respect to anchors $(1, 1)$ and $(2, 2)$.

Altogether, we propose to represent a picture language by a TEMPL $M = (T, E)$ with a map T and an automaton E. The map T uses a scanner M_r for producing a scanning sequence and a sequence dictionary D for mapping contents of the scanning window into a string. All parts of the model could be learned from a sample of a target picture language. In what follows, we will use a fixed scanner (row-by-row and left-to-right in each row), two types of sequence dictionaries, and two types of evaluator automata for E.

4   Experimental Results

To verify the suitability of our new representation of picture languages for their learning, we conducted a series of experiments with seven picture languages over the binary alphabet $\{\square, \blacksquare\}$ corresponding to white and black pixels, respectively (see Fig. 2 for sample pictures):

L_1 is the set of all white rectangles containing a black diagonal till the border of the picture. The diagonal can start in either of the top corners.

L_2 is the set of all white pictures of dimensions at least $(3, 3)$ with a black border of one-pixel width.

L_3 is the set of all pictures with a positive number of black rows followed by a positive number of white rows.

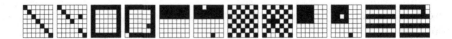

Fig. 2. Sample pictures from languages $L_1, \ldots, L_6$ showing crisp and noisy versions.

L_4 is the set of all pictures with a regular chessboard pattern of □-s and ■-s. The top-left corner of such picture can contain □ or ■, but the whole picture must have the chessboard pattern.

L_5 is the set of all pictures where the top left quadrant is black and the rest white.

L_6 is the set of all pictures with alternating black and white rows. The first row can be either black or white.

L_{mnist1} is based on the dataset of handwritten digits MNIST [9]. All pictures of digit one are treated as positive and all other digits as negative examples. We adapted the pictures from gray-scale to black and white.

We also added a noisy version for each crisp language $L_1, \ldots, L_6$. Each picture belonging to a noisy language must not differ from a positive crisp example by more than a given noise threshold – in our experiments, it was 5%.

4.1 Datasets

Experimental datasets for $L_1, \ldots, L_6$ were randomly selected from a randomly generated pool of pictures of various sizes. Pictures of larger sizes are more common in order to accommodate the larger space needed to be sampled. Positive and negative examples are handled separately. Positive examples with zero noise are always included in the pool.

The languages are sparse. Hence crisp positive examples can be obtained easily. Noisy positive examples are obtained by randomly flipping pixels. Negative examples are generated from the positive ones by flipping pixels selected uniformly randomly. We discard negative examples if they accidentally become positive. For noisy languages, the number of differing pixels is counted for each crisp positive picture and then tested whether it exceeds the noise threshold 5%.

From this pool, train and test sets are uniformly randomly chosen. If there is too few positive examples in the pool to match the number of negative examples, negative examples are added so that the sets have the same total sizes across all languages.

The specific training and testing sets required for our experiments were generated using scripts (available upon request) and are fully reproducible. Sample sets comprised 100, 200, 400, 800, 1600, and 3200 pictures for each sample language, ranging from dimensions $(5, 5)$ to $(10, 10)$. Each set contained the same number of positive and negative examples if possible. However, for training k-locally testable (string) languages, only positive examples in the training sets are used – see below.

4.2   Learning Setup

The experiment is set us as follows. First, a picture is rewritten into a string using a scanner and scanner dictionary. Then for each set S of sample strings, a deterministic finite-state automaton consistent with S is learned. In our experiments, we stick to a single scanning strategy – row-by-row and from left to right in each row using a scanning window of size 3-by-3. The first sequence dictionary simply rewrites the contents of a 3-by-3 scanning window into a string of 9 symbols. This is referred to as a one-to-one encoder.

The second sequence dictionary maps the contents of a scanning window into a single symbol from the alphabet $\{\Box, \blacksquare, \#\}$. As the alphabet is of size 19683, the symbol is represented as an integer between 0 and 19682. This sequence dictionary is referred to as a many-to-int encoder.

The training for k-locally testable languages consists in collecting the sets of possible prefixes, suffixes, and infixes of length k from all positive samples.

For learning from sets containing both positive and negative examples, we use the state merging algorithm `traxbar`, which is our python implementation of a version of breadth-first Trakhtenbrot-Barzdin's state merging algorithm [8].

Once the learning is finished, the resulting automaton is tested on an independent test set of pictures that are rewritten into strings in the same way. As our sample languages are rather sparse, we do not report accuracy as it usually considerably differs between positive and negative samples. Instead, we use F_1-score defined as $\dfrac{Precision \cdot Recall}{Precision + Recall}$, where $Precision = \frac{TP}{TP+FP}$, $Recall = \frac{TP}{TP+FN}$, TP, FP, and FN stand for the number of true positive, false positive, and false negative samples, respectively.

4.3   One-to-One Encoding of Window Contents

At first, we considered the same encoding of the contents of a scanning window into a string as in [7]. All symbols within the window are rewritten into a string row-by-row. That way, on each step of the scanning, the contents of the scanning window produced a string of length 9. Repeatedly, the window moved by one position to the right, and the current contents of the scanning window was rewritten into a string of length 9. In contrast to [7], we did not separate the consecutive contents of the scanning window by any separator. Further, we will call this encoding one-to-one.

Unsurprisingly, such encoding of the contents of the scanning window produces long strings. Training `traxbar` on a set of strings obtained from relatively small samples of pictures was prohibitively slow (see Fig. 3) with a terrible accuracy (F_1-score close to zero). From the plots, we can see that the combination of one-to-one encoding and `traxbar` is unusable.

Conversely, combining one-to-one encoding with learning k-locally testable languages is feasible (see Fig. 4). When using k-locally testable languages, we must also choose a proper order k for locally testable languages. In the figure, there are plotted results of experiments with $k = 2, 4, 6, \ldots, 20$ with the language

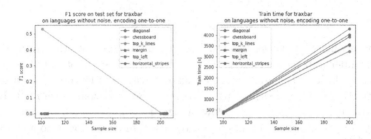

Fig. 3. Results of training finite automata using `traxbar` on 100 and 200 samples in train/test sets with the one-to-one encoding of the contents of the scanning window.

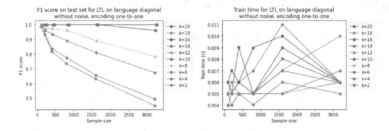

Fig. 4. Results of training finite automata using k-locally testable languages on sample sets with 100, 200, 400, 800, 1600, and 3200 samples in train/test sets with the one-to-one encoding of the contents of the scanning window. F_1-score on test sets on the left and time for training on train sets on the right. The markers for different values of k are shifted a little to show overlapping marks.

L_1 of diagonals. The best F_1-score was achieved for $k = 18$ and $k = 20$. The results for the other sample languages were also promising, and simultaneously, they were obtained in a very short training time.

4.4   Many-to-Int Encoding of Window Contents

As the train and test sets are generated randomly, the resulting F_1-score and training time is not constant. Therefore, in the following, we will plot the average F_1-score and train time from 10 randomly generated train and test sets for each size. To illustrate variance in the achieved results, we use error bars of length equal to the sample standard deviation of the measurements.

Using the many-to-int encoding, `traxbar` produced a reasonable F_1-score for all tested crisp languages and also for noisy languages (Figs. 5 and 6). Except for the noisy version of the language L_5, the training time for `traxbar` is not higher than 300 s up to the sample size of 3200. F_1-score is between 0.5 and 0.8, which is quite good for mostly sparse languages in our samples.

Next, we experimented with learning k-locally testable languages when using the many-to-int encoding. The obtained results are surprising. At first, we examined how the value of k influences the F_1-score. for the sample language L_1, we

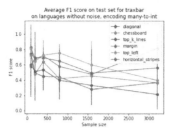
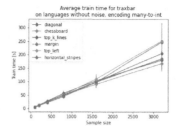

Fig. 5. Results of training finite automata for crisp sample languages using `traxbar` with the many-to-int encoding of the contents of the scanning window. Average F_1-score on test sets on the left and average time for training on train sets on the right. The length of the error bars is the sample standard deviation.

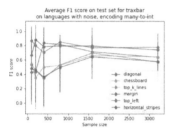
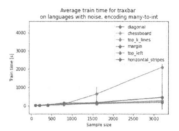

Fig. 6. Results of training finite automata for noisy sample languages using `traxbar` with the many-to-int encoding of the contents of the scanning window. Average F_1-score on test sets on the left and average time for training on train sets on the right. The length of the error bars is the sample standard deviation.

can see in Fig. 7 and Fig. 8 that the resulting average F_1-score is the highest for $k = 2$.

Similarly, for other sample languages, the value $k = 2$ together with the many-to-int encoding was the best combination. Interestingly, the value $k = 2$ with the many-to-int encoding corresponds exactly to $k = 18$ with the one-to-one encoding.

For the rest of our sample languages, the performance of learning crisp and noisy sample picture languages using 2-locally testable languages is plotted in Figs. 9 and 10. For all crisp sample languages, the F_1-score is close to 1, except for the sample language L_1 (of diagonals), for which it achieves 0.88 only. Probably it is caused by the very low number of positive samples. For all noisy sample languages, including the noisy version of L_1, the F_1-score converges to a value around 0.95 with the growing size of the training sample. The convergence is very stable, which can be seen from very short error bars.

Additionally, we can see a linear growth of the time required for training 2-locally testable languages with respect to the growing size of the training sample. Simultaneously, the training time is low. Lower than 0.03 s even for sample sets of size 3200.

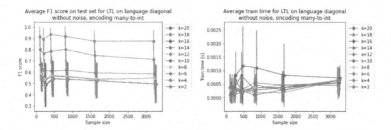

Fig. 7. Results of training finite automata using k-locally testable languages for the crisp sample language L_1 with the many-to-int encoding of the contents of the scanning window. Average F_1-score on test sets on the left and average time for training on train sets on the right. The markers for different values of k are shifted a little to show overlapping marks.

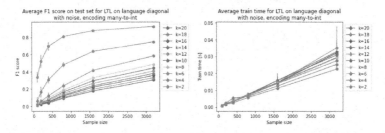

Fig. 8. Results of training finite automata using k-locally testable languages for the noisy sample language L_1 with the many-to-int encoding of the contents of the scanning window. Average F_1-score on test sets on the left and average time for training on train sets on the right.

4.5   Application on MNIST Dataset

Inspired by the high F_1-score obtained with many-to-int encoding, we decided to try the method on the known MNIST dataset[1] [9]. We prepared a training dataset consisting of 200 randomly selected pictures of handwritten digit one and 200 randomly selected pictures of other digits different from one. Similarly, we prepared a larger dataset consisting of 400 binarized pictures of digit one and 400 binarized pictures of other digits. Gray levels in the pixels of the original dataset were replaced by black and white pixels.

On these two datasets encoded with the many-to-int encoder, we applied traxbar and learning locally testable languages. While traxbar needed more than five hours to train on the second MNIST dataset, the training time for learning locally testable languages was under one second. In accordance with our experiments with the above sample languages, learning through locally testable languages achieved F_1-score more than 0.9 on the bigger MNIST dataset, while traxbar obtained F_1-score of only 0.45. The results are plotted in Fig. 11.

---

[1] The dataset is available from http://yann.lecun.com/exdb/mnist.

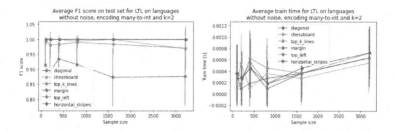

Fig. 9. Results of training finite automata using k-locally testable languages for the crisp sample language L_1 with the many-to-int encoding of the contents of the scanning window. Average F_1-score on test sets on the left and average time for training on train sets on the right. The markers for different values of k are shifted a little to show overlapping marks.

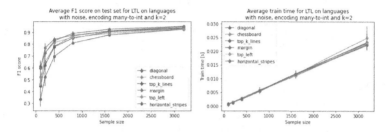

Fig. 10. Results of training finite automata using k-locally testable languages for the noisy sample language L_1 with the many-to-int encoding of the contents of the scanning window. Average F_1-score on test sets on the left and average time for training on train sets on the right.

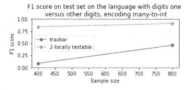

Fig. 11. Results of training finite automata using traxbar and 2-locally testable languages for the two subsets of binarized MNIST.

5   Conclusion

We have proposed and formalized the transcription-evaluation framework, including the Transducer-Evaluator and the four-way scanner automaton. These help us solidify the theoretical foundations for our experiments and formalize further options for picture scanning and processing.

In the experiments, the focus was set on several aspects. First was the comparison of different alphabet sizes in the scanning window transcription. We found out that with `traxbar` the many-to-int encoding clearly outperforms one-to-one encoding, while with LTL, the one-to-one encoding is perfectly sufficient.

The second aspect of interest was the behavior of the learning algorithms on our sample languages. The performance of `traxbar` with the many-to-int encoding on crisp languages varied significantly, but on languages with noise, it showed a decent performance with F_1-scores mainly in the 0.6 to 0.8 band.

As for the locally testable languages with the best performing value of $k = 2$, the learning algorithm succeeded for all of them, however, the diagonal language without noise turned out to be slightly more difficult to learn than others.

Lastly, we wanted to verify whether learning automata from picture languages has reached a stage where it can be used for practical problems – in our case learning numbers from the MNIST dataset. Clearly, we did not expect the learned automata to compete with deep neural networks, but the obtained results offer a promise for further research of this and other practical problems.

Acknowledgements. This research was supported by the Charles University Grant Agency (GAUK) project no. 1198519 and SVV-260588.

References

1. Costagliola, G., Deufemia, V., Ferrucci, F., Gravino, C.: On regular drawn symbolic picture languages. Inf. Comput. **187**(2), 209–245 (2003)
2. Freeman, H.: On the encoding of arbitrary geometric configurations. IRE Trans. Electron. Comput. **EC-10**(2), 260–268 (1961)
3. Giammarresi, D., Restivo, A.: Two-dimensional languages. In: Handbook of Formal Languages: Volume 3 Beyond Words, pp. 215–267. Springer, Berlin (1997). https://doi.org/10.1007/978-3-642-59126-6_4
4. De la Higuera, C.: Grammatical inference: learning automata and grammars. Cambridge University Press (2010)
5. Krizhevsky, A., Sutskever, I., Hinton, G.E.: Imagenet classification with deep convolutional neural networks. In: Advances in Neural Information Processing Systems, pp. 1097–1105 (2012)
6. Krtek, L.: Learning picture languages using restarting automata. Master thesis, Charles University, Faculty of Mathematics and Physics (2014)
7. Kubon, D., Mráz, F.: Learning picture languages represented as strings. In: Barták, R., Bell, E. (eds.) Proceedings of the Thirty-Third International Florida Artificial Intelligence Research Society Conference, pp. 529–532. AAAI Press (2020)
8. Lang, K.J.: Random DFA's can be approximately learned from sparse uniform examples. In: COLT 1992, pp. 45–52. ACM (1992)
9. LeCun, Y., et al.: Learning algorithms for classification: a comparison on handwritten digit recognition. In: Neural Networks: The Statistical Mechanics Perspective, pp. 261–276. World Scietific (1995)
10. Maurer, H.A., Rozenberg, G., Welzl, E.: Using string languages to describe picture languages. Inf. Control **54**(3), 155–185 (1982)
11. Pradella, M., Crespi Reghizzi, S.: A sat-based parser and completer for pictures specified by tiling. Pattern Recogn. **41**(2), 555–566 (2008)
12. Průša, D., Mráz, F., Otto, F.: Two-dimensional sgraffito automata. RAIRO-Theoretical Inf. Appl. **48**(5), 505–539 (2014)

Modelling Urban Traffic Configuration with the Influence of Human Factors

Ariadna C. Moreno Román[1], Mailyn Moreno Espino[1],
Cynthia Porras[1(✉)], and Juan Pavón[2]

[1] Facultad de Ingeniería Informática, Universidad Tecnológica de La Habana "José Antonio Echeverría", CUJAE, La Habana, Cuba
{amoreno,my,cporras}@ceis.cujae.edu.cu
[2] Instituto de Tecnología del Conocimiento, Universidad Complutense de Madrid, Madrid, Spain
jpavon@fdi.ucm.es

Abstract. Long vehicles queues at traffic signalized intersections are common elements on most urban roads. One of the causes of this problem is the configuration of installed traffic signals. The analysis of these configurations must consider human behavior, which is sometimes imprudent. Imprudence combined with poor signal configuration makes the waiting time on the road worse. Queuing theory is commonly used to represent traffic flow. This paper presents a queuing based model to evaluate traffic configurations with the inclusion of the parameters related to pedestrians and drivers simultaneously. An agent-based simulation is used to obtained results from the model with different human parameters. Comparisons show that when analyzing certain behaviors and characteristics of people, traffic performance, measured by waiting time, is affected.

Keywords: Agent based simulation · Human behaviors · Queuing based traffic model · Urban traffic light

1 Introduction

Achieving the proper functioning of urban traffic is a complex task for governments. An important indicator to consider is the occurrence of traffic jams [17]. These are long lines of vehicles waiting for the activation of a signal or for other vehicles to continue their route. This unfortunate situation has negative consequences such as environmental pollution due to gases, noise or excessive fuel consumption [1]. Additionally, drivers' activities are affected, delaying them in reaching their work or meetings, for instance.

One of the fundamental elements in traffic control is signal configuration. A correct configuration contributes to the reduction of congestion. Signal configuration is known as the selection of the times that make up the traffic lights cycle and the location of vertical road signs when a traffic light is not considered necessary [16]. Proper synchronization between placed vertical road signs and traffic light cycles contributes to uninterrupted vehicular flow [1].

A. C. Bicharra Garcia et al. (Eds.): IBERAMIA 2022, LNAI 13788, pp. 53–64, 2022.
https://doi.org/10.1007/978-3-031-22419-5_5

Poor signal configurations, far from optimizing traffic, obstruct it. For this reason, it is necessary to evaluate their performance before their implementation [15]. Other parameters that can determine the impact of a configuration are the decisions or actions taken by drivers or pedestrians. These parameters must be consider in order to analyze how influence the impact of the evaluated signal configuration.

Mathematical models are often used to measure the impact that a given traffic configuration will have prior to its installation [3]. There are several types of mathematical models, including linear complementary, system component calculations, and queuing theory, among others [20]. Table 1 summarizes a review of recent works that propose models to represent traffic and identifies those that consider the presence of human factor.

Table 1. Analyzed elements from literature reviewed

Reference	Type of model	Human Factor
[11]	Queuing Theory	–
[9]	Queuing Theory	–
[6]	Queuing Theory	–
[7]	Queuing Theory	–
[25]	Mixed Integer Linear Programming	Waiting time for pedestrian crossing
[5]	Queuing Theory	–
[22]	Queuing Theory	–
[23]	Macroscopic Dynamic	Waiting time for pedestrian crossing
[13]	Queuing Theory	Waiting time for pedestrian crossing
[14]	Queuing Theory	–
[12]	Queuing Theory	–
[24]	Green Wave Optimization	Waiting time for pedestrian crossing
[21]	Nagel-Schreckenberg (NaSch)	Pedestrian traffic lights
[18]	Markov Stochastic	Driver choices by environment
[19]	Dynamic Hybrid Choice	Driver choices by stress

This review of the state of the art shows that most of the models apply queuing theory to represent the waiting time in traffic. The human factor is analyzed in several proposals, where the main issues are the waiting time experienced by vehicles at crosswalks, the placement of pedestrian traffic lights and the variation in drivers' behaviors according to the characteristics of the environment.

However, the review has not identified a proposal that includes behaviors for combining both, pedestrians and drivers. In addition, among the proposals using queuing theory, only one was found that consider the human factor in traffic modeling. This proposal aims improving traffic volume and vehicle average speed for less queuing delays and better security for pedestrians [13].

The aim of this paper is to present a model based on queuing theory to represent traffic including behaviors and characteristics of drivers and pedestrians.

The model will be used to evaluate the influence of different types of factors and parameters in order to improve traffic performance. It is an extensible model that could include new characteristics and human behaviors that are considered of interest for the analysis of signal configurations.

The paper is structured as follows: Sect. 2 presents the mathematical model and queuing theory measurements to evaluate the performance of the signal configurations. Section 3 shows the results from an agent based simulation of the model. Finally, Sect. 4 presents the conclusions of the work.

2   Model Formulation Using Queuing Theory

Queuing theory is used to represent waiting situations [4], such as waiting for traffic signals. In this paper the traffic process is represented as a queuing system, where the clients are vehicles of all types that travel on the road. Vehicles must wait principally for the activation of the green light of traffic lights or for other vertical road sign on their way. These signals are considered service stations.

The queuing systems are represented by the notation $(A|B|C : D|E|F)$ where [6]: A is the distribution of arrivals or inter-arrival time, B is the distribution of service time, C is number of service channel, D is service discipline, E is maximum number of customers and F is the source of the units that will receive service.

Vehicle arrivals is assumed to be exponential [8]. Exponential distribution is used when modeling the time interval between independent events occurring at a fixed mean rate of change [6], as are the arrivals of the vehicles independent of each other. Since the exponential distribution satisfies the properties of the Markovian distribution, it is represented as (M), although other distributions such as uniform, general or Erlang can be used.

The number of servers will be the lanes on the roadway. The traffic follows a FCFS (First Come First Serve) discipline. There is no maximum number of vehicles for the system and the sources of arrival will be the intersection streets.

Service provision occurs at two types of stations: traffic lights and stop signs. Stop sign also have exponential distribution because the service time is short and not very variable so it will be directly proportional to the arrival distribution. So the model for these stations will be $(M/M/s)$. Traffic lights follow a general distribution [8] and the model will be $(M/G/s)$. This model accepts any data distribution for service times. It has been shown to fit the behavior of traffic lights [6].

At both signals there is an established waiting time, which vehicles must respect, otherwise it is considered an infraction. Particularly the service of traffic light is divided into three phases during the green light time of a traffic light [11]: (1) from the movement beginning of the first car in the queue until the flow of the recommended speed is reached; (2) from the moment the recommended speed is reached until the last car in the queue has passed; (3) from the moment of passing the queue to the end of the green traffic light.

The proposed mathematical model receives parameters from map, weather conditions, vehicles, people and the queuing system. These parameters are presented in Table 2. The possible values shown are specific to the data sources of this research then, they may be different if other sources are used.

Table 2. Model input parameters.

Group	Name	Expression	Possible Values in this research
Queuing Model	Average arrival rate	λ	–
	Average service rate	μ	–
	Variance of service time	σ^2	–
	Utilization factor	ρ	–
Map	Street Type	T_s	Primary, Secondary, Highway
	Surface condition	U_s	Good, Regular, Bad, Repair
	Width	D_s	> 0
	Maximum Allowed Speed	V_s	> 0
Vehicles	Type	T_v	Car, Bus, Articulated
	Technical condition	E_v	Good, Max or Min problem
	Vehicle length	l_a	> 0
People	Age	G_p	> 0
	Sex	S_p	Male, Female
	Knowledge of the area	K_p	$[1, 5]$
	Years of experience	X_p	≥ 0
Weather	Temperature	$Temp$	–
	Humidity	Hum	$[0, 100]$
	Rain	ll	$[1, 5]$
Probabilities	Red Light Violation	R_{prob}	$[0, 1]$
	Stop Violation	S_{prob}	$[0, 1]$
	Obstacles	O_{prob}	$[0, 1]$
Traffic	Minimum traffic light cycle	Min_c	> 0
	Maximum traffic light cycle	Max_c	> 0
	Standard speed of people	V_{ep}	> 0
	Speeds by T_v and T_s	V_{ev}	V value for each T_v and T_s
	Distance between vehicles	l_o	> 0

The equations below are used to calculate the performance measures of the queues, using the parameters obtained from Table 2.

$$a = \frac{d}{dt} V_{ev} \tag{1}$$

$$\mu_0(t) = \begin{cases} \frac{a\tau}{2(l_o+l_a)}, \tau \in [0; \tau^*] \\[2mm] \frac{V^2/2a+V(\tau-\tau^*)}{\tau(l_o+l_a)}, \tau \in [\tau^*; \tau^{**}] \\[2mm] \frac{V^2/2a+V(\tau^{**}-\tau^*)}{(l_o+l_a)+\lambda(\tau-\tau^{**})}/\tau, \tau \in [\tau^{**}; \tau^{***}] \end{cases} \tag{2}$$

$$\rho = \frac{\lambda}{s\mu} \tag{3}$$

$$P_0 = 1 - \rho \tag{4}$$

$$L_q = \begin{cases} \frac{\lambda^2 \sigma_s^2 + (\frac{\lambda}{s\mu})^2}{2(1 - \frac{\lambda}{s\mu})} \to (M/G/s) \\[2ex] \frac{(\frac{\lambda}{\mu})^s \rho \times P_0}{s!(1-\rho)^2} \to (M/M/s) \end{cases} \tag{5}$$

$$L = \begin{cases} \lambda W \to (M/G/s) \\ L_q + (\frac{\lambda}{\mu}) \to (M/M/s) \end{cases} \tag{6}$$

$$W_q = \frac{L_q}{\lambda} \tag{7}$$

$$W = W_q + \frac{1}{\mu} \tag{8}$$

Equation 1 corresponds to the instantaneous acceleration of vehicles at the start of the green light [11]. Equation 2 is used to calculate the average service rate of traffic light stations using the model presented in [11]. In this equation τ is the current time of the model and $\tau^*,\tau^{**},\tau^{***}$ the duration times of each subphase. Equation 3 calculates the factor of utilization of the queue [6] and Eq. 4 is the probability that exactly 0 clients will be in the queue. Equation 5 and 6 corresponds with number of vehicles in the every traffic signal (queues) and in the system respectively [6,10]. These measurements are calculated differently for each model used, so you will get one value for stop signs and another for traffic lights. Equation 7 and 8 get the waiting time in queue and system respectively [6, 10]. These measures are calculated in the same way for both models used.

Based on the behavior analyzed for queuing systems, simultaneous pedestrian and driver interactions are modeled according to the following equations:

$$ITH = 0.8 \times Temp + Hum \times (Temp - 14.4) + 46.4 \tag{9}$$

$$CF = ITH \times ll \tag{10}$$

$$Var_{CF-S_p} = 0.01 \times (CF - 70)^2 + (1.5 - S_p) \tag{11}$$

$$Var_{G_p-K_p} = \frac{35}{G_p} \times \frac{3}{K_p} \tag{12}$$

$$Var_{U_s-E_v-X_p} = \frac{3}{U_s + E_v + X_p} \tag{13}$$

$$V_p = [V_{ep} + Var_{CF-S_s}] \times Var_{G_p-K_p} \tag{14}$$

$$V_v = [\frac{V_{ev} + V_s}{2} + Var_{CF-S_p}] \times Var_{G_p-K_p} \times Var_{U_s-E_v-X_p} \tag{15}$$

Equation 9 obtains the humidity temperature index which is used to determine the level of comfort driver [2]. Equation 10 obtains a relationship between this index and rainfall. Equation 11–13 are the variations to be applied to the speed of people considering their characteristics and the external factors of the vehicle and the environment. Equation 11 depends on thermal comfort and gender, Eq. 12 on age and knowledge of the area, and Eq. 13 on the technical condition of the vehicles, roads and driving experience. Equation 14 and 15 calculate the speeds for drivers and pedestrians, based on the standard speeds and the variations applied to them.

Other behaviors to be analyzed are the violations committed by drivers, calculated by Eq. 16 and the waiting time of vehicles when a pedestrian stops in front of them, calculated by Eq. 17.

$$viol = R_{prob} \times S_{prob} \times \Sigma L_i \tag{16}$$

$$E_p = \frac{D_s}{V_p} \times O_{prob} \tag{17}$$

The variables that determine the performance of the model will be S and $Times$.

$$S = \begin{bmatrix} S_1 & S_2 & ... & S_n \end{bmatrix}$$

$$Times = \begin{bmatrix} GT_1 & RT_1 & A_1 \\ GT_2 & RT_2 & A_2 \\ ... & ... & ... \\ GT_k & RT_k & A_k \end{bmatrix}$$

The variable S represents the location of the signals where the values taken by the matrix positions are 0 if no signal, 1 if traffic light and 2 if stop sign. For each traffic light corner there will be a $Times$ matrix with the times of the lights (GT - green time, RT - red time) of the k phases of the traffic light.

$Times$ also stores the active phase in each traffic light. To determine the active phase, the Eq. 18 is applied. The active phase will have value 1 in the matrix, while the remaining phases will have value 0.

$$Active = k - \lfloor \frac{\Sigma GT_k}{\tau} \rfloor \tag{18}$$

Having the indices, variables and parameters of the model, the objective function of the proposed model is:

$$\mathbf{MIN} Z = \Sigma_{W_s} + \Sigma_{W_t} + E_p + viol \tag{19}$$

s.t.:

$$GT_n + RT_n + YT_n \leq Max_c \tag{20}$$

$$GT_n + RT_n + YT_n \geq Min_c \tag{21}$$

$$\Sigma_{i=1}^{k} A_i = 1 \tag{22}$$

The objective function of the model (Eq. 19) gets the total waiting time in the system, where W_t is the waiting time obtained for the traffic lights of the queuing system: $W \rightarrow (M/G/s)$ and W_s is the waiting time at Stop signs of the queuing system: $W \rightarrow (M/M/s)$. In addition, a penalization is applied for vehicles violations. Constraints 20 and 21 correspond to obtaining light cycles that are in the range set in the parameters. Constraint 22 ensures that each traffic light has only one active phase at each time τ.

3   Experimentation

To evaluate the presented model, an experiment with two test instances is performed. The objective is to show initially the influence of variable human factors on the model. To this end, we focus on the age of pedestrians and drivers as one of the factors that we consider most influential on traffic behavior. In addition, previous state-of-the-art scenarios where only pedestrians or drivers are included exclusively are modeled, and the results are compared to show the influence of both factors simultaneously. The model parameters are obtained from Open-StreetMap, OpenWeatherMap and set by the analyst[1]. Figure 1 shows the street and signal distribution for the two previously analyzed regions in the town of Plaza, Havana, Cuba.

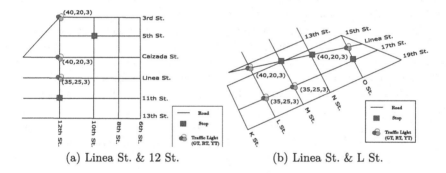

(a) Linea St. & 12 St.   (b) Linea St. & L St.

Fig. 1. Instances to evaluate the model

An agent-based simulation that reproduces the behavior of drivers and pedestrians on the road is used to obtain the values calculated by the model. For the execution of the simulations it is used a tool developed by the authors and documented at [16].

The environment where the agents interact is the map. Three main types of agents are identified: (1) Traffic Lights check if the light's time is finalized and change for the next one; (2) Vehicles decide whether to obey the signals or not. They can decide on the speed at which they travel according to their driver behavior. Finally, pedestrians decide the speed of travel, where to cross the streets, whether to commit infractions, and can walk in a group or alone.

---

[1] A complete parameters configuration can be downloaded from https://github.com/amoreno98/Mathematical-Model/blob/main/parameters.json.

Each instance was evaluated with two age configurations. Figure 2 shows the relationship of the number of people in each age range in each configuration.

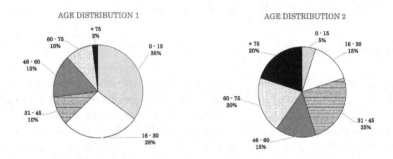

Fig. 2. Age range distributions

Through measurements and observation at the evaluated intersections, average arrival rates are $\lambda = 0.6$ for instance (a) and $\lambda = 0.77$ for instance (b)[2] The service time of the stop signal is $\mu = 0.8$ and for the traffic lights it is calculated with the model equations.

The main measures analyzed in the results are the average speed of pedestrians and vehicles and the waiting time of vehicles during a 6-hour simulation. The speed chosen by each agent to travel on the road is independent of the presence or behavior of other agents, so it is analyzed in general for all scenarios. Figure 3 shows a comparison between speed obtained by model for both instances and age distributions.

Pedestrian speed is faster with the first distribution. This is because 53% of the people on the road are in [0–30] ages. Young pedestrians have better physical condition and skills, so their movements are faster, while older people move slower due to illness, physical or motor problems. In the second distribution there are 40% of people over 60 years of age, which causes the average speed to decrease.

Vehicle speed for the first distribution is also faster in both instances, but at a lower rate. Younger drivers also drive at a higher speed, disobey speed limits more frequently and are more reckless, while older drivers are more careful when driving, sometimes even driving below the minimum speed limit.

Vehicle and pedestrian speed is one of the factors influencing the queue system performance measures. Figure 4 the performance measure analyzed is the waiting time obtained for each combination of instance and distribution in three different scenarios.

---

[2] Measurements and observations can be found in https://github.com/amoreno98/ Mathematical-Model/tree/main/Measurements.

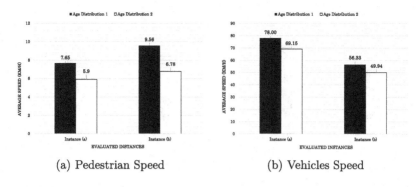

Fig. 3. Average speed of vehicles and pedestrians (km/h)

Analyzing the proposal showed in Fig. 4(a), it is found that in the first instance, the lowest waiting time is obtained with the first age distribution. This may be caused by the fact that pedestrians and vehicles travel faster. However, in the second instance the waiting times behave very similarly for both distributions. Although with the first instance vehicles and pedestrians will also travel faster, the intersection contained in this instance is high demand, and the variation in vehicle speed makes no difference in having long waits for signal activation.

Comparing with state-of-the-art scenarios it is evident that pedestrian behavior has lower influence than driver behavior, since in the simulations with only pedestrians, the waiting times for both age distributions are similar. When there are only drivers, the general trend seen in Fig. 4(a) is maintained.

It is important to note that the first age distribution remains generally with the lowest waiting time. Behavior according to age is maintained in all scenarios. For instance (a), when only pedestrians, the waiting time is the longest obtained because vehicles travel at approximately the same speed without violating speed limits due to age. In instance (b) waiting time increases in the combined evaluation, due to the slow pace of the elderly that makes the vehicles have to stop.

Generalizing results, for scenarios with young people (areas near schools, universities, amusement parks, etc.) the speeds are higher, while in areas with an aging population (near hospitals, grandparents' homes, etc.) the traffic is slower. In addition, it was shown that age is not the only factor that determines the behavior of the road, but also traffic demand and other parameters, so the model will be useful for analyses that integrate several factors and evaluate signal configurations that favor traffic performance.

The experiment shows how the model is able to capture the behaviors that are evident in real cases and with the support of the simulation visualize the traffic performance and generalize configurations for other instances with similar characteristics. In future work we can evaluate the influence of other factors.

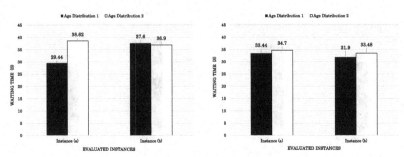

(a) Pedestrian and driver behaviors simultaneously

(b) Pedestrian-only behaviors

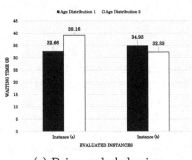

(c) Driver-only behaviors

Fig. 4. Waiting times obtained by the model (s)

4   Conclusions

There are numerous mathematical models built for traffic signal optimization that obtain efficient configurations in a short time. However, the literature review showed a low presence of an important factor such as the human behavior combining both, drivers and pedestrians.

In the proposed model these behaviors based on age are used to calculate their travel speed. The waiting time of vehicles is affected not only by the signals. Age proved to be an influential factor in the time spent in traffic. Other factors are also influential, such as the number of traffic violations, where the increase in waiting time is evident.

The analysis of a factor's influence on the model can be performed for gender, years of experience, comfort with the environment or knowledge of the traffic area. From each analysis, conclusions can be drawn and possible rules for signal configuration can be established. Speed variation equations can be extended by taking into account new traffic, vehicle and human parameters that are decisive.

The comparison with previous scenarios showed that driver behaviors are more influential than pedestrian behaviors. Possible reasons could be the difference in flow rates, a greater prudence of people when they are pedestrians, among other reasons that will be analyzed in future extensions of the work.

This model is extensible to new human behaviors such as physical disabilities, alcohol consumption or distractions and can be used to analyze different types of configurations and factors that can be considered to achieve better traffic performance. The proposal integrated to the multi-agent simulation constitutes a support tool for signal configuration. This integration can be used to evaluate real cases, where the inputs will be the traffic parameters and the region to be configured, in order to obtain an efficient configuration based on the data provided.

References

1. Afrin, T., Yodo, N.: A survey of road traffic congestion measures towards a sustainable and resilient transportation system. Sustainability **12**(11), 4660 (2020). https://doi.org/10.3390/su12114660
2. Barcia-Sardinas, S., Otero-Martín, M., Hernández-González, D., Gómez-Díaz, D., Gómez-Camacho, L.: Comparación de diferentes índices bioclimáticos en Cuba. Revista Cubana de Meteorología **26**(3) (2020). http://opn.to/a/uaURQ
3. Chiarello, F.A.: An overview of non-local traffic flow models. In: Puppo, G., Tosin, A. (eds.) Mathematical Descriptions of Traffic Flow: Micro, Macro and Kinetic Models. SEMA SIMAI Springer Series, vol. 12, pp. 79–91. Springer International Publishing (2021). https://doi.org/10.1007/978-3-030-66560-9_5
4. Ekeocha, R.J.O., Ihebom, V.I.: The use of queuing theory in the management of traffic intensity. Int. J. Sci. **4**(3), 56–63 (2018). https://doi.org/10.18483/ijSci.1583
5. Glushkov, A., Shepelev, V., Vorobyev, A., Mavrin, V., Marusin, A., Evtykov, S.: Analysis of the intersection throughput at changes in the traffic flow structure, vol. 57, pp. 192–199. Elsevier B.V. (2021). https://doi.org/10.1016/j.trpro.2021.09.042
6. Gunes, F., Bayrakli, S., Zaim, A.H.: Flow characteristics of traffic flow at signalized intersections and performance comparison using queueing theory. In: 4th International Symposium on Multidisciplinary Studies and Innovative Technologies (ISMSIT), pp. 1–9. IEEE (2020). https://doi.org/10.1109/ISMSIT50672.2020.9255044
7. Harahap, E., Darmawan, D., Fajar, Y., Ceha, R., Rachmiatie, A.: Modeling and simulation of queue waiting time at traffic light intersection. J. Phys. Conf. Ser. **1188**, 012001 (2019). https://doi.org/10.1088/1742-6596
8. Hillier, F., Lieberman, G.: Introduction to operations research. McGraw-Hill, 10th edn. (2015)
9. Jacyna, M., Zak, J., Golebiowski, P.: The use of the queueing theory for the analysis of transport processes. Logistics Transp. **41**(1) (2019). https://doi.org/10.26411/83-1734-2015-1-41-12-19
10. Joshi, P.K., Gupta, S., Rajeshwari, K.N.: Analysis and comparative study of various performance measures of M/G/1 and M/G/S queuing models. J. Comput. Math. Sci. **10**(1), 112–120 (2019). http://compmath-journal.org/dnload/Pradeep-K-Joshi-Shejal-Gupta-and-K-N-Rajeshwari-/CMJV10I01P0112.pdf
11. Kasatkina, E.V., Vavilova, D.D.: Mathematical modeling and optimization of traffic flows. J. Phys: Conf. Ser. **2134**(1), 012002 (2021). https://doi.org/10.1088/1742-6596
12. Keyvan-Ekbatani, M., Carlson, R.C., Knoop, V.L., Papageorgiou, M.: Optimizing distribution of metered traffic flow in perimeter control: Queue and delay balancing approaches. Control. Eng. Pract. **110**, 104762 (2021). https://doi.org/10.1016/j.conengprac.2021.104762

13. Li, W., Zhang, H., Huang, Z., Li, C.: Human-vehicle intersection traffic lights timing optimization research. J. Adv. Transp. 2022, 3 Mar 2022. https://doi.org/10.1155/2022/5549454

14. Ozerova, O., Lytvynenko, S., Sushchenko, R., Zapara, Y., Ovchar, P., Lavrushchenko, Y.: Factors influencing the modelling of transport flow dynamics in cities. Comptes Rendus de l'Academie Bulgare des Sciences **75**(2), 259–265 (2022). https://doi.org/10.7546/CRABS.2022.02.11

15. Qadri, S.S.S.M., Gökçe, M.A., Öner, E.: State-of-art review of traffic signal control methods: challenges and opportunities. Eur. Transp. Res. Rev. **12**(1), 1–23 (2020). https://doi.org/10.1186/s12544-020-00439-1

16. Román, A.M., Espino, M.M., Fraga, D.B.: Herramienta de simulación para evaluar configuraciones semafóricas. Revista Cubana de Transformación Digital **2**(1), 102–114 (2021), https://rctd.uic.cu/rctd/article/download/100/40

17. Sangaradasse, P., Eswari, S.: Importance of traffic and transportation plan in the context of land use planning for cities - a review. Int. J. Appl. Eng. Res. **14**(9), 2275–2281 (2019). https://www.ripublication.com/ijaer19/ijaerv14n9_33.pdf

18. Shin, J., Sunwoo, M.: Vehicle speed prediction using a Markov chain with speed constraints. IEEE Trans. Intell. Transp. Syst. **20**(9), 3201–3211 (2019). https://doi.org/10.1109/TITS.2018.2877785

19. Tarabay, R., Abou-Zeid, M.: A dynamic hybrid choice model to quantify stress in a simulated driving environment. IEEE Trans. Intell. Transp. Syst. (2021). https://doi.org/10.1109/TITS.2021.3056146

20. Tyagi, V., Darbha, S., Rajagopal, K.R.: A review of the mathematical models for traffic flow. Int. J. Adv. Eng. Sci. Appl. Math. **1**(1), 53–68 (2009). https://doi.org/10.1007/s12572-009-0005-8

21. Wang, Y., Zhang, C., Ji, P., Si, T., Zhang, Z.: Effect of pedestrian traffic light on traffic flow accompany with pedestrian crossing. Physica A **576**, 126059 (2021). https://doi.org/10.1016/j.physa.2021.126059

22. Yan, F., Li, P., Yan, G.W., Ren, M.F.: An iterative learning control strategy for traffic signals considering nonlinear characteristics of traffic flow. Zidonghua Xuebao/Acta Automatica Sinica **47**(9), 2238–2249 (2021). https://doi.org/10.16383/j.aas.c190183

23. Zeng, W., Chen, P., Yu, G., Wang, Y.: Specification and calibration of a microscopic model for pedestrian dynamic simulation at signalized intersections: a hybrid approach. Transp. Res. Part C-Emerg. Technol. **80**, 37–70 (2017). https://doi.org/10.1016/j.trc.2017.04.009

24. Zhang, J., Shang, H., Li, X., Yao, Y.: An integrated arterial coordinated control model considering green wave on branch roads and pedestrian crossing time at intersections. J. Manage. Sci. Eng. **5**(4), 303–317 (2020). https://doi.org/10.1016/j.jmse.2020.09.004

25. Zhang, Y., Gao, K., Zhang, Y., Su, R.: Traffic light scheduling for pedestrian-vehicle mixed-flow networks. IEEE Trans. Intell. Transp. Syst. **20**(4), 1468–1483 (2019). https://doi.org/10.1109/TITS.2018.2852646

Applying Anomaly Detection Models in Wastewater Management: A Case Study of Nitrates Concentration in the Effluent

Pedro Oliveira[1]([⊠]) , M. Salomé Duarte[2,3] , and Paulo Novais[1]

[1] ALGORITMI Centre, University of Minho, Braga, Portugal
pedro.jose.oliveira@algoritmi.uminho.pt, pjon@di.uminho.pt
[2] CEB - Centre of Biological Engineering, University of Minho, Braga, Portugal
salomeduarte@ceb.uminho.pt
[3] LABBELS - Associate Laboratory, Braga, Guimarães, Portugal

Abstract. With an increase in the diversity of data that companies in our society produce today, extracting insights from them manually has become an arduous task. One of the processes of extracting knowledge from the data is the application of anomaly detection models, which allows for finding unusual patterns in a given dataset. The application of these models in the context of Wastewater Treatment Plants (WWTPs) can improve water quality monitoring in these facilities, alerting decision-makers to act more quickly and effectively on anomalous events. Hence, this study aims to conceive and evaluate several candidate models based on Isolations Forest and Long Short-Term Memory-Autoencoders (LSTM-AE) to detect anomalies in the WWTP effluent, namely in the concentration of nitrates. Considering the obtained results, the best candidate was the LSTM-AE-based model, which had the best performance with an F1-Score of 97% and an AUC-ROC of 98%.

Keywords: Anomaly detection · Isolation forests · Long short-term memory-autoencoders · Nitrates · Wastewater treatment plants

1 Introduction

With the growth in the number of computing devices and their sensing capacity, the diversity of data available in different areas of our society is increasing [1]. Hence, an anomaly detection process, which aims to discover unusual deviations in a given dataset, has become arduous to perform manually [2,3]. By applying Machine Learning (ML) models to detect anomalies, it is possible to make this process automatic [4]. Within the scope of Wastewater Treatment Plants (WWTPs), anomaly detection models aim to help, in a better and more effective response, the decision-making process.

In the process of treatment in a WWTP, there are a variety of substances that can be taken into account in the context of anomaly detection. In this study, the

A. C. Bicharra Garcia et al. (Eds.): IBERAMIA 2022, LNAI 13788, pp. 65–76, 2022.
https://doi.org/10.1007/978-3-031-22419-5_6

focus was on the concentration of nitrates present in effluent. Nitrate contamination in water bodies is considered a serious environmental problem since nitrate can cause water quality deterioration and river eutrophication. Additionally, the presence of nitrite due to nitrate reduction is also a threat to humans' health [5]. Therefore, it is of extreme importance the detection of anomalies regarded to nitrate present in effluents, due to the impact that this compound can have on the ecosystems.

Therefore, this study aims to design, evaluate and tune different candidate models for each ML model to detect anomalies in a WWTP, specifically in the nitrate levels present in the effluent released by facilities. For this, we conceived models based on Isolation Forests and Long Short-Term Memory-Autoencoders (LSTM-AE). This manuscript is structured as follows: the next section presents the literature review carried out in the detection of water quality anomalies in WWTPs. The third section focuses on the description of the data exploration and preparation process carried out in this study and a brief explanation of the models and evaluation metrics used. The fourth section presents the experiments carried out throughout this study. The fifth section presents the obtained results and its respective discussion. The conclusions drawn from this study and future work are presented in the last section.

2   State of the Art

In comparison with other areas, in the context of WWTPs, the application of anomaly detection models is not yet fully explored. Although, some studies have already applied an anomaly detection process, especially concerning water quality [6–8].

A study by Mamandipoor et al. [6] focused on using a deep learning model, namely LSTMs, to detect anomalies regarding ammonia in a WWTP. To compare the performance of their model, the authors also used a statistical analysis method and a model based on the Principal Component Analysis-Support Vector Machine (PCA-SVM) for the anomaly detection process. The study was based on a WWTP in northern Italy, with data collection between January and December 2017, labelled as anomalous or non-anomalous by experts in the field. After that, the authors used a random search approach to select some of the best hyperparameters for the LSTM model. The data used were 70% and 30% for training and testing, respectively. To evaluate the performance of the models, the authors used different metrics, one of them being the F1-Score. Through the analysis of the results obtained, it was possible to verify that the LSTM model presented the best performance, compared to the others, with an F1-Score of 93%. The authors concluded that these results were obtained due to the high capacity that this model has to model temporal dependencies.

Another work carried out by Li et al. [7] aimed to detect contamination events, focusing on various water quality parameters in a water distribution system. The authors propose a stack-based learning model to detect anomalies in substances such as pH or turbidity. This model consisted of a first phase of predicting each parameter using a stacking model. In a second phase, anomalous and non-anomalous events are classified through the residuals between the predicted and measured data and the threshold obtained by the Sequential model-based optimization (SMBO) algorithm. The data used was based on the CANARY dataset, with data on water quality over a period of 4 months. These data were divided into 67% for training and 33% for tests, and cross-validation was used in the training set. The authors also used an Artificial Neural Network (ANN) to compare the designed model for anomaly detection. The F1-score value was higher for the stacked model than for the ANN model in all water quality parameters. For example, in the case of turbidity, the stacked model reached an F1-score of 75%.

Farhi et al. [8] carried out a study where they developed several models to detect anomalies in water quality in a WWTP, namely in ammonia concentration. The authors designed different ML models, such as LSTM-AE, LSTM or Gated Recurrent Units (GRUs). The data used in this study were based on a WWTP installed in Israel using the SCADA platform. The data were divided into 60% for training, 20% for testing and 20% for validation. Also, the data were normalized in an interval between 0 and 1. In the study, the authors performed only the number of epochs, the optimizer to be used, and the number of batch sizes considered in optimizing the hyperparameters. Before the anomaly classification process, the authors used the models designed to predict the values of substances for the next two days. After that, they defined a threshold, and if the predicted values were above the specified threshold, they were classified as anomalies. Concerning the evaluation metrics, the authors used the Accuracy and F1-Score. Through the results obtained, the LSTM-AE model obtained the best performance, in the case of ammonia concentration, with an F1-Score of 88%.

As mentioned, the use of anomaly detection models in WWTPs is still a topic that has not been much explored in the literature. From the analyzed studies, it is possible to verify that the process of searching for the best hyperparameters in the models used should be a more comprehensive topic. The studies that searched for the best hyperparameters were very contained in their scope, which may have achieved better results. The use of LSTM-AE in this context is also somewhat unexplored, and critical issues are not mentioned, such as the prevention of overfitting or the use of cross-validation appropriate to the cases of temporal sequences.

3   Material and Methods

Throughout this section, we will explain the steps developed in collecting, exploring and processing the data used in this study. The ML models used to detect anomalies are also presented as their evaluation metrics.

3.1   Data Collection

Concerning the data collection process, the data used in this study was provided by a multi-municipal Portuguese company responsible by the management of several WWTPs. The data provided based on one of its WWTPs, was collected between August 6th, 2018 and September 28th, 2019.

3.2   Data Exploration

The dataset used in the various experiments carried out in this study is based on the values of nitrates collected in a Portuguese WWTP. This dataset, containing 207 observations, presents a total of two features, namely the value of nitrates and the date on which this value was collected, which are described in Table 1. The data present in the dataset have a periodicity of every two days.

Table 1. Available features in the used datasets.

#	Features	Description	Unit
1	*date_time*	Timestamp	Date & time
2	*nitrates_value*	mg/L	nitrates concentration in the effluent

After verifying the number of features and observations in the dataset, the next step was to check for missing values in the entire dataset. When analyzing the data provided, it was possible to verify the existence of 3 missing timesteps, whose treatment is presented in the sub-section corresponding to the data preparation.

The next step focused on the statistical analysis of the column corresponding to the nitrate concentration. With this in mind, we analyzed different metrics such as the mean, the value of Kurtosis and Skewness. The mean value presented by nitrates throughout the dataset was 11.25. Regarding the Skewness, a value of 1.38 was obtained, thus affirming that the data related to nitrates present a positive asymmetric distribution. Finally, with a Kurtosis value of 1.08, it was possible to conclude that the nitrate data followed a leptokurtic distribution.

To understand the value of nitrates throughout the dataset, we conceived a graph with a set of boxplots by quarters. Through Fig. 1, it is possible to verify that the only quart that does not present outliers in the value of nitrates is the third quarter. However, in this quarter, the set of box plots illustrates the greatest dispersion of the data. In addition, the first and third quarters have a non-symmetrical distribution, unlike the second quarter, where this symmetry is evident. The highest outlier is presented in the first quarter, above 40 mg/L of nitrate, while in quantity, the second and fourth quarters have three outliers each.

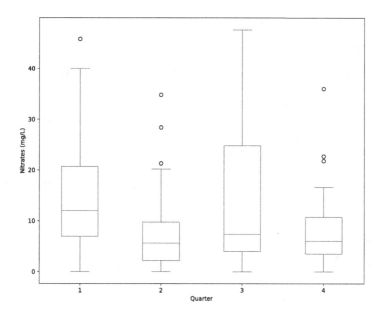

Fig. 1. Distribution of nitrate values per quarter.

The last step in data exploration was verifying if the data followed a Gaussian distribution. This is an essential process for deciding which data correlation analysis approach is used in the data processing phase. For this, the Kolmogorov-Smirnov test was used, with $p < 0.05$. Through the analysis of the obtained results, it was possible to conclude that the data did not follow a Gaussian distribution.

3.3 Data Preparation

The first step in data processing was the treatment of the missing timesteps identified earlier. In the exploratory analysis of the data, three missing timesteps were identified, and as in the case of the LSTM-AE model, the temporal sequence is essential. So it was necessary to insert these three observations into the data. The introduction of these missing timesteps resulted in missing values associated with these observations. To deal with these missing values, linear interpolation was used to fill them in.

Then, we applied a feature engineering process to create new resources to verify the possibility of correlation with the value of nitrates. From the *date_time* attribute, four new functionalities were created: the year, the month, the day of the month and the day of the week.

The next step developed was the correlation analysis. Considering the obtained results in the analysis of the Gaussian distribution of the data, as they did not follow a normal distribution, the nonparametric Spearman's rank correlation coefficient was chosen for correlation analysis. The correlation of the different features with the target was carried out, in this case, the value of nitrates. Through the results obtained, it was possible to verify that none of the features presented a strong correlation with the value of nitrates. Consequently, we removed all attributes, except the target, with the final dataset having 210 observations, sorted ascending by the index (*date_time*). Then, to label the data as anomalous and non-anomalous, to evaluate the performance of the developed models, the data from the final dataset were labelled by specialists. It should be noted that the feature with data labelling only served to evaluate the performance of the models, thus not being used in their training.

Finally, taking into account the use of a model based on LSTMs, the data that feed the LSTM-AE model were normalized considering the *MinMaxScaler*, between −1 and 1.

3.4 Evaluation Metrics

Two evaluation metrics were considered to assess the performance of the different candidate models conceived. Considering that we face a classification problem, the metrics chosen to evaluate the performance were the F-Score and the Area Under the Curve-Receiver Operating Characteristics (AUC-ROC).

The first metric, the F-Score, is a measure used to evaluate binary classification systems. The F-Score is a metric that combines two metrics, in this case, Precision and Recall, and is, therefore, a weighted average of Precision and Recall. When the F-Score value is 1, there is a perfect Precision and Recall. On the other hand, when it has a value of 0, it indicates that the other two metrics are 0 [9]. In this study, the F1-Score was determined according to the following equation:

$$\text{F1-Score} = 2 \times \frac{Precision \times Recall}{Precision + Recall} \tag{1}$$

The second metric, the AUC-ROC, is used to evaluate the performance of classification models. This metric helps to determine the ability of a given model to distinguish classes, where ROC is the probability curve, and AUC represents the degree of separability. A model with an excellent separability measure has an AUC-ROC close to 1, whereas a value close to 0 means worse separability. If the value assigned to the AUC-ROC is 0.5, the model could not separate the classes present in the data [10].

3.5 Isolation Forests

Isolation Forests are a type of anomaly detection model based on decision trees, showing similarities with Random Forests. This model is based on the fact that the anomalies in a given dataset are generally found in small numbers, which are different from most of the data. Unlike other anomaly detection models, Isolation Forests isolate anomalies rather than profiling what non-anomalous instances are [11].

The isolation process carried out in an Isolation Forest model is based on creating several Isolation Trees (iTrees) for a given data set. The instances defined as anomalies are those that are not found in the depths of the iTrees because it was easier for the tree to separate it from the other instances. On the other hand, the instances with greater depth in iTrees are those where there were more cuts to isolate them, thus becoming less likely to be classified as anomalies [11]. Figure 2 presents an example of the architecture of the Isolation Forest model.

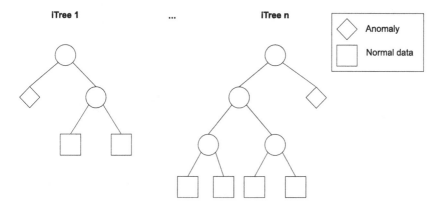

Fig. 2. Example of an isolation forest architecture.

3.6 LSTM-Autoencoder

A LSTM-AE is a network that implements an autoencoder for a given sequence of data using LSTMs. An autoencoder aims to reduce the dimensions of the data without changing the main information present in the structure of the data. This type of network is characterized by an input and output layer, encoding and decoding neural network, and a latent space. Through the encoding network, the objective is to compare the data in the latent space. On the other hand, the decoding network focuses on decompressing the encoded representation at the output layer [12].

Using a LSTM-AE, both the encoder and decoder are part of an LSTM network. Taking into account the ability of LSTMs to learn the temporal sequences existing in the data, the combination of an autoencoder with an LSTM makes it possible to perceive patterns of sequential data and recreate the input sequence. The performance of these models is evaluated on the ability of the trained model to recreate the input sequence. In the case of using these models for an anomaly detection process, the critical point is to determine the threshold of the reconstruction error. This threshold can be defined as the maximum mean absolute error loss value. For a given point in the test data, if its reconstruction error is greater than the defined threshold, it is labelled an anomaly [12]. Figure 3 presents an example of a LSTM-AE. In our study, at the end of the LSTM encoder there is a Dropout layer, as well as at the end of the LSTM decoder.

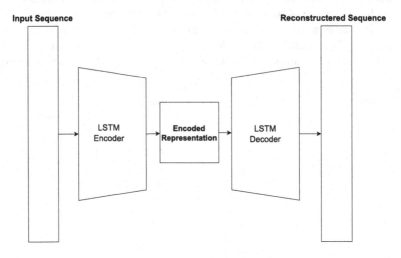

Fig. 3. Example of a LSTM-AE architecture.

4   Experiments

Several experiments were developed following a uni-variate approach to obtain the best candidate model for detecting anomalies in the nitrates present in the effluent of a WWTP. All experiments carried out were evaluated by considering the previously mentioned metrics to discover the best combination of hyperparameters to select the best one. For this search, the grid search technique was used.

Table 2 describes the hyperparameters, and the different values tested, with various combinations thereof, in the experiments carried out.

In all the experiments, the models were trained without the labels that identify if an observation is an anomaly. These labels were only used in the test data to evaluate the performances of the different candidate models. Data were divided into 70% for training and 30% for testing.

In the case of candidate models based on LSTM-AE, the learning curves were analyzed so that the models did not suffer from overfitting or underfitting. For the models not to go through an overfitting process, the epoch value used was 100. In addition, specific cross-validation for time series was used, namely the *TimeSeriesSplit*, with a value of k equal to 3. The threshold used in the LSTM-AE was defined based on the maximum value of the Mean Absolute Error obtained in the training phase.

Regarding the technologies used in the development of this study, Python 3.9 was the programming language selected for the exploration and processing of

Table 2. Isolation Forests vs LSTM-AE hyperparameters' searching space.

Parameter	Isolation forest	LSTM-AE
Neurons	–	[32,64,128]
Batch size	–	[10,20]
Drop-Out	–	[0.0, 0.2, 0.5]
LSTM layers	–	[1,2,3]
Activation	–	[ReLU, tanh]
№Estimators	[100,150,200]	–
Contamination	[0.02, 0.04, 0.06, 0.08, 0.1]	–
Bootstrap	[True, False]	–
Max Samples	[80, 100, 120]	–

data and the conception of the different candidate models. Different libraries were used, such as pandas, scikit-learn and TensorFlow v2.0. The experiments carried out were developed on the hardware made available to Google's Collaboratory.

5   Results and Discussion

With all the experiments carried out, the next step was to analyze their results. In Table 3 and Table 4, it is possible to observe the top-3 of the best candidate for the models based on Isolation Forest and LSTM-AE, respectively. In these tables, it is possible to verify the value of each hyperparameter for each candidate model and the respective value of the two evaluation metrics taken into account. In addition, the training time of each candidate model is also illustrated.

Table 3. Isolation Forest top-3 candidate models. Legend: a. n_estimators; b. contamination; c. max_samples; d. bootstrap; e. F1-Score; f. AUC-ROC; g. time (in seconds).

a.	b.	c.	d.	e.	f.	g.
100	0.02	80	True	**0.91**	**0.92**	0.243
100	0.08	80	True	0.87	0.88	0.248
100	0.08	80	False	0.83	0.87	0.196

Analyzing the results obtained, expressed in the previous tables, it is possible to verify that the best candidate model is a model based on LSTM-AE with an F1-Score of 0.97 and an AUC-ROC of 0.98. Compared with the other two best candidate models, it is possible to verify that this one needs a smaller number of layers and fewer neurons per layer. On the other hand, the best model based on LSTM-AE required a higher drop-out value than the others. Regarding the

Table 4. LSTM-AE top-3 candidate models. Legend: a. layers; b. neurons; c. activation function; d. dropout-rate; e. batch size; f. F1-Score; g. AUC-ROC; h. time (in seconds).

a.	b.	c.	d.	e.	f.	g.	h.
2	64	ReLu	0.5	10	**0.97**	**0.98**	12.787
3	128	ReLu	0.2	20	0.95	0.97	23.469
3	128	tanh	0.2	20	0.94	0.96	24.436

activation function, in comparing the three models, there is a higher prevalence of the ReLu function. Also, the best model needed a lower value than the other two models in terms of batch size.

Regarding models based on Isolation Forests, the best candidate model obtained an F1-Score of 0.91 and an AUC-ROC of 0.92. It is, therefore, possible to verify a certain homogeneity in the value of the hyperparameters present in the three best candidate models, mainly in terms of *n_estimators* and *max_samples*. Considering the best model, it is possible to verify that it needs a lower dataset contamination value, in this case, 0.02, than the other two that need a value of 0.08. In terms of the bootstrap hyperparameter, there was a prevalence of the value True, as far as the three best candidate models are concerned.

When comparing the two types of models, as expected, in terms of training time, LSTM-based models have a higher value, as this type of model has a higher computational cost. Considering the evaluation metrics (F1-Score and the AUC-ROC), it is possible to verify that the candidate models based on LSTM-AE are always superior when compared to the Isolation Forests-based models. Focusing on the best candidate models of both models, there is a 6% improvement over the F1-Score and the AUC-ROC. It is also important to note a more pronounced decrease of both metrics in Isolation Forest-based models.

Considering the best candidate model, LSTM-AE-based, Fig. 4 illustrates the anomalies detected in nitrate concentration by this model. It is possible to verify that the model detected as anomalies the values above 20 mg/L. At the bottom of the graph, it is clear that the anomalies detected are at values close to 0 mg/L.

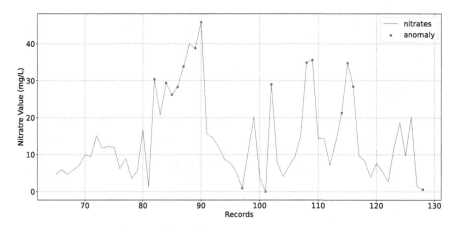

Fig. 4. Anomalies detected by the best candidate model.

6   Conclusions

Automating an anomaly detection process aims to help and alert decision-makers more quickly. In the case of WWTPs, the alert of a possible anomaly in water quality has the consequence that it is possible to act in advance on it, which can lead decision-makers to prevent events throughout the decision process., Therefore, through this study, the objective was to design ML models to detect anomalies in terms of nitrates in the effluent of a WWTP, namely through Isolation Forests and LSTM-AE.

To understand which candidate model had the best performance, considering the focus of the study, we developed several experiments in this sense. The results verified that the best candidate model was LSTM-AE-based, with an F1-Score of 0.97 and an AUC-ROC of 0.98. As expected, the LSTM-AE-based models had a higher training time than the Isolation Forests-based ones. Another conclusion is that in the three best candidate LSTM-AE-based models, there is no marked decrease in their performance, taking into account the two evaluation metrics. On the contrary, this decrease is more evidently verified in the Isolation Forest-based models.

Regarding the following steps to be taken, as future work, the objective is to design more anomaly detection models, such as the One-Class Support Vector Machines, in addition to the development of hybrid models, such as the conjunction of the LSTM-AE with the Isolation Forest. In the case of LSTM-AE, in which the threshold is static in this study, the objective is to apply a threshold moving technique to verify the model's performance with this new approach. In addition, the next step is to use the anomaly detection models for other aspects of WWTPs, such as energy consumption or volumetric flows.

Acknowledgments. This work is financed by National Funds through the Portuguese funding agency, FCT - Fundação para a Ciência e a Tecnologia within project DSAIPA/AI/0099/2019.

References

1. Pang, G., Shen, C., Cao, L., Hengel, A.V.D.: Deep learning for anomaly detection: a review. ACM Comput. Surv. (CSUR) **54**(2), 1–38 (2021). https://doi.org/10.1145/3439950
2. Zenati, H., Romain, M., Foo, C. S., Lecouat, B., Chandrasekhar, V. (2018, November). Adversarially learned anomaly detection. In 2018 IEEE International conference on data mining (ICDM) (pp. 727–736). IEEE. https://doi.org/10.1109/ICDM.2018.00088
3. Cook, A.A., Mısırlı, G., Fan, Z.: Anomaly detection for IoT time-series data: a survey. IEEE Internet Things J. **7**(7), 6481–6494 (2019). https://doi.org/10.1109/JIOT.2019.2958185
4. Nassif, A.B., Talib, M.A., Nasir, Q., Dakalbab, F.M.: Machine learning for anomaly detection: a systematic review. IEEE Access (2021). https://doi.org/10.1109/ACCESS.2021.3083060
5. Yun, Y., Li, Z., Chen, Y.H., Saino, M., Cheng, S., Zheng, L.: Elimination of nitrate in secondary effluent of wastewater treatment plants by Fe0 and Pd-Cu/diatomite. J. Water Reuse Desalination **8**(1), 29–37 (2018). https://doi.org/10.2166/wrd.2016.122
6. Mamandipoor, B., Majd, M., Sheikhalishahi, S., Modena, C., Osmani, V.: Monitoring and detecting faults in wastewater treatment plants using deep learning. Environ. Monit. Assess. **192**(2), 1–12 (2020). https://doi.org/10.1007/s10661-020-8064-1
7. Li, Z., Zhang, C., Liu, H., Zhang, C., Zhao, M., Gong, Q., Fu, G.: Developing stacking ensemble models for multivariate contamination detection in water distribution systems. Sci. Total Environ. **828**, 154284 (2022). https://doi.org/10.1016/j.scitotenv.2022.154284
8. Farhi, N., Kohen, E., Mamane, H., Shavitt, Y.: Prediction of wastewater treatment quality using LSTM neural network. Environ. Technol. Innov. **23**, 101632 (2021). https://doi.org/10.1016/j.eti.2021.101632
9. Tharwat, A.: Classification assessment methods. Appl. Comput. Inform. (2020). https://doi.org/10.1016/j.aci.2018.08.003
10. Muschelli, J.: ROC and AUC with a binary predictor: a potentially misleading metric. J. Classif. **37**(3), 696–708 (2019). https://doi.org/10.1007/s00357-019-09345-1
11. Al Farizi, W. S., Hidayah, I., Rizal, M.N.: isolation forest based anomaly detection: a systematic literature review. In: 2021 8th International Conference on Information Technology, Computer and Electrical Engineering (ICITACEE), pp. 118–122. IEEE, September 2021. https://doi.org/10.1109/ICITACEE53184.2021.9617498
12. Tran, P.H., Heuchenne, C., Thomassey, S.: An anomaly detection approach based on the combination of LSTM autoencoder and isolation forest for multivariate time series data. In Developments of Artificial Intelligence Technologies in Computation and Robotics: Proceedings of the 14th International FLINS Conference (FLINS 2020), pp. 589–596 (2020). https://doi.org/10.1142/9789811223334_0071

Optimal Architecture Discovery
for Physics-Informed Neural Networks

Taco de Wolff$^{(\boxtimes)}$ ⓘ, Hugo Carrillo ⓘ, Luis Martí ⓘ, and Nayat Sanchez-Pi ⓘ

Inria Chile Research Center, Av. Apoquindo 2827, Santiago, Chile
{taco.dewolff,hugo.carrillo,luis.marti,nayat.sanchez-pi}@inria.cl
https://inria.cl

Abstract. Physics-informed neural networks allow the neural network to be trained by both the training data and prior domain knowledge about the physical system that models the data. In particular, it has a loss function for the data and the physics, where the latter is the deviation from a partial differential equation describing the system. Conventionally, both loss functions are combined by a weighted sum, but this leaves the optimal weight unknown. Additionally, it is necessary to find the optimal architecture of the neural network. In our work, we propose a multi-objective optimization approach to find the optimal value for the loss function weighting, as well as the optimal activation function, number of layers, and number of neurons for each layer. We validate our results on the Burgers and wave equations and show that we are able to find accurate approximations of the solution using optimal hyperparameters.

Keywords: Physics-informed neural networks · Multi-objective optimization · Evolutionary algorithms

1 Introduction

Physics-informed neural networks (PINNs) [21] constitute a novel (deep) neural network learning method that consists of the joint use of data accompanied with a model coming from a physical law, typically in form of a partial differential equation (PDE), in order to train a neural network that will be a simulation of the physical phenomenon. Recently, this technique has gained a broad attention for being applicable in science and engineering. This general concept has been previously explored and is known as data assimilation [25], but PINNs bring a novel and sound approach to consolidate the existing models and sampled data. As a matter of fact, it can be said that they could represent an stepping stone towards machine learning explainability and, eventually, the connectionism-symbolism dichotomy.

In contrast, in classical numerical schemes, a discretized version of the PDE with an associated error with respect to the exact PDE is used. For this case, there are results establishing the existence and uniqueness of the exact solution

A. C. Bicharra Garcia et al. (Eds.): IBERAMIA 2022, LNAI 13788, pp. 77–88, 2022.
https://doi.org/10.1007/978-3-031-22419-5_7

for this discretized PDE [1]. Most classical numerical methods in PDEs are finite differences, finite elements and finite volumes, and they have been studied and have good properties, see for example [1]. However, these classical methods are not suitable for efficiently approximating PDEs with high-dimensional state or parameter spaces. An advantage of PINNs with respect to classical simulation methods is the fact that PINNs present a differentiable, mesh-free approach and capable of avoiding the curse of dimensionality [11]. Therefore, problems that are intractable with classical methods because of their computational requirements become viable with PINNs.

A natural approach for PINNs is to define a total loss which is a weighted sum of each task/objective loss function. In the seminal paper [21], the weights are the same for each loss term, but such weighting is not efficient for several models, so it is necessary to obtain a better choice of the weights.

The 'natural' approach to PINNs is to define an aggregated loss which is a weighted sum of each task/objective loss function $\ell = (1-\lambda)\ell_{\text{data}} + \lambda\ell_{\text{physics}}$ that aggregates the loss of the data (ℓ_{data}) and the loss of the physical law (PDE) component(s) (ℓ_{physics}). λ is commonly set as an *a priori* hyperparameter. This ℓ loss is then used to learn the neural network parameters (weights). In the seminal paper [21], the weights are the same for each loss term, but such weighting is not efficient for several models, so it is necessary to obtain a better choice of λ.

This poses an overlooked and -in our opinion- important issue: as the optimization process as both model and data can lead to conflicts between ℓ_{data} and ℓ_{physics} depending on the initialization of the neural network. This is not at issue in classical methods for PDEs as in that case the initial and boundary conditions do not lead to conflicts with the actual PDE, and there are results establishing the existence and uniqueness of an exact solution for the PDE exactly satisfying the boundary/initial conditions [10].

Even if such conflict does not occur, setting the correct value of λ is not an intuitive process. At first glance, it can be expected that λ expresses a preference between the two loss terms. For instance, in some problems where data is particularly noisy or not representative one would trust more the physical model. In other cases, where there is a big set of measured data, more attention would be paid to the data loss.

A reason for weighting the loss terms is that, in general, PDE and data losses have different physical units, and it is possible that they have different magnitudes, which can cause that in a stochastic initialization of the neural network, and hence the loss functions can considerably differ in their magnitudes. This fact can cause that PINNs suffer problems for the convergence and efficiency. As ℓ_{data} and ℓ_{physics} will likely have very different numerical characteristics (range, absolute values of the gradients, etc.), the role of λ overloaded as it is not only to expresses a preference but is also responsible of compensating the numerical characteristics of both losses. For example, a way in which PINNs fail to converge is the presence of gradient pathologies arising from imbalanced loss terms, as reported in [26].

Furthermore, there is a huge number of choices for free hyperparameters such as the configurations of the neural networks and types of activation functions and the neuron weight initialization. The choice of activation function is particularly critical because of its numerical impact in the computation of ℓ_{physics}. Some methods have been proposed in order to balance the contribution of the terms of the PINNs loss and their gradients, for example, Learning Rate Annealing [26], GradNorm [6], SoftAdapt [13], ReLoBRaLo [3], among others.

For this reason, we decided to propose a different approach to attempt to overcome this issue: instead of *a priori* guessing what are the values of λ we split this process in two parts:

- first find the set of trade-off values for ℓ_{data} and ℓ_{physics} as a Pareto front (and the values of λ that generate those trade-offs) as well as activation functions then
- allow practitioners (decision makers) to determine what point of the Pareto front best expresses their preferences between ℓ_{data} and ℓ_{physics}.

We tackle this by formulating the PINNs learning problem as a multi-objective optimization problem (MOP). MOPs consist on jointly optimizing multiple objectives, often competing between them under their constraints and thus would be capable of representing those trade-offs. This seems viable taking into account the current state of the art of evolutionary multi-objective optimization (EMO) [7], multi-objective AutoML [19,20,23], and MOP applications in multi-task learning problems [5,12,15–18,28].

In a related work [22], the Pareto optimal set is estimated by varying the loss weight for a given architecture and hyperparameters. However, a manual tuning of these scaling factors requires arduous grid search and becomes intractable as the number of terms grows due to the sensitivity and interdependence of these hyperparameters.

In this paper we report our preliminary results on multi-objective physics-informed neural networks (MOPINNs). MOPINNs uses an EMO algorithm to find the set of trade-offs of ℓ_{data} and ℓ_{physics} and the corresponding values of λ. Because of time and computational resources constraints we have limited our search space to individuals that represent a given value of λ and the activation function used by the neural network.

To the best of our knowledge, this is the first work relating MOPs and PINNs via evolutionary algorithms, so our contribution is to provide an exploratory analysis of this technique, in order to determine its feasibility. With that purpose, we apply MOPINNs on PDEs of particular interest: the heat equation and the following hyperbolic PDEs: waves, and Burgers equations. The motivation for the selection of these problems is derived by our interest on applying the results here described in ocean modelling problems where it is necessary to deal with the approximation of ocean waves.

This paper is organized as follows. In Sect. 2, we provide the necessary concepts used in our approach for solving PINNs. After that, in Sect. 3, we give a rationale and state-of-the-art in the problem of balancing the loss terms in PINNs, and we present our method. Then, in Sect. 4, we show the setting of our

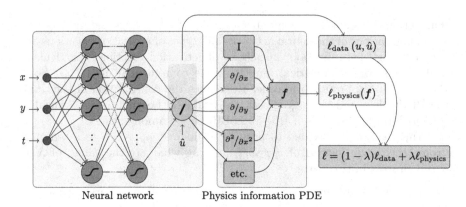

Fig. 1. Schematic representation of a physics-informed neural network with inputs x, y, and t; output $\hat{u}$. In this work we applied metaheuristics for determining the optimal hyperparameters for each case. Using automatic gradient calculation we can differentiate the neural network by its input variables and construct a physics error function f. The loss function involves a loss term for the data and a loss term for the physics function.

numerical experiments and present the results of them. Finally, in Sect. 5, we discuss the results and show our conclusions.

2   Physics-Informed Neural Networks (PINNs)

The above-mentioned approaches have the limitation of high computational costs for creating such data sets. Physics-informed neural networks (PINNs) were recently proposed [21] with the goal of overcoming those high computational costs. They are defined as a hybrid approach that considers a process where a source of model knowledge in the form of PDEs is available. They specify the problem of training the neural networks as a multi-objective learning task, where we want to minimize the error with the data as well as the error with a physical law.

Now we have three problems instead of one: minimizing the error with the data, minimizing the error with the physical law, and the combination of training for both objectives. In [21] this is solved by incorporating a physics loss term to the loss function, including a relative weight between both loss terms.

PINNs are neural network models that are trained to obey laws of physics described by PDEs. They are used to solve supervised tasks in which we both minimize the error with respect to the data and to the physics law. Their schematic representation is summarized in Fig. 1.

Let F be a differential operator on $\Omega \times (0, T)$. Let us consider a PDE for the unknown function $u = u(x, t)$, such that the physical law can be written as $F(u)(x, t) = 0$ for all $(x, t) \in \Omega \times (0, T)$. Let us assume that we have measurements of $u(x, t)$ for a discrete set of points (x, t), usually on the boundary and in the initial condition $\in \partial\Omega \times (0, T) \cup \Omega \times \{t = 0\}$, or even including some points in the interior of the spatio-temporal domain $\Omega \times (0, T)$.

Definition 1 (Weighted Physics-Informed Neural Network Formulation). *Given $\lambda \in (0,1)$ and a neural network $u_\theta(x,t)$, find $\theta = (W,b)$ the set of weights and biases that minimizes the loss function*

$$\ell(\theta, \lambda) = (1 - \lambda)\ell_{data}(\theta) + \lambda\ell_{physics}(\theta), \tag{1}$$

where ℓ_{data} is the loss with respect to the data, $\ell_{physics}$ is the loss with respect to the PDE expressing a physics law, and $\lambda \in [0,1]$ is the relative trade-off between the losses.

The losses are

$$\ell_{data}(\theta) = \frac{1}{N_u} \sum_{i=1}^{N_u} \left| B(u)(x_i^u, t_i^u) - B(u_\theta)(x_i^u, t_i^u) \right|^2, \tag{2}$$

$$\ell_{physics}(\theta) = \frac{1}{N_f} \sum_{j=1}^{N_f} \left| F(u_\theta)(x_j^f, t_j^f) \right|^2, \tag{3}$$

where $\{(x_i^u, t_i^u)\}_{i=1}^{N_u}$ are the points where the data is collected, and $\{(x_j^f, t_j^f)\}_{j=1}^{N_f}$ are the points where the PDE holds.

3   Evolutionary Multi-objective Optimization

Let $\mathcal{X}$ be a nonempty set and let $F : \mathcal{S} \to \mathbb{R}^M$ be a function $F = (f_1, \ldots, f_m)$, we define

$$x^* \in \arg\min_{x \in S} F(x), \tag{4}$$

as the optimal solution.

We state that x solves the multi-objective problem if $x \in \mathcal{S}^*$, with $\mathcal{S}^*$ the *Pareto-optimal set* defined by

$$\mathcal{S}^* = \{x \in \mathcal{S} | \; (\forall \hat{x} \in \mathcal{S}) \; (\exists i \in \{1, \ldots, m\}, \; f_i(\hat{x}) > f_i(x))$$
$$\vee (\forall j \in \{1, \ldots, m\}, \; f_j(\hat{x}) \geq f_j(x))\}$$

The set $\mathcal{O}^* = F(\mathcal{S}^*)$ is called the *Pareto-optimal front* [4].

As finding the explicit formulation of $\mathcal{S}^*$ is often impossible, generally, an algorithm solving (4) yields a discrete non-dominated set, $\mathcal{P}^*$, that approximates $\mathcal{S}^*$. The image of $\mathcal{P}^*$ in objective set, $\mathcal{PF}^*$, is known as the *non-dominated front*.

A broad range of heuristic and metaheuristic approaches has been used to address MOPs [4]. Among these, Evolutionary multi-objective optimization algorithms (EMOAs) [7] have been found to be a competent approach in a wide variety of application domains. Their main advantages are ease of use, inherent parallel search and lower susceptibility to the shape or continuity of the image of the efficient set, compared with traditional mathematical programming techniques for multi-objective optimization [4].

For this work we have selected three representative EMOAs: the non-dominated sorting genetic algorithm (NSGA-II) [9], the reference-point-based selection NSGA (NSGA-III) [8], and the multi-objective evolutionary algorithm by decomposition (MOEA/D) [27].

NSGA-II is one of the classical algorithm in the field. It is an improvement over the non-dominated sorting genetic algorithm (NSGA) [24]. NSGA-II incorporates two key operations: fast non-dominated sorting of the population and crowding distance computation with the aim of promoting diversity in the population. We have included in the analysis as it represents a solid well-understood baseline.

The crowding distance considers the size of the largest cuboid enclosing each individual without including any other member of the population. This feature is used to keep diversity in the population, where solutions belonging to the same rank and with a higher crowding distance are assigned a better fitness than those with a lower crowding distance, avoiding the use of the fitness sharing factor.

Similarly to NSGA-II, NSGA-III employs the Pareto non-dominated sorting to partition the population into a number of fronts. In the last front however, rather than using the crowding distance to determine the surviving individuals, a novel niche-preservation operator is applied. This niche-preservation operator relies on reference points organised in a hyper-plane in order to promote a diverse population. As a result, solutions associated with a smaller number of crowded reference points are more likely to be selected. Finally, we note that a sophisticated normalisation scheme is incorporated into the NSGA-III, which is aimed to effectively handle objective functions of different scales.

MOEA/D decomposes the objectives into subproblems with only one objective. This characteristic makes it particularly appealing for addressing the PINNs with a multi-objective approach as can be directly associated with the loss function weight λ. The population is split into subpopulations where each individual is associated with a subproblem. Each subproblem is linked to a certain number of neighboring subproblems, which contribute to the optimization process of the subproblem itself. The fitness function of each subproblem is the decomposition function, implemented to encompass all the objectives of the original problem. The Tchebycheff decomposition is widely used because is less parameterized, e. g., when compared to Penalty-based Boundary Intersection (PBI) [27], another well-known decomposition function [14].

4   Experiments

We consider the case of weighted PINNs where we look for the solution of a neural network u_θ, which approximates the PDE by estimating the solution of the multi-objective problem by minimizing

$$\min_{\theta,\lambda} \left((1 - \lambda)\ell_{\text{physics}}(\theta), \ \lambda\ell_{\text{data}}(\theta) \right). \tag{5}$$

Here, the minimum is understood as the Pareto efficient set. The individuals for the EMO algorithms are composed of $\lambda \in [0,1]$ from (1), the number of neurons per layer, as well as the activation function used by the network, including: LeakyReLU, ReLU, Tanh, Sigmoid, Softplus, Softsign, TanhShrink, CELU, GELU, ELU, SELU, and LogSigmoid. This allows us to investigate the two main dimensions of interest: the feasible intervals of λ and the ideal activation function and architecture per problem.

The MOEA/D multi-objective algorithm is run with a population size of 25 individuals and run for 20 generations. Sampling, crossover, and mutation were implemented for the mixed variable case of both optimizing a continuous variable λ and discrete variables for the activation function and the number of neurons per layer. Simulated binary crossover was used with $\eta = 3.0$, and polynomial mutation with $\eta = 3.0$. A set of 12 uniformly distributed reference directions were generated.

We select a sample subset $(X_i, Y_i) \; \forall i \in [0, N_u - 1]$ at random from the entire solution space for training, with $X_i \in \mathbb{R}^D$ the D features and $Y_i \in \mathbb{R}^P$ the labels of dimension P. Note that for the physics loss the solution is not needed, allowing applications to be able to train even when few data are available of the solution.

4.1 Burgers Equation

We consider the Burgers equation

$$\frac{\partial u}{\partial t} + u \frac{\partial u}{\partial x} = \nu \frac{\partial^2 u}{\partial x^2} \,, \tag{6}$$

where the unknown is $u = u(x,t)$, and ν is the diffusion coefficient. This equation usually appears in the context of fluid mechanics, and more specifically models one-dimensional internal waves in deep water. It represents a hyperbolic conservation law as $\nu \to 0$ and it is the simplest model for analyzing the effect of nonlinear advection and diffusion in a combined way.

We simulated the Burgers equation in the spatio-temporal domain $x \in [-1, 1]$ and $t \in [0, 1]$ using a Fourier spectral method [2] with 512 spatial points and 100 points in time in uniform grids. The diffusion coefficient is taken as $\nu = 0.01/\pi$ m^2/s. We consider the initial condition

$$u(x, 0) = -\sin(\pi x) \quad \forall \, x \in [-1, 1] \,, \tag{7}$$

and the boundary condition

$$u(1, t) = u(-1, t) = 0 \quad \forall \, t \in [0, 1] \,, \tag{8}$$

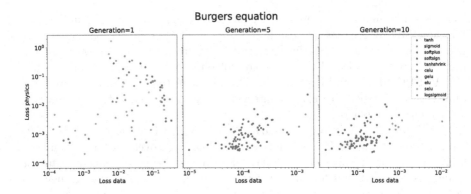

Fig. 2. The individuals for three generations for the Burgers equation, each trained with 25000 epochs.

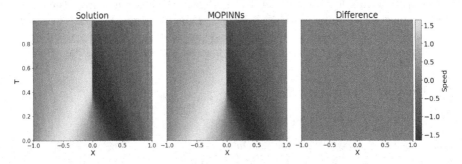

Fig. 3. Burgers equation solution with $50 \times 50 \times 40$ neurons, $\lambda = 0.15$, using the tanh activation function for 25000 epochs.

Therefore, the simulation produced $512 \times 100 = 51200$ spatio-temporal points. From them, for PINNs we randomly selected $N_u = 46080$ and $N_f = 46080$ points.

The evolution of the multi-objective optimization is shown in Fig. 2, where we observe how the first generation selects a wide range of activation functions, but quickly converges to the best performing activation function. The last five generations have an additional objective to optimize the number of neurons in the neural network. On the Pareto front only the tanh activation function and a configuration of $50 \times 50 \times 40$ survived.

The solution of MOPINNs compared to the solution simulated by Fourier spectral method (which is understood as a sufficiently accurate solution) is shown in Fig. 3. The largest deviation from the exact solution is 0.0162 m/s, validating that MOPINNs is able to accurately find the solution of the Burgers equation.

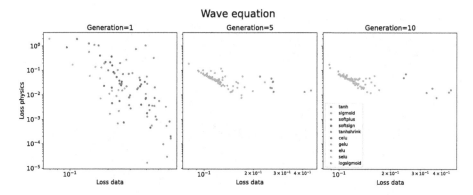

Fig. 4. The individuals for three generations for the wave equation, each trained with 2500 epochs.

4.2 Wave Equation

We consider the wave equation for $\Omega \in \mathbb{R}^2$, defined by

$$\frac{\partial^2 \eta}{\partial t^2} = \nabla \cdot (H \nabla \eta),$$

(9)

where the unknowns are η the displacement and H the depth.

We simulate the wave equation on a rectangular spatial domain $\Omega = (0,1)^2$ in the time interval $(0, T_f)$, with null Dirichlet boundary condition for η on $\partial \Omega \times (0, T_f)$ with initial conditions

$$\eta(x, y, 0) = \exp\left(-10 \cdot \left((x - 0.5)^2 + (y - 0.75)^2\right)\right),$$
$$\frac{\partial \eta}{\partial t}(x, y, 0) = 0.$$

(10)

The depth H is taken to be

$$H(x, y) = (1 - x)(2 - \sin(3\pi y)).$$

(11)

We implement FEM for this equation considering Lagrange finite elements of degree 1, the spatial domain is represented by an unstructured rectangular mesh with 218 nodes and 384 triangles. The time scheme is explicit and $T_f = 1.0$, $n = 100$. Therefore, the simulation produced $218 \times 100 = 21800$ spatio-temporal points. From them, we randomly selected $N_u = 19620$ and $N_f = 19620$ points for PINNs.

In Fig. 4 we observe how the multi-objective optimization quickly converges to the SELU activation function. Again, from the 5th generation onwards we additionally optimize for the number of neurons.

The solutions of the wave equation are shown in Fig. 5, where we have optimized for both the displacement η and the height H fields. The first row, called

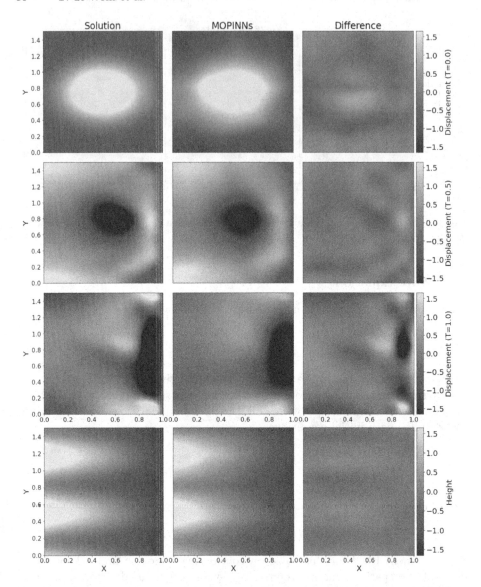

Fig. 5. Wave equation solution of η (top three rows) and H (bottom row) with $60 \times 50 \times 50$ neurons, $\lambda = 0.76$, using the SELU activation function for 25000 epochs.

"Solution" is the one simulated by FEM, since it is understood as a sufficiently accurate solution. The wave equation is a significantly harder problem than the Burgers equation, but we are still able to replicate the large features of the original solution. The first three rows show the solution for η at different values of $t \in \{0.0, 0.5, 1.0\}$. The last row shows the solution for H which is constant in time.

5   Conclusions

We have shown that using a multi-objective approach to training physics-informed neural networks allows to find the optimal architecture for each equation by optimizing the value for λ, number of hidden layers, neurons per layer, and the activation function.

The evolutionary nature of the optimization algorithms are effective at selecting and mixing successful individuals on the Parent front, and allow the survival of the optimal neural network architecture. We have shown that the optimal architecture is highly dependent on the problem, and that an automatic search using multi-objective optimization is required.

Our method is able to recover an accurate solution with respect to the exact solution, validating that the physics-informed neural network is able to learn the physical model behind the solution. It is important to mention that the amount of points for the data loss is large, so future work will need to explore the ability of PINNs for predicting the solution in a larger set of points, where it will be interesting to see the role of λ. It will also make MOPINNs be an alternative solver as FDM, FEM and FVM, and at the same time a corrector of models in the case data is taken for predicting the solution for a given PDE model that is not fully known, which would give a special ability to this technique and will place it as a useful technique for applications.

Acknowledgements. Funded by CORFO/ANID International Centers of Excellence Program 10CEII-9157 Inria Chile, Inria Challenge OcéanIA, STICAmSud EMISTRAL, CLIMATAmSud GreenAI and Inria associated team SusAIn.

References

1. Allaire, G.: Numerical Analysis and Optimization: An Introduction to Mathematical Modelling and Numerical Simulation. Oxford University Press, Oxford (2007)
2. Basdevant, C., et al.: Spectral and finite difference solutions of the burgers equation. Comput. Fluids **14**(1), 23–41 (1986)
3. Bischof, R., Kraus, M.: Multi-objective loss balancing for physics-informed deep learning (2021)
4. Branke, J., et al. (eds.): Multiobjective Optimization, Lecture Notes in Computer Science, vol. 5252. Springer, Heidelberg (2008). https://doi.org/10.1007/978-3-540-88908-3
5. Cai, T., et al.: The multi-task learning with an application of Pareto improvement. In: The 2nd International Conference on Computing and Data Science. ACM, January 2021. https://doi.org/10.1145/3448734.3450463
6. Chen, Z., Badrinarayanan, V., Lee, C., Rabinovich, A.: GradNorm: gradient normalization for adaptive loss balancing in deep multitask networks. In: Proceedings of Machine Learning Research, ICML, pp. 794–803 (2018)
7. Coello Coello, C., Lamont, G., van Veldhuizen, D.: Evolutionary Algorithms for Solving Multi-Objective Problems. Genetic and Evolutionary Computation, 2nd edn. Springer, New York (2007)

8. Deb, K., Jain, H.: An evolutionary many-objective optimization algorithm using reference-point based non-dominated sorting approach, part I: solving problems with box constraints. IEEE Trans. Evol. Comput. **18**, 577–601 (2014)

9. Deb, K., Pratap, A., Agarwal, S., Meyarivan, T.: A fast and elitist multiobjective genetic algorithm: NSGA-II. IEEE Trans. Evol. Comput. **6**, 182–197 (2002)

10. Evans, L.C.: Partial differential equations. Grad. Stud. Math. **19**(4), 7 (1998)

11. Grohs, P., Hornung, F., Jentzen, A., Von Wurstemberger, P.: A proof that artificial neural networks overcome the curse of dimensionality in the numerical approximation of Black-Scholes partial differential equations. arXiv preprint arXiv:1809.02362 (2018)

12. Gupta, S., Singh, G., Lease, M.: Scalable uni-directional pareto optimality for multi-task learning with constraints. arXiv preprint arXiv:2110.15442 (2021)

13. Heydari, A., Thompson, C., Mehmood, A.: SoftAdapt: techniques for adaptive loss weighting of neural networks with multi-part loss functions. arXiv preprint arXiv:1912.12355 (2019)

14. Li, K., Deb, K., Zhang, Q., Kwong, S.: An evolutionary many-objective optimization algorithm based on dominance and decomposition. IEEE Trans. Evol. Comput. **19**(5), 694–716 (2015)

15. Lin, X., Yang, Z., Zhang, Q., Kwong, S.: Controllable Pareto Multi-Task Learning, October 2020. https://openreview.net/pdf?id=5mhViEOQxaV

16. Lin, X., et al.: Pareto multi-task learning. arXiv:1912.12854, December 2019

17. Ma, P., Du, T., Matusik, W.: Efficient continuous Pareto exploration in multi-task learning. arXiv:2006.16434, June 2020

18. Mahapatra, D., Rajan, V.: Exact Pareto optimal search for multi-task learning: touring the Pareto front. arXiv:2108.00597, August 2021

19. Mondal, I., Sen, P., Ganguly, D.: Multi-objective few-shot learning for fair classification. arXiv:2110.01951, October 2021

20. Pfisterer, F., et al.: Multi-objective automatic machine learning with AutoxgboostMC. CoRR. arXiv:1908.10796 (2019)

21. Raissi, M., Perdikaris, P., Karniadakis, G.: Physics-informed neural networks: a deep learning framework for solving forward and inverse problems involving nonlinear partial differential equations. J. Comput. Phys. **378**, 686–707 (2019). https://doi.org/10.1016/j.jcp.2018.10.045, https://www.sciencedirect.com/science/article/pii/S0021999118307125

22. Rohrhofer, F., Posch, S., Geiger, B.: On the pareto front of physics-informed neural networks. arXiv preprint arXiv:2105.00862 (2021)

23. Ruchte, M., Grabocka, J.: Efficient multi-objective optimization for deep learning. arXiv:2103.13392, March 2021

24. Srinivas, N., Deb, K.: Muiltiobjective optimization using nondominated sorting in genetic algorithms. Evol. Comput. **2**(3), 221–248 (1994)

25. Vetra-Carvalho, S., et al.: State-of-the-art stochastic data assimilation methods for high-dimensional non-Gaussian problems. Tellus A: Dyn. Meteorol. Oceanogr. **70**(1), 1–43 (2018). https://doi.org/10.1080/16000870.2018.1445364

26. Wang, S., Teng, Y., Perdikaris, P.: Understanding and mitigating gradient flow pathologies in physics-informed neural networks. SIAM J. Sci. Comput. **43**(5), A3055–A3081 (2021)

27. Zhang, Q., Li, H.: MOEA/D: a multiobjective evolutionary algorithm based on decomposition. IEEE Trans. Evol. Comput. **11**, 712–731 (2007)

28. Zhou, D., et al.: Multi-task multi-view learning based on cooperative multi-objective optimization. IEEE Access: Pract. Innov. Open Solut. **6**, 19465–19477 (2018). https://doi.org/10.1109/access.2017.2777888

An AI–Based Approach for Failure Prediction in Transmission Lines Components

Alberto Reyes[1]([⊠])(iD), Ramiro Hernández[1], Alberto Hernández[1],
Leonardo Rejón[1], Karla Gutiérrez[1], Ricardo Montes[2], and Alejandro Valverde[2]

[1] Instituto Nacional de Electricidad y Energías Limpias, 62490 Cuernavaca, Mexico
albertoreyesballesteros@gmail.com
[2] Comisión Federal de Electricidad, 01780 Mexico City, Mexico

Abstract. In this paper, a novel AI method for failure prediction in transmission lines components is presented. The method combines machine learning and deep learning capabilities. The approach was tested using degradation simulated data of a composite insulator exposed to different levels of environmental pollution. The failure model was constructed using historical real data of a Mexican utility. Preliminary experimental results shows that the joint use of deterministic forecasting and probabilistic diagnosis methods help determine the future failure of a transmission line component for different time horizons with very acceptable precision rates.

Keywords: Transmission line components · Composite polymeric insulators · Failure prediction · Machine learning · Deep learning

1 Introduction

Transmission lines are the physical assets through which the electrical energy is transported, and they are made up of: conductors, towers, insulators, fittings accessories between insulators and towers, ground wires, etc. Due to their importance, it is necessary to have a permanent and reliable diagnosis that allows to adequately plan their maintenance actions. The components of a transmission line have an average service life of about 30 years, but with a proper maintenance they can last up to 50 years. By the general and preventive overhaul can be possible to achieve with the goal of making the components 100% available, in order to hold a high reliability

Currently, the Mexican public electric utility, Comisión Federal de Electricidad (CFE), evaluates its transmission lines using a reliability index. This indicator provides an idea of the state of a transmission line. Some of the parameters for calculating this index are obtained based on the experience of the inspection personnel. In order to strengthen the way to assess the condition of transmission lines (TL) and schedule the maintenance of their assets effectively in the Mexican transmission network, in the PE-A-11 project financed by the Mexican Center

A. C. Bicharra Garcia et al. (Eds.): IBERAMIA 2022, LNAI 13788, pp. 89–100, 2022.
https://doi.org/10.1007/978-3-031-22419-5_8

for innovation in Intelligent Electrical Networks and Microgrids (CEMIE-Redes), which is developed by the National Institute of Electricity and Clean Energies (INEEL), have redefined the way of evaluating the TL through the implementation of maintenance techniques based on the condition, and for which now a condition index will be used.

There are important and inspiring related work found in literature such as [1], in which a proposed framework of health index of the transmission line using a condition-based method is presented. An inspection concept for maintenance of overhead power transmission lines is also described in [7]. Irfan and Handika [3] further explain the method of calculating the health index taking into account possible parameters, components, and weighting criteria that have not been clearly discussed in previous research. In [4], a knowledge-based Fuzzy Inference System (FIS) is designed using Matlab's Fuzzy Logic Toolbox as part of the methodology and its application is demonstrated on utility visual inspection practice of porcelain cap and pin insulators.

With the idea of anticipating the possible failure of a component, a transmission line component degradation model based on deep learning techniques will be proposed. This model will be able to show the evolution of a component in the future based on its historical evaluations. The degradation model is based on recurrent neural networks (RNN) and its implementation in Python language. From historical data provided by the CFE, a probabilistic failure model is obtained. It will determine the probability of failure from on-site evaluations of the components of the TL that have historically presented failures over the years. The failure model is based on Bayesian networks (BN) and its implementation in the OpenMarkov tool. In this way, the assembly of Recurrent Neural Networks and Static Bayesian Networks will allow to predict failures in transmission lines components with a good degree of certainty.

2   Transmission Lines Components and Common Failures

Considering a transmission line as a system, it could be described as a set of subsystems interrelated in such a way that works as a whole. A subsystem is an indicator that mainly determine the operation availability or continuity of the electric power supply. Each subsystem is conformed by components whose functional condition contributes to the own indicator or subsystem availability. The subsystems might be: a component, a set of components, or a environmental aspect. The main subsystems in this document are: founding, structures, lightning, and contamination. Table 1 shows a sample of components of a transmission line classified by subsystems.

One of the main components of an overhead transmission line is the insulator, which can be made out of glass, ceramic and polymer matrix composite. Composite insulators have been increasingly accepted by utilities as suitable substitutes for porcelain and glass insulators since their introduction in the early 1970s s due to their hydrophobic property providing better performance in areas of high contamination. This is not the case of glass and porcelain insulators where the

Table 1. Sample of components of a transmission line classified by subsystems

Component	Description	Subsystem
AB	Shield angle	Lightning (DA)
ST	Grounding systems	Lightning (DA)
AP	Line lightning surge arrester	Lightning (DA)
EHG	Ground wire/splices for ground wire/ground wire damper	Lightning (DA)
CHCH	Fittings for protection against ground wire impact	Lightning
PCI	Impact protection	Foundation (CIM)
CIM	Steel or concret foundation	Foundation (CIM)
EST	Structure	Structure (EST)
LC	Vertically	Structure (EST)
FA	Insulators	Contamination (CON)
CHC	Fittings for conductor	Contamination (CON)
SEP	Conductor damper	Contamination (CON)
CECN	Splice for conductor	Contamination (CON)
CHH	Insulator pin	Contamination (CON)
COND	Conductor	Contamination (CON)
TOF	Type of fault	

combination of contamination with moisture increases the risk of failure due to the presence of dry bands and consequent flashover.

Unfortunately, the composite insulator could present degradation of its fiberglass rod and therefore it can suffer from mechanical failure that can cause the risk of the conductor falling. When some fibers are fragile, all the mechanical stress is supported by other fibers, over-stressing them to the point of rupture, as shown in Fig. 1. This failure process is known as Brittle Fracture and can cause a destructive failure.

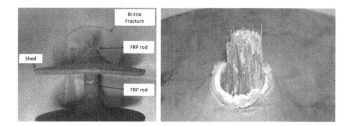

Fig. 1. Mechanical failure of an insulator due to the process of Brittle Fracture.

Based on this, it is very relevant to consider the composite insulator as an element whose main properties must be monitored in real time and taken into account as something important to determine the state and risk of a transmission line in order to avoid a failure.

The term potential failure is the process at which it is possible to detect that the failure is occurring or is about to occur. The condition of the fiberglass rod indicates its potential to fail i.e. break. And because conditions on an insulator rod can deteriorate with time, the relationship between the rod condition and the time at which the insulator is put in service can be represented by the P-F curve shown in Fig. 2. P is the moment at which the rod could fail (potential failure) and F is the point at which it really fails.

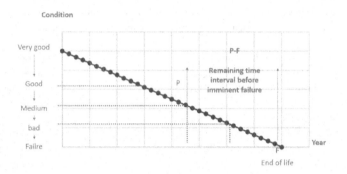

Fig. 2. Characteristic P-F curve of useful-life in a synthetic insulator.

3 Fundamentals

3.1 Bayesian Networks

A Bayesian network or belief network [5] is a probabilistic graphical model (acyclic and directed) represented by a set of variables and their conditional dependencies. Nodes represent random variables and edges conditional dependencies. Nodes might be observable amounts, unknown parameters, or hypothesis. Non–interconnected nodes represent variables that are conditionally independent of others. Each node has a probability function associated that takes a set of particular values as input from the parent nodes and that returns the variable probability represented by the node.

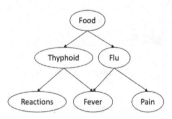

Fig. 3. A Bayesian network representing probabilistic relationships between failures and observations.

As an example, a Bayesian network might represent probabilistic relationships between diseases and symptoms. Given certain symptoms, the network might be used to compute the probability of one or more diseases. In the Bayesian network of Fig. 3, nodes represent diseases, symptoms and other causal factors. The variable pointed by the arrow is dependent conditionally of the variable that originates the arrow. For instance, fever is dependent of typhoid and flu.

3.2 Recurrent Networks

Recurrent networks [6] are dynamic systems also known as space-time networks. The calculation of an input in one step depends on the previous step, and in some cases on the future step. They are an attempt to establish a correspondence between input and output sequences that are nothing more than temporal patterns.

The tasks that can be performed with this type of networks are:

- Sequence recognition: A particular output pattern is produced when an input sequence is specified.
- Stream playback: The network must be able to generate the rest of a stream when it sees part of it.
- Temporal association: In this case, a particular output sequence must be produced in response to a specific input sequence.

One way to classify recurrent networks is depending on the number of their hidden layers and the way of back-propagation, which is a way of training a neural network. They are of three types: Simple recursive networks, LSTM networks (Long Short Term Memory), GRU networks (Gated Recurrent Unit).

LSTM networks [2] have a feedback connection, which consists of adapting the network to the "hidden" information about the data it analyzes so that it learns.

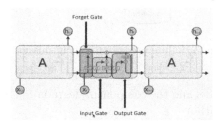

Fig. 4. A LSTM network and its gates.

The way of processing the data is the same as that of standard recurrent neural networks. However, the information propagation operation is different. As information passes through it, the operation decides which information to process further and which information to let go. The main operation consists of cells and gates. The cellular state functions as a pathway to transfer information. Cells can be considered as memory.

There are several gates in the LSTM process. When the cellular state carries information, these gates help new information flow. The gates will indicate what data is useful to save and what data is not useful. So only the relevant data is passed through the sequence chain for easy prediction. Figure 4 shows the gates of a LSTM unit and three steps of the network.

4　Failure Prediction of Transmission Lines Components

The estimation of future probability of failure is based on a current failure probability model that is instantiated with the evaluation of the future transmission line component. The current failure probability is estimated using a Bayesian model trained with historical evaluation-failure pair data. For this part, the future evaluation of the elements is estimated by means of an LSTM-type recurrent network that estimates the damage or degradation in the components over time. This network is trained with historical data of evaluations of each component to know its behavior in the future. In this way, current component evaluation data is fed into the degradation model to obtain an estimate of what would be expected in the future. Figure 5 shows a block diagram of the process for estimating the probability of future failures.

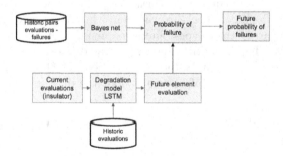

Fig. 5. Block diagram for failure prediction of transmission lines components.

Both the failure model and the component degradation model are built using machine learning algorithms.

4.1　Failure Model Using Bayesian Networks

The training data set for obtaining the probabilistic failure model is composed of historic pairs evaluations-failures with the form {C0, C1, C2,..Cm}, {F}, where components C0, C1, C2, C3..,Cm are the attributes and the failure F is the class. The evaluation of a particular component can be expressed as the state condition s0, s1, s2, ..., sn, where sn means the best condition and state s0 the worst condition. Table 2 shows a sample of a hypothetical training data set, and Table 3 shows the nomenclature of the different types of failures considered. The resulting model is a static Bayesian network that can determine a failure given the state condition of one or more components.

Table 2. Hypothetical training data set

	C0	C1	C2	...	F
1	s0	s0	s0		C0
2	s0	s1	s0		A1
3	s1	s1	s1		A1
4	s1	s1	s2		C1
..					
i	s4	s3	s5	...	C2

Table 3. Nomenclature of the different types of failures considered.

ID	Failure	Type
A0	Structure	Ordinary
A1	Fittings	Ordinary
A2	Insulators	Ordinary
A3	Conductor	Ordinary
A4	Ground thread	Ordinary
A5	Gap	Ordinary
A6	Contamination	Ordinary
B0	Lightning	Ordinary
C0	Strong winds	Ordinary
C1	Cane burning	Ordinary
C2	Vegetation burning	Ordinary
D0	Unknown failures	Ordinary
E0	Equipment failures	Extraordinary
E1	System disturbance	Extraordinary
E2.I	Vandalism	Extraordinary
E2.II	Natural phenomena	Extraordinary
E2.III	Accidents	Extraordinary
E2.IV	Other failures	Extraordinary
E3	Technical failures	Extraordinary
E4	Energy failure	Extraordinary

4.2 Degradation Model Using a LSTM Recurrent Network

A failure model is a mechanism that allows to diagnose a failure, while a degradation model allows to predict the state of a component in a certain time horizon. This is why, the training data set must be given as a time-series.

Consider then a set of historic evaluations of the state condition of a component Ci within a fixed time-step TS, i.e. $s10$ $s10$ $s9$ $s9$ $s9$ $s8$ $s8$ $s8$ $s7$. In order to train a recurrent network, the sequence of evaluations must be divided into input/output patterns of u time steps and v time steps. The input pattern are the u historic observations of the component Ci. If $v = 1$ then the model will predict one-step ahead but if $v = 3$ it will predict multiple steps in the future.

The equivalent training set for a $u = 3$ and $v = 2$ is then converted into a matrix of the form:

$$
\begin{array}{cc}
x & y \\
s10\ s10\ s9 & s9\ s9 \\
s10\ s9\ s9 & s9\ s8 \\
s9\ \ s9\ s9 & s8\ s8 \\
\end{array}
$$
...

where x is the input vector and y the output vector.

The resulting model is a recurrent network for uni-variate time-series that predicts two-step ahead in periods of TS time steps, given 3 observations.

5 Experimental Results

In this section, an evaluation of the quality of the lifetime estimation model and the failure model is presented. The predicted failure results of first predict the potential evaluation in a certain horizon and then use this value to instantiate the failure model in the interest element to obtain a probability distribution of failure.

5.1 Evaluation of the Probabilistic Failure Model

The test scenario consisted of using historical pairs evaluations-failure in order to build a probabilistic failure model capable to determine the distribution of probabilities around the set of failures given the evaluation of a particular element in the transmission line. The emphasis was given on the subsystem of contamination. A Bayesian network was used to represent the failure model.

In order to train alternative models, 357 examples of historical evaluations presenting some type of failure and 180 cases with no failure were used. The failure data corresponded to years 2004 to 2012 from different transmission areas in Mexico. The learning process used the hill climbing algorithm with a K2 metrics and an alpha parameter=5. The software tool utilized was OpenMarkov and a python package called pgmpy.

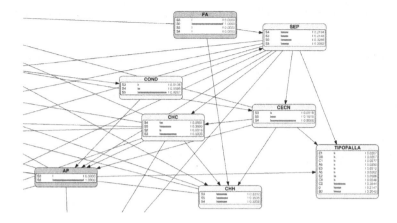

Fig. 6. Bayesian failure model. TIPOFALLA = Type of failure.

To select the best model, 5 alternative models were generated with different data sets. The selected network was repeated in 2 of 5 tests carried out, being the selected structure the one shown in Fig. 6.

Table 4 shows the precision of the failure model for the most likely value and the second most likely value in terms of true positives (TN) and false positives (FP) of the confusion matrix. Preliminary results for the failure model using the most likely value are not very acceptable. However, when using the second most likely value are much better. The reason behind the decision to present the second most like values is that during the experiment the modeled suffered overfitting when selecting a non-appropriated number of training records with no failure.

Table 4. Precision of the failure model.

Test	TP	TN	precision
Most likely value	252	177	0.58741259
2nd most likely value	372	57	0.86713287

5.2 Evaluation of Lifetime Estimation with LSTM Models

The test consisted of training a prediction model using simulated data based on real experiments of the lifetime of a polymer composite insulator under different pollution conditions (see Table 5). The idea is that in the future, after collecting real data, this initial model can be updated progressively (historical evaluations).

For this experiment, it was assumed that the inspections were performed each 6 weeks (1.5 months) to the same insulator element during 4 years. This is, a total of 33 observations (or inspections) for each contamination condition (see Fig. 7).

Table 5. Lifetime table for a silicone rubber insulator with alumina T under different levels of contamination

Pollution	Wetting	Leakage distance mm/kV	Lifetime (years)
Industrial and salty (very high)	Sea ($> 50m$), factories emanations	39	0–5
Industrial and salty (medium)	Sea ($> 50m$), factories emanations	39	5–10
Industrial and salty (light)	Sea ($> 100m$), factories emanations	39	10–30

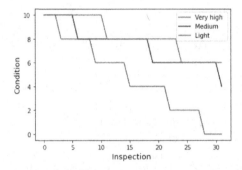

Fig. 7. Simulated data set for a Rubber-Silicone with alumina insulator under three contamination scenarios (very high, medium and light).

A vanilla LSTM model shown in Fig. 8 was used to perform uni-variate forecast for one-step ahead. In this way, the model was capable of predicting one evaluation ahead given a set of 5 sequential inspections (observations). A comparison of real and predicted data for different contamination environments are shown in Fig. 9.

Table 6 shows the mean absolute percent error (MAPE) for the three models constructed under different levels of environmental pollution (very high, medium and light). The MAPE for models of insulators exposed to medium and light degree of pollution are very acceptable. However, the MAPE for insulators with very high degree of pollution are numerically not good. One reason, is that the output of the model provide a continuous value and not a category. Visually, the results are quite good (see Fig. 9 left).

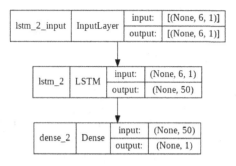

Fig. 8. Vanilla LSTM model.

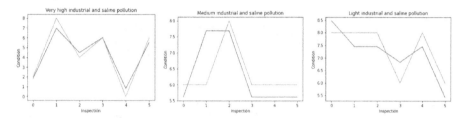

Fig. 9. Comparison of real and predicted data for different contamination environments (left: very high, center: medium, right: light). Orange line corresponds to the real value and blue line to the predicted value. (Color figure online)

Table 6. Mean absolute percent error (MAPE) for the degradation models constructed

Degree of pollution	MAPE (%)
Very high	33.5981241
Medium	7.248735698
Light	8.3264011

6 Conclusions and Future Work

In this paper, a novel AI-based method for failure prediction in transmission lines components was presented. The proposal combines machine learning and deep learning capabilities. The approach has been tested using simulated data of a composite polymeric insulator exposed to different degrees of ambient contamination. Preliminary results shows that the joint use of probabilistic and deterministic methods help determine the future failure of a transmission line component based on an assembly of diagnosis and forecast models.

As future work, we plan to: i) Collect historical data from real inspections of composite insulators and other TL components such as: conductors, towers, fittings accessories between insulators and towers, ground wires. ii) Pre-process more existing failure historical data. iii) Re-parameterize and tune up both degradation and failure models. iv) Build degradation models for multiple-steps ahead.

Acknowledgment. The authors thank CFE-GL staff for the support given during the development of project PE-A-11, which is still in progress by the National Institute of Electricity and Clean Energies (INEEL). This project has been founded by CONA-CyT through the Mexican Center for innovation in Intelligent Electrical Networks and Microgrids (CEMIE-Redes).

References

1. Hashim, R., Usman, F., Baharuddin, I.N.Z.: Determining health index of transmission line asset using condition-based method. Resources **8**(2) (2019). https://doi.org/10.3390/resources8020080,https://www.mdpi.com/2079-9276/8/2/80
2. Hochreiter, S., Schmidhuber, J.: Long short-term memory. Neural Comput. **9**(8), 1735–1780 (1997)
3. Irfan, J., Handika, P.: Comprehensive calculation of overhead transmission line health index. In: 2019 11th International Conference on Information Technology and Electrical Engineering (ICITEE), pp. 1–6 (2019). https://doi.org/10.1109/ICITEED.2019.8929988
4. Noor, M.J.M.: Application of knowledge-based fuzzy inference system on high voltage transmission line maintenance. Master's thesis, Queensland University of Technology (2004), https://eprints.qut.edu.au/16050/
5. Pearl, J.: Bayesian networks, causal inference and knowledge discovery. UCLA Cognitive Systems Laboratory, Technical Report (2001)
6. Rumelhart, D.E., Hinton, G.E., Williams, R.J.: Learning representations by back-propagating errors. Nature **323**(6088), 533–536 (1986). https://doi.org/10.1038/323533a0,http://www.nature.com/articles/323533a0
7. Thongchai, P., la or, P.P., Kulworawanichpong, T.: Condition-based health index for overhead transmission line maintenance. In: 2013 10th International Conference on Electrical Engineering/Electronics, Computer, Telecommunications and Information Technology, pp. 1–4 (2013)

Ethics and Smart City

Crowdsensing on Smart Cities:
A Systematic Review

Rui Miranda⬭, Vasco Ramos⬭, Eduarda Ribeiro⬭, Carla Rodrigues⬭,
António Silva⬭, Dalila Durães⁽✉⁾⬭, César Analide⬭, António Abelha⬭,
and José Machado⬭

ALGORITMI Research Center, University of Minho, Braga, Portugal
{rui.miranda,antonio.silva,dalila.duraes}@algoritmi.uminho.pt,
{pg42852,a85412,a84710}@alunos.uminho.pt,
{analide,abelha,jmac}@di.uminho.pt

Abstract. With the rise of the internet and the Internet of Things
(IoT), the concept of a Smart City began to materialise. Crowdsensing is the process of using portable sensing devices to gather information
about people's surroundings. Furthermore, the research on these domains
has pivoted from solely technology concepts to now including how they
improve the quality of life of citizens and their utility.

This paper presents a systematic review aiming to identify the role
and purpose of crowdsensing and the improvement of citizens' lives in
a smart city through this technique. Using the SCOPUS citation and
abstract database, six papers were picked out as relevant for discussion
in the review.

Keywords: Smart cities · Crowdsensing · Smart notifications ·
Internet of Things

1 Introduction

The study and development around Smart Cities has been a topic of great interest for several years, however its first phase of research initially focused on strictly
technological concepts. More recently, the focus of interest has shifted towards
a holistic perspective on the information systems associated with them, with
a particular focus on the quality of life of citizens and the impact of the utility of smart technologies on their lives, as well as on the social, economic and
environmental sustainability of cities [4].

The crowdsensing field is highly diverse and has experienced a lot of evolution, either by integrating sensory data collection through users' mobile devices,
which consists of a citizen contributing to the collective, or by providing intelligent information tailored to each citizen, in which a citizen is taking advantage
of the collective.

Thus, the aim of this systematic review to search for recent scientific publications that present an outlook of the current state of development in the fields

© The Author(s), under exclusive license to Springer Nature Switzerland AG 2022
A. C. Bicharra Garcia et al. (Eds.): IBERAMIA 2022, LNAI 13788, pp. 103–106, 2022.
https://doi.org/10.1007/978-3-031-22419-5_9

of crowdsensing in smart cities. In this context, a Research Question was proposed: "What is the role and purpose of crowdsensing and how does it impact the citizen's lives in smart cities?".

The following document is structured in four sections. Next section describes the research and review process. On Sect. 3, the obtained results are presented and discussed, and the last section aims to summarise the main conclusions and contributions obtained through the review, including recommendations for future work.

2   Methodology

This systematic review is built on the PRISMA[1] (Preferred Reporting Items for Systematic Reviews and Meta-Analyses) statement and respective checklist.

The preliminary research was conducted on 25 April 2022 and the used data source was SCOPUS[2], due to its size, quality assurance and wide coverage in terms of publication subjects.

In order to carry out the bibliographic research, some keywords were identified as a starting point. These keywords were applied in the title, abstract and keywords fields, and organized into two groups, which are combined with a conjunction. Keywords in each group are combined with disjunctions. This choice fulfils the purpose of each group selecting all documents that include at least one of its keywords and then ensuring that only documents containing one or more terms from each of the selected groups. The first group is related to the areas and technical subjects directly related to the research topic (*"Smart Cities"*, *"Crowdsensing"*, *"Smart Notifications"* and *"Location-based Notifications"*) and the second group aims to filter by broader areas of the technological scope (*"Information Communication Technology"*, *"Internet of Things"*, *"Information Systems"* and *"Mobile Computing"*), in order to focus the results in the context of information systems and agents.

To screen the articles and studies collected, some eligibility criteria (in the form of exclusion criteria) were define. As such, all documents that matched any of the following criteria were excluded: i) not accessible in open access mode; ii) Were not produced in the last 4 years (from 2018) or have not yet been fully published; iii) Do not come from the field of *Computer Science* or *Engineering*; iv) Are not an Article or a Review/Survey and are not written in English; v) Were not written by relevant authors or don't have a relevant number of citations; vi) Were not produced inside the European Union (due to similar policies regarding extraction, manipulation and exploitation of data for the production of knowledge, as well as data protection); and vii) Do not focus on the variables studied or are out of context.

---

[1] http://www.prisma-statement.org.
[2] https://www.scopus.com.

3   Results and Discussion

Of the 21362 studies identified, the application of the PRISMA methodology resulted in a final set of 8 studies that were relevant to previously defined research questions. This section presents and discusses the findings of the review against the research question.

Amaxilatis et al. [1] present a solution for deploying and managing crowdsensing campaigns and experiments across cities. As a result, the authors concluded that the system allows developers to reverse course if they notice that their campaign design is underperforming through the system.

Foschini et al. [3] presented an edge-enabled mobile crowdsensing technology that uses edge nodes to compute potentially dangerous crowd scenarios. Yang et al. [6] present a perspective which displays crowdsensing as a technique in which a large number of people using mobile devices with sensors share sensory data to measure, analyse, or infer any issue of common interest.

Ismagilova et al. [4] present a discussion on the important findings from existing research on issues connected to smart cities from an Information Systems perspective. The authors outlined existing advancements' shortcomings as well as prospective future possibilities.

Capponi et al. [2] provided a comprehensive review of the challenges, solutions and opportunities of mobile crowdsensing systems and proposed a four-layered architecture to characterise the works in mobile crowdsensing.

Nizetic et al. [5] reviewed and discussed the IoT and its prospects, challenges, and concerns in the context of a smart and sustainable future.

In terms of the role and purpose of crowdsensing in smart cities, Amaxilatis et al. [1] and Foschini et al. [3] suggest that it enables the collection and sharing of large volume of data, which can be used to track citizen habits and movements in urban environments. Furthermore, crowdsensing enables the development of cost-effective and high-quality monitoring systems for urban infrastructures, services, and the environment, and could help speed up the deployment of smart city projects [3,6].

Crowdsensing has the potential to significantly improve citizens' daily lives while also providing urban civilizations with new perspectives [2,5]. In smart cities, citizens can not only interact and engage with services but also provide data for these services via crowdsensing [4]. More information leads to more comprehensive conclusions and, as a result, high-quality data sets [3].

Emergency management and prevention, environmental monitoring, health care and wellbeing, e-commerce, indoor localization, intelligent transportation systems, mobile social networks (MSNs), public safety, unmanned vehicles, urban planning, waste management, and Wi-Fi characterization are among the most promising application domains where crowdsensing can operate to improve citizens' quality of life in a smart city, according to Capponi et al. [2].

4   Conclusion

The concept of smart cities is a cornerstone to the sustainability of cities and places with high population density.

In this paper, a systematic review was conducted to analyse the literature produced in the fields of smart cities and its application using crowdsensing. The review was based in the PRISMA model using the SCOPUS database as the source. In the studies analysed, all the authors agree that the implementation of crowdsensing initiatives can improve the citizens' lives in a smart city and, the most promising application domains where crowdsensing can play an important role were identified.

This work was developed as part of a research project whose goal is to study the application of crowdsensing, in the smart cities' infrastructure and framework. As such, and given all the compiled information, future work includes the study and development of a platform to integrate these concepts into a viable and helpful information system that can bring value to the smart city paradigm and the citizen's lives where it is applied.

Acknowledgements. This work has been supported by FCT Fundação para a Ciência e Tecnologia within the RD Units Project Scope: UIDB/00319/2020. and the FCT - Fundação para a Ciência e Tecnologia within the R&D Units Project Scope: NORTE-01-0145-FEDER-000086.

References

1. Amaxilatis, D., Mylonas, G., Diez, L., Theodoridis, E., Gutiérrez, V., Muñoz, L.: Managing pervasive sensing campaigns via an experimentation-as-a-service platform for smart cities. Sensors (Switzerland) **18** (2018). https://doi.org/10.3390/s18072125

2. Capponi, A., Fiandrino, C., Kantarci, B., Foschini, L., Kliazovich, D., Bouvry, P.: A survey on mobile crowdsensing systems: Challenges, solutions, and opportunities. IEEE Commun. Surv. Tut. **21**, 2419–2465 (2019). https://doi.org/10.1109/COMST.2019.2914030

3. Foschini, L., Martuscelli, G., Montanari, R., Solimando, M.: Edge-enabled Mobile Crowdsensing to Support Effective Rewarding for Data Collection in Pandemic Events. J. Grid Comput. **19**(3), 1–17 (2021). https://doi.org/10.1007/s10723-021-09569-9

4. Ismagilova, E., Hughes, L., Dwivedi, Y.K., Raman, K.R.: Smart cities: advances in research — an information systems perspective. Int. J. Inf. Manag. **47**, 88–100 (2019). https://doi.org/10.1016/j.ijinfomgt.2019.01.004,https://linkinghub.elsevier.com/retrieve/pii/S0268401218312738

5. Nižetić, S., Šolić, P., de-Ipiña González-de Artaza, D.L., Patrono, L.: Internet of things (iot): Opportunities, issues and challenges towards a smart and sustainable future. J. Cleaner Prod. **274** (2020). https://doi.org/10.1016/j.jclepro.2020.122877

6. Yang, M., Zhu, T., Liang, K., Zhou, W., Deng, R.H.: A blockchain-based location privacy-preserving crowdsensing system. Future Gener. Comput. Syst. **94**, 408–418 (2019). https://doi.org/10.1016/j.future.2018.11.046

Sentiment Gradient, An Enhancement to the Truth, Lies and Sarcasm Detection

Fernando Cardoso Durier da Silva(✉) , Ana Cristina Bicharra Garcia ,
and Sean Wolfgand Matsui Siqueira

Federal University of the State of Rio de Janeiro - UNIRIO, Rio de Janeiro, Brazil
{fernando.durier,cristina,sean}@uniriotec.br

Abstract. Information sharing on the Web has also led to the rise and spread of fake news. Considering that fake information is generally written to trigger stronger feelings from the readers than simple facts, sentiment analysis has been widely used to detect fake news. Nevertheless, sarcasm, irony, and even jokes use similar written styles, making the distinction between fake and fact harder to catch automatically. We propose a new fake news Classifier that considers a set of language attributes and the gradient of sentiments contained in a message. Sentiment analysis approaches are based on labelling news with a unique value that shrinks the entire message to a single feeling. We take a broader view of a message's sentiment representation, trying to unravel the gradient of sentiments a message may bring. We tested our approach using two datasets containing texts written in Portuguese: a public one and another we created with more up-to-date news scrapped from the Internet. Although we believe our approach is general, we tested for the Portuguese language. Our results show that the sentiment gradient positively impacts the fake news classification performance with statistical significance. The F-Measure reached 94%, with our approach surpassing available ones (with a p-value less than 0.05 for our results).

Keywords: Fake news · Gradient · Sentiment analysis · Machine learning · NLP

1 Introduction

Taking advantage of the overwhelming amount of information on the Web, big corporations, governmental organizations, and ill-intended people may use technology to spread propaganda, manipulate the information people will consume and misguide the beliefs of the masses. Worsening the situation, psychology theories have shown that people prefer fake information even in the presence of the known truth [10]. The reasons for this preference include the low cognitive effort for understanding the fake message, the high social acceptance of the lies, and even the old saying that repeating a lie many times turns it into a truth [14].

From the automatic detection process, we find out that sentiment analysis plays an important role in detecting the urgency feeling that boosts the fake

© The Author(s), under exclusive license to Springer Nature Switzerland AG 2022
A. C. Bicharra Garcia et al. (Eds.): IBERAMIA 2022, LNAI 13788, pp. 107–118, 2022.
https://doi.org/10.1007/978-3-031-22419-5_10

news spread, or the serious tone of a message. Works that use such features rely on discrete values that imply either happiness, neutrality, or sorrow, while when in reality, we can see a varying spectrum of those in a simple single sentence [11].

We developed a model capable of handling the multi-classification aspect of a text and the sentimental variance. We trained and tested on a set of sociopolitical Brazilian Portuguese news articles we outsourced and on a public dataset.

This paper presents our gradient sentiment method for detecting fake news that outperforms current fake news detection approaches. Next section presents the related works that provides an overview of the state-of-the art in fake news detection. Then we present the Sentiment Gradient method followed by the datasets' descriptions and the experiments. Last, we present the discussion of the results reflecting on the reasons our method outperform the others and the conclusion.

2   Related Works

For fake news detection, the literature defines different strategies to solve the problem of labeling data as truth or lie. Shu et al. [16], in an extensive review, pointed that the trend of sentiment analysis in fake news Detection is funda- mental for completing this task. Wang et al. [20] analyzed the different feature choices available in this research area, one of them the sentiment analysis of fake news, that they divided into four possible classes (Factual, Manipulative, Hoax, and Incomplete) and with pure sentiment analysis (provided by Linguis- tic Inquiry and Word Count - LIWC technique) achieved 94.2% of accuracy for a known English set the PoliFact. Bhutani et al. [2] relied upon the traditional term frequency-inverse document frequency(TF-IDF) vectorizer modeling of fake news, in order to be compliant to neural networks models, achieving 84.7% of accuracy.

Monteiro et al. [11] proposed a new Portuguese Dataset with 7200 news (3624 True News and 3576 fake news) called Fake.BR. They tested classifiers upon their dataset and got 89% for their best classifier, an Support Vector Machine(SVM), for all the features they worked with (POS-Tagging, Word Embedding, Senti- ment Analysis, etc.). Focusing only on sentiment analysis, they got the best score of 56% accuracy.

We observe in the literature that the classical machine learning models are used as baselines of comparison against neural networks models [9], which are modern and less feature engineering oriented approaches (as their inner architec- ture is capable of automatic feature extraction) [8]. However, we should not forget that the classical models can be as effective as the neural ones; the main differ- ence is that they require more manual effort on feature engineering. Manjusha and Raseek [8] obtained a winning 79.7% mean F-Measure for classifying articles into satire, humor, and irony with a Convolutional Neural Network(CNN) that competed against SVM, Decision Tree, K-Nearest Neighbors(KNN), and Gaus- sian Naive Bayes(GNB). On the other hand, de Morais et al. [12] obtained 80% F-Measure with classical models only for a Brazilian set of Portuguese News.

According to the literature, the datasets are outsourced from social networks and microblogs originated from polemic subjects as elections, polarizing discussions, and events. Few works deal with the Portuguese language, such as: [13]; [5]; [15]; [12]. They all have in common the same outsourcing strategy of web scraping articles from the news portals available in Brazil, and relying upon these portals' reputation to label them as true or fake news. The trend we might notice is that they follow the same strategy of English oriented works, and were implemented almost simultaneously, therefore having no direct reference or intersection with one another. They chose as metrics f-measure, accuracy, precision, and recall. Moreover, they are also based on the classical machine learning models.

The works that deal with sentiment analysis usually focus on a sum of sentiment scores. Luo et al. [7] try to change the perspective for financial news analysis by configuring an Long short-term memory neural net(LSTM) model to have attention over sentences instead of words, but, in the end, it still aggregates the sentiment score. The works of Wang et al. (e.g., [19]) explore the same change of perspective on images by trying to get the sentiment of filtered regions of the image to understand the bigger picture, but they also aggregate in the end. Finally, Abburi et al. [1] try to decompose music into parts (beginning and end) to understand the sentiment variance better. All these decomposed attempts had great results for their specific tasks, compared against their own baselines, even though in the end they aggregated the final product and only considered the singular values of the parts when measuring.

3   Sentiment Gradient

Our proposed method changes the sentiment analysis representation. Instead of summarizing a message by a holistic sentiment (a number), we represent the message with its full nuances. The message is represented by a vector of sentences.

The traditional sentiment analysis representation works as the "Bag-of-Features" (BoF), which is the set of descriptors extracted from a textual message, denoted by $A = $ bag of a_k, $k \in \{1, \dots, N\}$, in which a_k is a feature and N is the total number of features in a message. For traditional sentiment analysis, each feature is a word that will have a sentiment score according to a function f, given by pre-defined mapping of words to a sentiment score. The average of those scores is the output of the Sentiment Analysis function (see Eq. 1). Then, Sentiment(A) is one more feature considered to train the fake news classifier.

$$\text{Sentiment}(A) = \frac{1}{N} \sum_{k=1}^{N} f(a_k) \qquad (1)$$

Instead of modeling the BoF of the message by their word components, we chose to model it by sentences because each sentence is an utterance about a target subject. Then, we propose a novel technique of applying derivatives into the array of features (see Eq. 3), like what we would do on a time series (Eq. 2), that way we would be able to capture the information we need, i.e., the rise, the fall and the stability of a sentimental gradient. Allowing classifiers to comprehend the sentiment variance of figurative languages such as Sarcasm. We named this new interpretation and technique Sentiment Gradient (Algorithm 1).

$$S(A) = (Y_t : t \epsilon N) \tag{2}$$

$$Y_t = f(x) = \begin{cases} Y_1 = f(a_1), & i = 1 \\ Y_i = g(\frac{\partial f(a_{i-1})}{\partial sentiment}), & i > 1 \end{cases} \tag{3}$$

The sentences can be understood as the phases of our text signal, the sentiment gradient as the amplitude (which mathematically would indicate the intent of upward or downward change in sentiment), and the frequency is fixed by 20 sentences as it is the average in the dataset, as can be seen in Fig. 1. To follow the trend, like the other works, we also have an aggregated average additional to our time-series. This measure is the average sentiment gradient of the series.

In the same figure, we can observe that the three classes of news fall in a specific spectrum of the signal, the True news maintain mostly the neutral to positive overtone, the fake news maintain mostly the negative impact overtone, and the sarcastic ones vary along the neutral point showing the subtle imbalance game to make us laugh.

Any given message that is not truncated by the communication channel is continuous on time, for word or sentence level. Therefore, by deriving a sentence's sentiment, we are not simply getting a singular value, but, depending on the mathematical signal, the direction to which the signal is flowing, because the derivative of a point 'a' shows us the tangent inclination of the variation of that point in relation to the linear function [3].

However, why choose to derive the sentences' time-series? Because the derivative function shows us the areas of increase, and decrease of a function as well as the magnitude of such change, passing more information than the singular value [6].

Algorithm 1: Sentiment Gradient Algorithm

Result: Sentiment Gradient of the News
sentiment_timeseries = empty array;
sentence_array = SentenceTokens(News);
if *Length(sentence_array) > 1* **then**
    for *each sentence in sentence_array* **do**
        sentiment_rate =
        sentence[sentiment_charge]\Length(sentence[tokens])
        sentiment_timeseries.append(sentiment_rate)
    end
    return mean(getGradients(sentiment_timeseries))
else
    return sentence_array[0][sentiment_charge]
end

---

4 Dataset

We want the machine to understand the difference between ironic humorous critic, true facts and false statements. We expect the machine to perceive the

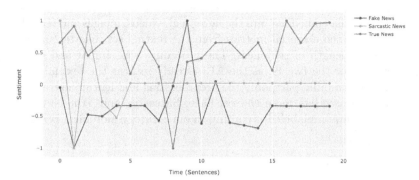

Fig. 1. Example of sentiment gradient extracted from three different pieces of news with different tones: a fact (sentiment avg.: 0.5), a lie (sentiment avg.: -0.3), and a sarcastic (sentiment avg.: -0.1).

nuances in sentiment charge across the sentences, and the style of writing (too dense text, too wordy, etc.).

Due to the lack of open datasets in Portuguese for both the fake and sarcastic news, we needed to leverage upon known popular sources of sarcastic and fake news in Brazil, E-Farsas' fake news session[1] and Sensacionalista[2] Furthermore, for the True news, we chose Folha de São Paulo[3] a journalistic source of news known to the Brazilian population. Although we have a full set of other sources for True News in Brazil, their sites have peculiarities and blockers that make the scraping process harder than Folha de São Paulo. Also, we took advantage of a produced set of 2000 pair-wise true and fake news from the works of Silva et al. [18] and appended it into our main raw set to be preprocessed following our proposed methodology.

We considered news agencies and governmental accounts to outsource the truthful tweets. For the sarcastic ones, we chose popular sarcastic accounts, e.g., Não Salvo (@naosalvo), Sensacionalista (@sensacionalista) and O Criador (@OCriador). Finally, for fake news, we considered a set of reported known fake news spreader accounts, such as @opavao and current politics involved in the fake news scandal in Brazil from 2020 on-wards, which nonetheless, due to the extreme political propaganda material on its own, they can be classified as fake news as well as its unverified disseminated texts.

In order to extract the training set for our models, we created a web scraper for each of the sources. To do so, we relied upon Python programming language and its libraries, such as Beautifulsoup (lib for webscraping)[4] NLTK (natural

---

[1] E-Farsas - http://www.e-farsas.com/secoes/falso-2.

[2] Sensacionalista - https://www.sensacionalista.com.br/.

[3] Folha de São Paulo - https://www.folha.uol.com.br/.

[4] BeautifulSoup - https://www.crummy.com/software/BeautifulSoup/.

language toolkit)[5] and re(for regular expression)[6] Since our focus is on the textual content, we extracted only the textual news content from those sources and ignored extra media embedded into the news, e.g., videos, images and recordings.

We extracted the essential metrics from the textual set, like textual length and the average length of sentences. The procedure was similar and yet more straightforward for the tweets, as the API already returns the textual content, not needing the scraping part, only the preprocessing and feature extraction. In the end, our dataset has three classes, Fake, Sarcastic, and True News. With 76,782 rows of news prelabeled according to our strategy aforementioned.

5   Experiment

The most used features by the literature for the sentiment analysis [4] are Word Sentimental Score and Sentence Sentimental Score. However, different from our novelty, this step is usually focused only on the singular synthesized metric of sentiments (sum or average) from the entire text or sentence. For the basic features the literature recommend word count, sentence count, space count, and POS Tag count.

We investigated the difference between the basic sentiment analysis and the Sentiment Gradient combined with basic features (results in Table 3). In order to fairly compare against the related works, we trained models with all the features, analogously they did in their respective works (results in Table 4). The only public dataset in Portuguese we found available was the Fake BR, that we used to train the same baseline SVM model of [11], which we obtained the same 0.89% score to compare against our own.

We ran the experiment of cross validating (using 5 folds, repeated 10 times) each one with each set of the features we engineered as well as combinations of the basic and each approach for sentiment analysis. After this process we obtained distributions of 10 repetitions of 5, i.e., 50 registries to which we compared against each other through Mann Whitney U Hypothesis test to check for statistical significance in our results (obtaining p-values lower than 0.05 for all comparisons).

5.1   Machine Learning Algorithms and Hyperparameters

The state of the art in fake news Detection indicates the following techniques as most used ones [17]: K-Nearest Neighbors (KNN), Naive Bayes (Gaussian, GNB, and Multi-nominal, MNB), Decision Tree, Random Forest (R.For.), Adaboost Tree, Gradient Boosting, Support Vector Machine (SVM), Linear Regression Classifier (LNR), LSTM and Multi-Layer Perceptron (MLP).

The following algorithms used the hyperparameter tuning process of 5 fold cross validation, using f1-weighted as the comparison metric to decide the best

---

[5] NLTK - https://www.nltk.org/.
[6] Regular Expression - https://docs.python.org/3/library/re.html.

hyperparameter configuration. This tuning process used the existing Gridsearch of Python's SKLearn Library.

- **KNN**: Number of neighbors as 5, weighting strategy of euclidean distance, and deciding algorithm of ball tree.
- **Decision tree**: Max depth of 10, and both minimum leaf samples, and minimum split samples of 2. With a random state of zero, for replication purposes.
- **Random Forest**: 500 estimators in the forest, each base classifier with max depth of 40, min samples in the leaf nodes of 1, and min samples of split of 2. With a random state of zero, for replication purposes.
- **Gradient Boosting**: 500 base estimators, 0.01 learning rate, max depth of 10, min samples in leaf of 1, and min samples of split as 2. For the Adaboost classifier, we got as best configuration set: 500 base estimators, and learning rate of 0.01.
- **Multi-layer Perceptron**: Adam as solver, alpha of 1-e5, constant learning rate, activation function of relu, and the following architecture tuple (10,20,30,40,50,40,30,20,10).
- **LSTM**: Activation function of hyperbolic tangent, optimizer adam, loss of sparse categorical crossentropy, and the following architecture tuple (12 Dense, 4 Recurrent Units, Flatten, and Softmax, each hidden layer with a subsequent dropout of 20%).

For the **Linear Regression** algorithm, **Support Vector Machine**, **Gaussian Naive Bayes**, and **Multinomial Naive Bayes** we configure as studied from literature because of the simplicity of naive bayes algorithms and SVM and Linear Regression algorithms' complexity when using the probability functions (causing computer resource overload).

- **Linear Regression Classifier** with multi-class parameter set to multinomial.
- **SVM** with probability option set to True, and divided in 6 jobs.
- **Gaussian Naive Bayes** with standard parameters.
- **Multinomial Naive Bayes** with standard parameters.

5.2 Study Case in Twitter

We wanted to experiment how our approach would behave on messages coming from Twitter. In order to do this, what we did was to repeat the data gathering part of our experiment, now this time for Twitter.

In order to preemptively obtain the label values of each tweet, we took profiles known for spreading messages of each category (Fake, Sarcastic and True messages) and composed our dataset from there. We ran the exact same preprocessing we ran for the news set on the twitter-set, obtaining the same features.

The key point of analysis here is that the sentiment gradient on tweets would operate in a short number of sentences, due to Twitter's post size restriction (240 characters only). The average number of sentences in a tweet is 2. And following our algorithm the remaining pads (18 right pads) would be attenuated with 0. This is a limitation of the classical models, in the sense that all features should be fixed, linear, and normalized.

Then, we repeated the cross-validation experiment with the same parameters, 5 folds, repeating 10 times. In Table 1 we can see better the metrics we got.

Table 1. Twitter Dataset Cross-Validation Metrics.

Model	Approach	Score
GNB	Basic + Sentiment	0.864 (+/−0.004)
GNB	Basic + SentimentGradient	0.715 (+/−0.004)
LSTM	Basic + Sentiment	0.887 (+/−0.004)
LSTM	Basic + SentimentGradient	0.890 (+/−0)
R.For.	Basic + Sentiment	0.890 (+/−0)
R.For.	Basic + SentimentGradient	0.890 (+/−0)
SVM	Basic + Sentiment	0.890 (+/−0)
SVM	Basic + SentimentGradient	0.890 (+/−0)

From the results in this subsection we can see that more studies and experimentation is needed to make sentiment gradient as effective to tweets as it was for news.

Also for the joint of all attributes, we can see it's metrics in Table 2.

Table 2. Comparison between our approach and the related works, compared against the Tweets classification score.

Work	Winning Model	Score
Our Approach with News	Grad. boosting classifier	0.949
Our Approach with Tweets	R.For.	0.893
Wang et al. 2018	Logistic regression	0.942
Bhutani et al. 2019	CNN	0.847
Manjushaa and Raseek, 2018	CNN	0.797
Monteiro et al., 2018	SVM	0.890
de Morais et al., 2019	LP	0.800

6   Experiment Results Discussion

As shown in Table 3, the Sentiment Gradient feature positively impacted most of the results of the model trained. We can observe an increase of 14% for Random Forests, and 6% for Gradient Boost Classifiers for the best models.

Table 3. F1 Measure comparing the usage of regular sentiment analysis and sentiment gradient. As seen in bold, the best performing models were the Random Forest and Gradient Boosting Classifiers setup with Sentiment Gradient.

Model	Feature Choice	F1(+/-Stdv)
Adaboost	Basic + Sentiment	0.736(+/−0.007)
Adaboost	Basic + SentimentGradient	0.739(+/−0.007)
DecTree	Basic + Sentiment	0.757(+/−0.007)
DecTree	Basic + SentimentGradient	0.754(+/−0.008)
GNB	Basic + Sentiment	0.612(+/−0.019)
GNB	Basic + SentimentGradient	0.594(+/−0.011)
GradientBoost	Basic + Sentiment	0.778(+/−0.005)
GradientBoost	**Basic + SentimentGradient**	**0.832(+/−0.008)**
KNN	Basic + Sentiment	0.748(+/−0.007)
KNN	Basic + SentimentGradient	0.661(+/−0.008)
LNR	Basic + Sentiment	0.551(+/−0.003)
LNR	Basic + SentimentGradient	0.632(+/−0.007)
LSTM	Basic + Sentiment	0.656(+/−0.016)
LSTM	Basic + SentimentGradient	0.677(+/−0.011)
MLP_ADAM	Basic + Sentiment	0.756(+/−0.013)
MLP_ADAM	Basic + SentimentGradient	0.769(+/−0.012)
MNB	Basic + Sentiment	0.24(+/-0-.000)
MNB	Basic + SentimentGradient	0.24(+/−0.000)
R.For.	Basic + Sentiment	0.788(+/−0.007)
R.For.	**Basic + SentimentGradient**	**0.846(+/−0.006)**
SVM	Basic + Sentiment	0.554(+/−0.005)
SVM	Basic + SentimentGradient	0.577(+/−0.008)

We experimented our model on tweets, wondering whether the data type would influence the results. Employing our scrapers, we were able to outsource tweets about the same topics of our news from known reference profiles on the classes (Fake, True, and Sarcastic), obtaining 80,843 tweets. We observed that due to the lack of sentences in Tweet's microblog structure the sentiment gradient is unable to get the nuance because the granularity would need to change to world level maybe, as our results for that kind of data were of 89% F-Measure.

Even though the MLP and LSTM are modern approaches from the neural networks class of algorithms, they were not able to beat the ensemble methods, and this may be explained by the lack of data that we have nowadays in Portuguese just yet. However, analyzing on the contribution of the sentiment gradient, those models were impacted positively as well.

On the other hand, we see that GNB, Decision Tree, and KNN models got inverted results, benefiting more from the traditional sentiment analysis than the sentiment gradient. We attribute this behavior to the increase in the data dimensionality as we consider twenty sentences for representing the sentiment gradient, but KNN and Decision Tree rely on their linear separation of observations' space, a strategy that can be harmed by higher dimension sets.

We ran the cross-validation training to perceive the real impact of sentiment gradient. We defined a 10 folded cross-validation setup on F1 Measure
and applied the Mann Whitney U Test to check if the results from the cross
val distributions were statistically significant. We got P-Values inferior to the
threshold of 0.05, i.e., our results have a confidence interval of 95%.

Comparing our results with the related works (see Table 4), we understand
that our approach strongly contributes to the Brazilian Portuguese differentia-
tion of True, Fake, and Sarcastic News, as we achieved more than 5% in relation
to similar works. Although we are not in the same language context, we sur-
passed the two CNN works and got pretty close to the best English detection
work with a difference of only 0.7%. We attribute this success to the fact that our
classifiers were able to comprehend the nuances of sentences' sentiment charge
different from the related works handling a singular fixed value (concentrated
by an average, rather than the expansion of the series), and also due to the
information that the derivative function brings us, i.e., the direction of change
and its magnitude.

Table 4. Comparison between our approach and the related works.

Work	Winning Model	Score
Our Approach	Gradient Boosting Classifier	0.949
Wang et al. 2018	Logistic Regression	0.942
Bhutani et al. 2019	CNN	0.847
Manjusha and Raseek, 2018	CNN	0.797
Monteiro et al. 2018	SVM	0.89
de Morais et al., 2019	LP	0.80

Comparing the results for the application of our approach over tweets, we
can see that the solution was not as effective since we could see that for some
algorithms such as the Random Forest and SVM the inclusion of it seems not
to make difference, and for GNB and LSTM the traditional sentiment analysis
worked better than sentiment gradient. Also in Table 2 we can see that even
though we got competitive metrics against the related works, it is not as effective
as the joint set for news, implying that the sentiment gradient didn't impact the
classification of tweets.

We credit this for the fact that tweets have a much different structure if
compared to news, they are smaller, and with more slang and abbreviations.
For the sentiment gradient to work with tweets, we may need to change the
algorithm for taking words in consideration rather than sentences, however, this
can be challenging as well, since there tweets that don't rely much on writing
and instead on links, or medias.

Our results show a benefit of using Sentiment Gradient on the fake news
detection process. However, more experimentation on other language datasets is
recommended, as well as exploring other format of news such as microblogs that

might require mathematical modeling experimentation in the sentiment gradient approach (from sentences to words) in order to get results closer to the ones in news analysis.

7 Conclusions

From the results we got in our experiments, we were able to test our proposed novelty of the sentiment gradient being effective to help differentiate what is fake, true, or sarcastic as it can provide to the machine the sentimental imbalance that occurs in sarcastic cues.

The current limitation of the novelty is the fact that it is not as effective for tweets due to their different writing style, which relies on faster, smaller, and more direct sarcastic cues. Furthermore, more experiments on other language datasets are suggested, as we restricted our experiments to the Brazilian Portuguese Scenario.

We see as contributions of this work the sentiment gradient technique we propose and the dataset which allowed us to do such study, as Portuguese sets of labeled news are hard to find due to the commodity of using known English sets.

For future works, we intend to expand the sentiment gradient concept to other datasets, such as the tweets, and also explore other machine learning algorithms besides the classic models.

References

1. Abburi, H., Akkireddy, E.S.A., Gangashetti, S., Mamidi, R.: Multimodal sentiment analysis of telugu songs. In: SAAIP@ IJCAI (2016)
2. Bhutani, B., Rastogi, N., Sehgal, P., Purwar, A.: Fake news detection using sentiment analysis. In: 2019 Twelfth International Conference on Contemporary Computing (IC3), pp. 1–5. IEEE (2019)
3. Bos, H.J.M.: Differentials, higher-order differentials and the derivative in the leibnizian calculus. Arch. Hist. Exact Sci. **14**(1), 1–90 (1974)
4. Charalampakis, B., Spathis, D., Kouslis, E., Kermanidis, K.: A comparison between semi-supervised and supervised text mining techniques on detecting irony in greek political tweets. Eng. Appl. Artif. Intell. **51**, 50–57 (2016)
5. CORDEIRO, P.R.D., Pinheiro, V., Moreira, R., Carvalho, C., Freire, L.: What is real or fake?-machine learning approaches for rumor verification using stance classification. In: IEEE/WIC/ACM International Conference on Web Intelligence, pp. 429–432. ACM (2019)
6. Górecki, T., Łuczak, M.: Using derivatives in time series classification. Data Min. Knowl. Disc. **26**(2), 310–331 (2013)
7. Luo, L., et al.: Beyond polarity: Interpretable financial sentiment analysis with hierarchical query-driven attention. In: IJCAI, pp. 4244–4250 (2018)
8. Manjusha, P., Raseek, C.: Convolutional neural network based simile classification system. In: 2018 International Conference on Emerging Trends and Innovations In Engineering And Technological Research (ICETIETR), pp. 1–5. IEEE (2018)

9. Manzoor, S.I., Singla, J., et al.: Fake news detection using machine learning approaches: A systematic review. In: 2019 3rd International Conference on Trends in Electronics and Informatics (ICOEI), pp. 230–234. IEEE (2019)
10. Marchi, R.: With facebook, blogs, and fake news, teens reject journalistic "objectivity". J. Commun. Inq. **36**(3), 246–262 (2012)
11. Monteiro, R.A., Santos, R.L.S., Pardo, T.A.S., de Almeida, T.A., Ruiz, E.E.S., Vale, O.A.: Contributions to the study of fake news in portuguese: new corpus and automatic detection results. In: Villavicencio, A., et al. (eds.) PROPOR 2018. LNCS (LNAI), vol. 11122, pp. 324–334. Springer, Cham (2018). https://doi.org/10.1007/978-3-319-99722-3_33
12. de Morais, J.I., Abonizio, H.Q., Tavares, G.M., da Fonseca, A.A., Barbon Jr, S.: Deciding among fake, satirical, objective and legitimate news: A multi-label classification system. In: Proceedings of the XV Brazilian Symposium on Information Systems, p. 22. ACM (2019)
13. Pinto, M.R., de Lima, Y.O., Barbosa, C.E., de Souza, J.M.: Towards fact-checking through crowdsourcing. In: 2019 IEEE 23rd International Conference on Computer Supported Cooperative Work in Design (CSCWD), pp. 494–499. IEEE (2019)
14. Polage, D.C.: Making up history: False memories of fake news stories. Eur. J. Psychol. **8**(2), 245–250 (2012)
15. Reis, J., Correia, A., Murai, F., Veloso, A., Benevenuto, F.: Explainable machine learning for fake news detection. In: Proceedings of the 10th ACM Conference on Web Science, pp. 17–26. ACM (2019)
16. Shu, K., Sliva, A., Wang, S., Tang, J., Liu, H.: Fake news detection on social media: a data mining perspective. SIGKDD Explor. **19**(1), 22–36 (2017). https://doi.org/10.1145/3137597.3137600,https://dl.acm.org/doi/10.1145/3137597.3137600
17. Cardoso Durier da Silva, F., Vieira, R., Garcia, A.C.: Can machines learn to detect fake news? a survey focused on social media. In: Proceedings of the 52nd Hawaii International Conference on System Sciences (2019)
18. Silva, R.M., Santos, R.L., Almeida, T.A., Pardo, T.A.: Towards automatically filtering fake news in portuguese. Expert Systems with Applications **146**, 113199 (2020). https://doi.org/10.1016/j.eswa.2020.113199,http://www.sciencedirect.com/science/article/pii/S0957417420300257
19. Wang, J., Fu, J., Xu, Y., Mei, T.: Beyond object recognition: Visual sentiment analysis with deep coupled adjective and noun neural networks. In: IJCAI, pp. 3484–3490 (2016)
20. Wang, L., Wang, Y., De Melo, G., Weikum, G.: Five shades of untruth: Finer-grained classification of fake news. In: 2018 IEEE/ACM International Conference on Advances in Social Networks Analysis and Mining (ASONAM), pp. 593–594. IEEE (2018)

Green and Sustainable AI

Selection of Acoustic Features for the Discrimination Between Highly and Moderately Transformed Colombian Soundscapes

Fernando Martínez-Tabares[1]($\boxtimes$)(iD) and Mauricio Orozco-Alzate[1,2](iD)

[1] Universidad Nacional de Colombia - Sede Manizales, Grupo de Control y Procesamiento Digital de Señales, Km 7 vía al Magdalena, Manizales 170003, Colombia
{fmartinezt,morozcoa}@unal.edu.co

[2] Universidad Nacional de Colombia - Sede Manizales, Departamento de Informática y Computación, Km 7 vía al Magdalena, Manizales 170003, Colombia

Abstract. Several acoustic features have been proposed in the literature as useful indexes to characterize natural soundscapes. Such a characterization can be applied, for instance, to study the effect of the land transformation on the audible properties of a place. The collection of available features is relatively large and, therefore, requires an examination to decide which ones are really informative and discriminative for the problem at hand. In this paper, we pursue an empirical study for the selection of acoustic features in the problem of discriminating between highly and moderately transformed versions of two Colombian soundscapes. Classical supervised feature selection methods were used, along with exploratory tools such as correlation matrices and scatter plots. Results reveal that a small number of acoustic features are enough to discriminate between the classes, typically those that estimate either the acoustic complexity via the intrinsic variability of the sound intesities or the biodiversity through the species richness or abundance in particular frequency bands.

Keywords: Acoustic features · Classification · Feature selection methods · Soundscapes

1 Introduction

The Artificial Intelligence (AI) community has recently focused efforts on the application of AI for social good (AI4SG). According to [15], such a social good is understood in relation to the 17 United Nations Sustainable Development Goals, from which a number of them are related to ecological and environmental concerns. Therefore, a branch of AI4SG directs its attention to the application of AI techniques for the conservation of biodiversity, ecosystems and landscapes [9].

A. C. Bicharra Garcia et al. (Eds.): IBERAMIA 2022, LNAI 13788, pp. 121–132, 2022.
https://doi.org/10.1007/978-3-031-22419-5_11

Conservation has been traditionally based on visual assessments, either remote or *in situ*, of the environment; however, an emerging field called Ecoacoustics [10] is intersecting methods from Ecology, Signal Processing and Pattern Recognition (PR[1]) for the remote and, hopefully, continuous monitoring of the acoustic nature of the habitats: the so-called *soundscapes*.

An ecoacoustic monitoring system is composed by the traditional blocks of a PR system [7]; namely: sensing, segmentation, feature extraction/selection and classification. The first block corresponds to the usage of electronic devices for the recording of audio files; the second one consists in cutting long audio recordings into shorter signals, by typically applying energy-based thresholds or just a length-based splitting criterion; the third one provides measurements of informative and discriminative properties of the the segmented signals, in order to facilitate their vectorial representation for the subsequent step; finally, the last block corresponds to the application of a rule that, based on a set of examples and typically operating in a vector space, allows to discriminate among different classes of soundscapes or audible events.

The step of feature extraction—or, more in general, called *representation*—is of utmost importance for the PR pipeline; in fact, according to [6], any ill-definition in this step cannot be improved later in the PR process. Many different acoustic features[2] have been proposed in Ecoacoustics, ranging from traditional signal processing measurements to specialized acoustic attributes. Nevertheless and in contrast to our human intuition, considering a large number of features does not necessarily leads to a better representation; indeed, the opposite may occur, i.e. that the representation—and, therefore, the whole PR system—deteriorates when considering a large number of features. Such a deterioration might be in terms of reduction of classification accuracy, difficulty for the human interpretation of the results or prohibitive speeds and costs for operating the system. That is why selecting a proper subset of features is required. Depending on the application, different feature selection methods are available either for supervised [12, Chap. 7] or unsupervised problems [14].

Assessing the state of conservation in a place is a frequent task in Ecology, often defining it as a supervised two-class problem for categorizing an ecosystem into either healthy (class#1) or poorly conserved (class #2). Even though such an evaluation, in many cases, can be directly inferred from the visual landscape, the audible component may contribute with additional insights as well as to provide a non-intrusive alternative to the human visits that might change the behavior of the vocally-active species. Moreover, exploratory studies about soundscapes are of interest to characterize habitats whose acoustic properties are barely known or just qualitatively assessed. That is precisely the case of dis-

---

[1] PR is, according to [1], a branch of AI. Other authors highlight practical and philosophical differences between them, see https://bit.ly/3slL6Ht. Anyway, both disciplines are closely related, showing a significant overlap in their techniques.

[2] In Ecoacoustics, authors prefer to use the name *Acoustic indexes* instead of *Acoustic features*. However, in this paper we use the latter in order to adhere to the standard PR terminology. Other synonyms are *Acoustic variables* and *Acoustic attributes*.

tinguishing between different states of conservation of the ecosystems through the analysis of their corresponding soundscapes.

Recently, the authors of [13] discussed the importance of Ecoacoustics to study Colombian ecosystems and presented examples of highly and moderately transformed Colombian soundscapes, namely to distinguish between i) pure coffee plantations (monoculture) or mixed with forest patches; ii) "páramos"[3] suffering low or high transformations and iii) tropical dry forests also exhibiting either low or severe transformations of the landscape. Those examples were presented with both spectrograms and audio files but, to the best of our knowledge, have not yet been analyzed in terms of neither quantitative properties nor to select which acoustic features are the most discriminative to distinguish between the two categories of conservation for each ecosystem.

The aim of this paper is, therefore, to consider the variety of acoustic features available in the ecoacoustic literature and evaluate their discrimination power to distinguish between the two above-mentioned conservation categories of two Colombian soundscapes. Several supervised feature selection methods are used for both selecting the best individual features and finding the best sets of them. The remaining part of the document is as follows. A summary of acoustic features and supervised feature selection methods is presented in Sect. 2. Afterwards, the results of their application to the examples of Colombian soundscapes are presented and discussed in Sect. 3. Finally, our concluding remarks, along with future work directions, are given in Sect. 4.

2   Background and Methods

2.1   Acoustic Features

As stated above, a variety of acoustic features have been proposed in the literature. The main ones, along with their Python implementations, were recently outlined in [8] by Eldridge and Guyot. They grouped the acoustic features into three categories, namely: i) ecoacoustic features, which are particularly designed for characterizing soundscapes and mainly focused on studying complexity from an ecological point-of-view; ii) spectral features, which are directly computed from the spectrogram and are generic to characterize any audio signal and iii) temporal features, which are directly computed from the signal waveforms.

In this work we consider the acoustic features surveyed by Eldridge and Guyot; in addition, the recently proposed Ecoacoustic Global Complexity Index (EGCI) [4] was also considered and computed by using the implementation available in [3]. A list of the acoustic features considered in our study is shown in Table 1. Notice that several features correspond to summary statistics of the same attributes. As a result, many of them are potentially redundant. Therefore, applying feature selection methods is required in order to avoid such a redundancy as well as to reduce the dimensionality of the feature space.

---

[3] A "Páramo" is an Andean high mountain biome, roughly delimited between 2700 and 3700 m above the sea level.

Table 1. Acoustic features considered in this study.

ID number	Feature names
1 to 7	Acoustic Complexity Index (main, min, max, mean, median, std, var)
8, 37	Acoustic Diversity Index (main, NR main)
9, 38	Acoustic Evenness Index (main, NR)
10, 39	Bioacoustic Index(main, NR)
11	Normalized Difference Sound Index
12 to 17	RMS energy (min, max, mean, median, std, var)
18 to 23	Spectral centroid (min, max, mean, median, std, var)
24, 40	Spectral Entropy (main, NR)
25	Temporal Entropy
26 to 31	Zero Crossing Rate - ZCR (min, max, mean, median, std, var)
32 to 35	Wave Signal to Noise Ratio (SNR, Acoustic activity, Acoustic events, Average duration)
36	Number of Peaks
41	Normalized Entropy
42	Ecoacoustic Global Complexity Index

2.2 Feature Selection Methods

Feature selection is aimed at finding an optimal feature subset, such that the risk of the so-called curse of dimensionality [11, Sec. 3] is avoided. The simplest options to select features consist in removing or fusing those that are highly correlated or judging their individual ability to discriminate between the classes. However, these simple methods are often inconvenient to be applied because, firstly, non-linear correlations are difficult to assess and, secondly, a well-known fact in pattern recognition is that the best individual features does not necessarily compose the best subset [5].

Finding an optimal feature subset is a combinatorial search problem and, therefore, may turn computationally very costly. As a solution, A number of suboptimal feature selection methods are available for the supervised case which differ in the applied evaluation criteria and the chosen search algorithms. Among the evaluation criteria, those based on the ratio of inter-intra class distances and the classification performance of a parameterless classifier—typically the nearest neighbor (NN) rule—are often preferred.

In this study we considered the main selection methods whose implementations are available in PRTools[4]. A brief description of them is provided below (further details can be found in [12, Chap. 7]).

- *Individual selection*: it provides a ranking of the features according to their independent ability to separate the classes.

---
[4] http://37steps.com/prtools.

- *Forward selection and backward selection*: The search of a subset of features can be done either by adding features (named as forward search) or sequentially removing them (backward search). The latter requires more computations that the first one [11, Table 5]; thereby, the forward selection is typically preferred over the backward one.
- *Float selection*: A refined version of the forward selection, known as its floating variant. It provides the best tradeoff between a nearly optimal result and an affordable computational cost to perform the search.
- *Branch-and-bound*: this method is recursive but computationally tractable; moreover, it guarantees that the optimal subset is found, provided that the evaluation criterion—i.e. the objective function of the optimization procedure—is monotonous. However, this monotonicity requirement is difficult to be satisfied; indeed, classification performance while including/excluding features is far from being monotonously increasing or decreasing. The implementation also requires the specification of a desired cardinality of the feature subset.

3   Results and Discussion

For the experiments, we considered four 10-minute audio recordings[5] provided by Instituto Humboldt, which correspond to each one of the four soundscape categories: a) pure coffee plantations (monoculture) or b) mixed with forest patches; c) "Páramo" with low transformations or d) "Páramo" highly transformed. Soundscapes a) and b) are associated to the first classification problem and, similarly, soundscapes c) and d) constitute the second one. The spectrograms of the four audio recordings are shown in Fig. 1.

The audio signals for "Páramo" were recorded at Chingaza Natural Park and those of the coffee plantations were acquired in locations near Belén de Umbría. The four recordings were sampled at 48 kHz and cleaned by using a 1 kHz high-pass filter to remove wind noise perturbations [13]. In order to build a dataset, the 10-min length files were segmented into observations (windows) lasting for 1 min each and with a 50% of overlap (i.e. an overlap of 30 s) Afterwards, the 1-minute segments were characterized by computing all the acoustic features listed in Table 1. As a result, each soundscape class is exemplified by 19 points in the feature space. Consequently, the data matrices corresponding to the two-class problems contain 38 observations and 42 features each.

The very first inspection consists in computing the correlation matrices for both classification problems. They are shown in Fig. 2. Notice that the absolute values of the correlation coefficients are shown instead of their signed versions. This is because, even though the sign allows to distinguish the direction of the relation (either positive or negative), here we are only interested in its strength: the darker the entry, the stronger the relationship between the pair of acoustic features. Some groups of acoustic features are clearly highly correlated; see for

---

[5] The audio files are available at: http://colecciones.humboldt.org.co/rec/sonidos/publicaciones/ret2019/.

instance the top-left corner in Fig. 2(a), which correspond to all the summary statistics of the Acoustic Complexity Index.

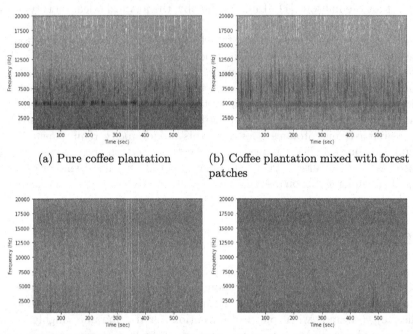

(a) Pure coffee plantation

(b) Coffee plantation mixed with forest patches

(c) "Páramo" with high transformation (d) "Páramo" with low transformation

Fig. 1. Spectrograms of the four Colombian soundscapes considered in this study.

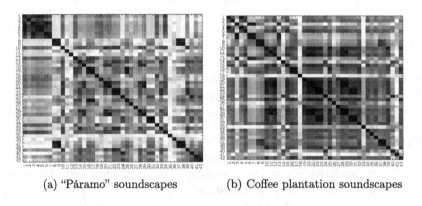

(a) "Páramo" soundscapes

(b) Coffee plantation soundscapes

Fig. 2. Correlation matrices for the two studied classification problems.

After inspecting the correlation matrices, we proceeded to apply four of the above-mentioned feature selection methods, namely: individual selection, forward selection, float selection and branch-and-bound. When a desired dimensionality (size of the feature subset) was not specified, we only use float selection due to its appropriate tradeoff between quality of the result and computational cost. Moreover, considering the limited amount of available observations per dataset, we also restrict ourselves to apply the leave-one-out NN classification performance as evaluation criterion for all the methods; indeed, according to [11], the NN performance is a common exemplar criterion. Moreover, other methods such as those based on either the inter-intra distances or the sum of Mahalanobis distances tend to be ill-posed for small-sample problems due to the inversion of almost singular matrices.

3.1 Individual Feature Selection

The individual selection returns a ranking of the features, in descending order of importance to discriminate between the classes. Below we report the rankings for both soundscape classification problems. In order to save space, we refer to the ID number of the features instead of their names, as they were codified in Table 1.

- *Páramo:* In this problem, the ranking of the features is the following one: 5, 1, 4, 35. 18, 3, 2, 11, 40, 34, 13, 19, 24, 6, 7, 26, 27, 32, 38, 8, 37, 9, 12, 20, 30, 31, 42, 14, 16, 17, 25, 41, 10, 15, 21, 29, 33, 39, 22, 23, 36 and 28. Notice that the best individual feature to discriminate between the two classes of Páramo soundscape is the median of the Acoustic Complexity Index, followed by two features corresponding to the main value and the mean of the very same acoustic attribute. Therefore, all in all, the main information to discriminate in this problem is provided by the Acoustic Complexity Index.
- *Coffee:* In this problem, the ranking of the features is the following one: 10, 11, 12, 15, 19, 20, 21, 28, 29, 18, 25, 5, 22, 23, 27, 30, 31, 32, 13, 16, 17, 34, 35, 36, 39, 42, 2, 24, 9, 26, 1, 4, 14, 3, 38, 8, 37, 6, 7, 33, 40 and 41. In this case, the best individual feature corresponds to the main value of the Bio Acoustic Index, followed by the Normalized Difference Sound Index and, afterwards, by the minimum of the RMS Energy. Notice that, in contrast with the behavior observed for the Páramo problem, in the case of the Coffee soundscape classification problem there is no a dominant attribute among the first ones of the individual ranking. Moreover, it is interesting that the minimum of the RMS Energy appears among the best ranked ones, indicating that the presence/absence of an energetic acoustic source is a key factor for the discrimination.

The individual selection provides some insights but, as stated above, it is well-known that the best individual features do not necessarily form the best subset; so, the performance when using combinations of features must be explored. In particular, we looked for the best pair, the best triple and the best group of features. The first two cases were explored for the sake of a direct visualization of the corresponding scatter plots, the last case was considered in order to remove any conditioning on the resulting dimensionality of the feature vectors.

3.2   Selection of the Best Pair of Acoustic Features

The best pair of features was searched by using all the selection methods described in Sect. 2.2—except backward search—and using the NN performance as evaluation criterion. The selected features per method are reported below. The corresponding scatter plots are shown in Fig. 3.

Páramo:
- *Float* and *Forward:* 5 (Median of the Acoustic Complexity Index) and 9 (Main value of the Acoustic Evenness Index)
- *Branch-and-bound:* 35 (Average duration of the Wave SNR) and 5 (Median of the Acoustic Complexity Index)

Coffee:
- *Float*: The method only returns feature 10 (Main value of the Bio Acoustic Index) as the selected one.
- *Forward*: 10 (Main value of the Bio Acoustic Index) and 2 (Minimum of the Acoustic Complexity Index)
- *Branch-and-Bound*: 28 (Mean of the ZCR) and 29 (Median of the ZCR)

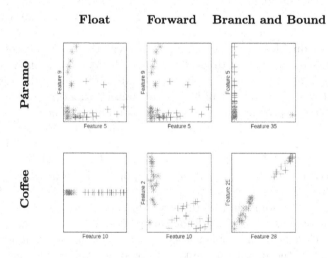

Fig. 3. Scatter plots for the best pairs of acoustic features per classification problem and selection method.

3.3   Selection of the Best Triple of Acoustic Features

Similarly, the same methods and evaluation criterion were used for selecting a subset of three features. The selected triples of features per method are reported below. Their associated scatter plots are shown in Fig. 4.

Páramo:

- *Float*: This method reports that only the following two features are needed: 5 (Median of the Acoustic Complexity Index) and 9 (Main value of the Acoustic Evenness Index)
- *Forward*: 5 (Median of the Acoustic Complexity Index), 9 (Main value of the Acoustic Evenness Index) and 8 (Main value of the Acoustic Diversity Index)
- *Branch-and-bound*: 4 (Mean of the Acoustic Complexity Index), 35 (Average duration of the Wave SNR) and 5 (Median of the Acoustic Complexity Index)

Coffee:

- *Float*: A single feature is required, namely feature 10 (Main value of the Bio Acoustic Index)
- *Forward*: 10 (Main value of the Bio Acoustic Index), 2 (Minimum of the Acoustic Complexity Index) and 4 (Mean of the Acoustic Complexity Index)
- *Branch-and-Bound*: 21 (Median of the Spectral Centroid), 28 (Mean of the ZCR) and 29 (Median of the ZCR)

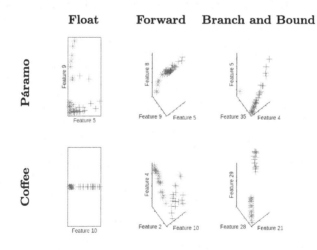

Fig. 4. Scatter plots for the best triples of acoustic features per classification problem and selection method.

In the case of the Coffee classification problem, it is noteworthy that Branch-and-bound selected features 28 (Mean of the ZCR) and 29 (Median of the ZCR); that is, two summary statistics of the same acoustic feature (ZCR) that are clearly highly correlated as can be observed in Fig. 2(b).

An interesting observation in both classification problems, as well as for results in Sects. 3.2 and 3.3, is that float and forward tend to coincide in the composition of the selected acoustic features. Remember that the float method is a modified version of the forward one and, thereby, it is expected that their results are similar. However, notice that float consistently selects subsets smaller than the specified dimensionality; that is, a singleton for Coffee even though either a pair or a triple of features was desired and, similarly, a pair of features for Páramo when a subset of size three was looked for.

3.4   Selection of the Best Subset of Acoustic Features

Lastly, no dimensionality was specified when using float selection, such that the resulting subset is the best one found by that method with the NN criterion. For each classification problem, the best subsets were the following ones:

Páramo: The best subset is composed by the following three features: 5 (Median of the Acoustic Complexity Index), 11 (Main value of the Normalized Difference Sound Index) and 2 (Minimum of the Acoustic Complexity Index)
Coffee: The best feature subset is a singleton: 10 (Main value of the Bio Acoustic Index)

Since the best found subsets are three-dimensional and one-dimensional, respectively, the corresponding scatter plots can be visualized; see Fig. 5.

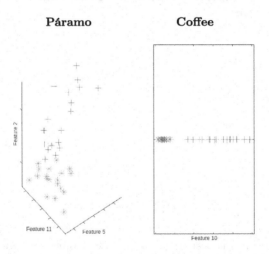

Fig. 5. Scatter plots for the best subsets of acoustic features, per classification problem, found by float selection.

Notice that, in the case of Páramo, the best subset selected by the float method is a triple that includes two statistics of the Acoustic Complexity Index, namely its median (feature 5) and minimum (feature 2) values. Indeed, the correlation coefficient between those features is high, as can be confirmed with a visual inspection of the top-left corner in Fig. 2(a). Therefore, in this case, the method fails in finding a subset of informative but independent acoustic features. Regarding this observation, remember that the optimal solution is only guaranteed for float selection if a monotonic function is optimized.

4 Conclusion

In this work, a supervised selection of acoustic features was undertaken in the context of discriminating between two classes for, in turn, two classification problems of Colombian soundscapes. A collection of well-documented acoustic features was used as the initial set, finding that many of them correspond to summary statistics of the very same attributes (such as mean, mode, standard deviation, min/max, among others), implying that a strong redundancy is expected as confirmed by the entry values in the correlation matrices. Agreements—and some discrepancies—among the feature subsets selected by the methods were observed; such disagreements are explained, at least partially, by the non-monotonicity of the objective function involved in the selection procedure, which does not allow to guarantee that an optimal subset of acoustic features is found.

A notorious fact was that the acoustic complexity index was preferred by the selection methods for the Páramo problem and the bio acoustic index for the Coffee problem; the first one quantifies the inherent irregularity in biophony and the second one estimates the richness or abundance of vocally-active species by inspecting how much energy is concentrated between 2 and 11 kHz [2]. Indeed, by listening to the audio recordings we noticed that the Páramo soundscape with low transformation contains more bird sounds than the amount that are heard in the one with high transformation and, therefore, a lowly-transformed Páramo is a more complex soundscape. Similarly, an examination of the Coffee audio recordings confirm that the pure soundscape includes a persistent background sound (a water stream), which is absent in the mixed soundscape. This difference might be crucial for the discrimination.

In addition to the two-class classification problem, a quantification of the degree of transformation might be derived as a measure of the proximity of a feature point (that represents a query soundscape) to the decision boundary. Future work includes the application of alternative supervised strategies, for instance those using evolutionary computation (e.g. genetic algorithms), as well as the application of unsupervised feature selection methods [14] and non-negative matrix factorization algorithms.

Acknowledgements. The authors acknowledge support from Universidad Nacional de Colombia - Sede Manizales and Ministerio de Ciencia, Tecnología e Innovación - *Minciencias*, República de Colombia [Programa de excelencia doctoral del Bicentenario, 2019].

References

1. Boesch, G.: What is Pattern Recognition? A Gentle Introduction, July 2021. https://viso.ai/deep-learning/pattern-recognition/

2. Bradfer-Lawrence, T., et al.: Guidelines for the use of acoustic indices in environmental research. Methods Ecol. Evol. **10**(10), 1796–1807 (2019). https://doi.org/10.1111/2041-210X.13254

3. Colonna, J.G.: EGCI: A Python library (2020). https://github.com/juancolonna/EGCI

4. Colonna, J.G., Carvalho, J.R.H., Rosso, O.A.: Estimating ecoacoustic activity in the Amazon rainforest through information theory quantifiers. PLOS ONE. **15**(7), 1–21 (2020). https://doi.org/10.1371/journal.pone.0229425

5. Cover, T.M.: The best two independent measurements are not the two best. IEEE Trans. Syst. Man Cybern. SMC. **4**(1), 116–117 (1974). https://doi.org/10.1109/tsmc.1974.5408535

6. Duin, R.P.W., Pękalska, E.: Open issues in pattern recognition. In: Kurzyński, M., Puchala, E., Woźniak, M., Żolnierek, A. (eds.) Computer Recognition Systems: Proceedings of the 4th International Conference on Computer Recognition Systems CORES 2005. Advances in Soft Computing, vol. 30, pp. 27–42. Springer Verlag, Berlin (2005). https://doi.org/10.1007/3-540-32390-2_3

7. Duin, R.P.W., Roli, F., de Ridder, D.: A note on core research issues for statistical pattern recognition. Pattern Recogn. Lett. **23**(4), 493–499 (2002). https://doi.org/10.1016/s0167-8655(01)00181-7

8. Eldridge, A., Guyot, P.: Acoustic_Indices: A Python library (2019). https://github.com/patriceguyot/Acoustic_Indices

9. Fang, F., Tambe, M., Dilkina, B., Plumptre, A.J. (eds.): Artificial Intelligence and Conservation. Artificial Intelligence for Social Good, Cambridge University Press, Cambridge, UK (2019). https://doi.org/10.1017/9781108587792

10. Farina, A.: Ecoacoustics: a quantitative approach to investigate the ecological role of environmental sounds. Mathematics. **7**(1), 1–9 (2019). https://doi.org/10.3390/math7010021

11. Jain, A., Duin, R., Mao, J.: Statistical pattern recognition: a review. IEEE Trans. Pattern Anal. Mach. Intell. **22**(1), 4–37 (2000). https://doi.org/10.1109/34.824819

12. Lei, B., et al.: Classification, Parameter Estimation and State Estimation: An Engineering Approach Using MATLAB, 2 edn. Wiley, Chichester, (2017). https://doi.org/10.1002/9781119152484

13. Rodríguez-Buriticá, S., et al.: Paisajes sonoros de Colombia: La otra dimensión de la biodiversidad. In: Moreno, L.A., Andrade, G.I., Gómez, M.F. (eds.) Biodiversidad 2018: Estado y tendencias de la biodiversidad continental de Colombia, chap. 103, p. Ficha 103. Instituto de Investigación de Recursos Biológicos Alexander von Humboldt, Bogotá, Colombia, September 2019. http://reporte.humboldt.org.co/assets/docs/2018/1/103/biodiversidad-2018-103-ficha.pdf, ficha metodológica disponible en: http://reporte.humboldt.org.co/assets/docs/2018/1/103/biodiversidad-2018-103-fichametodologica.pdf

14. Solorio-Fernández, S., Carrasco-Ochoa, J.A., Martínez-Trinidad, J.F.: A review of unsupervised feature selection methods. Artif. Intell. Rev. **53**(2), 907–948 (2019). https://doi.org/10.1007/s10462-019-09682-y

15. Tomašev, N., et al.: AI for social good: unlocking the opportunity for positive impact. Nat. Commun. **11**(1), 2468 (2020). https://doi.org/10.1038/s41467-020-15871-z

A Multi-objective Hyperparameter Optimization for Machine Learning Using Genetic Algorithms: A Green AI Centric Approach

André M. Yokoyama[1] , Mariza Ferro[1,2(✉)] , and Bruno Schulze[1]

[1] National Laboratory for Scientific Computing, Petrópolis, RJ, Brazil
{yokoyama,mariza,schulze}@lncc.br
[2] Institute of Computing, Fluminense Federal University, Niterói, RJ, Brazil

Abstract. This work proposes, develops, and evaluates an approach to improve the efficiency of ML models. This approach is centered on a Green AI, and the models' efficiency is a trade-off of accuracy, time to solution, and energy consumption. This leads to a multi-objective optimization problem implemented through the Genetic Algorithms (GA). We present the GA scheme and operators designed for this work focused on the architecture and hyperparameter optimization of ML pipeline, developed to be part of an AutoML solution. GA was evaluated for the XGBoost algorithm and results show the effectiveness of the GA for this multi-objective optimization. Also, it was possible to reduce energy consumption with minimal losses of predictive performance.

Keywords: Genetic algorithms · Green AI · Auto machine learning

1 Introduction

Artificial Intelligence (AI), especially its subarea, Machine Learning (ML), touches almost every part of our lives. It has attracted considerable research interest. We repeatedly hear about AI's benefits to our society, helping us solve society's challenges. On the other side, much has been debated about the negative impacts of AI on ethical and environmental issues. This negative view is intimately related to AI data-driven approaches, like ML, which has been trained using increasingly large datasets and requires significant computation times and capacities to fine-tune these models, leading to high energy consumption and carbon emissions.

A more pressing concern about how modern AI directly creates amounts of emissions from the data-intensive training of ML algorithms came after the study published in June 2019 by the College of Information and Computer Sciences at the University of Massachusetts [28]. The study pointed out that a common Natural Language Processing (NLP) model training and tuning pipeline using Deep Learning (DL) produced the same amount of carbon dioxide equivalent (CO_2e)

A. C. Bicharra Garcia et al. (Eds.): IBERAMIA 2022, LNAI 13788, pp. 133–144, 2022.
https://doi.org/10.1007/978-3-031-22419-5_12

as five cars during their lifespan. In sequence, [27] demonstrated through case studies that ML systems can significantly contribute to CO_2 emissions and introduced the term Green AI, in which our work is centered, an area of *"AI research that is more environmentally friendly and inclusive"* [27]. The recent work [1] made a fuss involving AI and one Big Tech, arguing that the current trajectory of AI, with NLP models trained on increasingly large data sets, is unsustainable and harmful in several ways, including its massive carbon footprint.

One approach that has shown promise in reducing these costs is the development of Automated Machine Learning (AutoML). In AutoML [13], the goal is to automate the development of parts, or even the entire pipeline, of an ML model. There are already several AutoML frameworks and tools available [7]. However, the most proposed approaches only seek to maximize the predictive performance in a given task without considering aspects related to energy efficiency.

Thus, the main contribution of this paper is to propose, develop and evaluate an approach to improve the efficiency of ML models. The term efficiency encompasses the accuracy of the models, the time to reach a solution, and its energy consumption. This leads to a multi-objective optimization problem implemented through the Genetic Algorithms (GA) [11] for the model generation part of the ML pipeline focusing on the architecture and hyperparameter search. Finally, it is being developed to be part of an AutoML solution to increase ML pipeline automation's efficiency. We describe implemented GA scheme and operators. Experiments are performed for the XGBoost ML model, and the results show the effectiveness of the GA for multi-objective optimization. Also, it was possible to reduce energy consumption with minimal losses of predictive performance.

2   Background and Related Works

This section briefly introduces the domain of Automated Machine Learning and Green AI with some background information and its state of the art, presenting the main works related to our proposal. The aim is not to do a literature review, which can be found in [6,7,13,21] about AutoML and about GreenAI in [8] and in the aforementioned references [1,27,28].

Schwartz et al. [27] introduce the concept of Red AI, which refers to AI research that seeks to improve accuracy through massive computational power while being both environmentally unfriendly and prohibitively expensive, raising barriers to participation. The costs of the actual state-of-the-art AI research limit the ability of many researchers to study it and practitioners to adopt it. The use of massive data and large amounts of computation in tuning hyperparameters of computationally intensive models creates barriers for many researchers to reproduce these models' results and train their models on the same setup. Red AI research has yielded valuable scientific contributions to the field, but it is time to increase the prevalence of Green AI. Creating efficiency in AI research will decrease its carbon footprint and increase its inclusivity as AI study should not require large budgets and computational power. Green AI refers to research that yields novel results while considering the computational cost [8].

AutoML is an approach that has shown great promise and has been among the AI trends in the coming years[1] since ML model trained to one particular dataset would not work well in another one. So there is a need to create separate ML models for each dataset, including steps of data preparation, feature engineering, algorithm selection, and hyperparameter tuning. Most of these steps require trial and error approaches, making the process inefficient and costly (economic, time and environmental). This has motivated the development of AutoML [13], which aim to automate parts or even the entire pipeline of building an ML model. The pipeline can involve several processes: data preparation, attribute engineering, model construction and evaluation. The model building process consists of the selection of algorithms and optimization methods. The optimization methods are further divided into hyperparameter optimization (HPO) and architecture optimization (AO) or Neural Architecture Optimization (NAS) when it involves only neural network models [13]. In HPO the former indicates the parameters related to training (e.g., learning rate and batch size), and the latter indicates the parameters related to the model (for example, the number of layers for neural network architectures or the number of trees in XGBoost).

The works found in the literature are distinguished mainly by the optimization technique used in their formulation, the most common being the Bayesian [9], Reinforcement Learning [14], Random Search [19] and Evolutionary Algorithms (EA) [25]; by the type of input data being tabular, text, images and time series; and by the *pipeline* processes that are involved [7]. However, what they all have in common is that the formulation of the optimization process most only involves maximizing the predictive performance (Red AI approach), without considering the energy consumption as our proposal, which is Green AI centric.

Among the optimization techniques used in the formulation of AutoML approaches, the methods based on EA, more specifically the Genetic Algorithms [15], have shown good results to optimize different ML models [10,12, 16,17,20,29,31]. Several works with AutoML are for DL models and using GA as optimizer [5,12,20,26,29,30]. This is justified by the fact that, as already mentioned, these algorithms have achieved surprising results for unstructured data. Furthermore, these models involve a large number of hyperparameters to be adjusted. At the same time, eXtreme Gradient Boosting (XGBoost [3]) is very effective for structured data. It is among the most used ML algorithms for all kinds of data science problems and responsible for solving and winning most of Kaggle's challenges [18].

There is a growing number of tools for AutoML: Auto-Weka, Auto-sklearn [9], TPOT [23], Autokeras, Auto PyTorch, H2O [19] and rminer[2] [7], and others [22].

Multi-objective optimization is implemented in some existing AutoML solutions and various criteria can be involved in its design. AutoxgboostMC [24] proposes Bayesian approach for optimize predictive performance, fairness and interpretability in HPO process for XGBoost. [25] proposes an EA as a model

---

[1] Gartner https://www.gartner.com/smarterwithgartner/top-trends-on-the-gartner-hype-cycle-for-artificial-intelligence-2019.

[2] https://cran.r-project.org/web/packages/rminer/rminer.pdf.

design to be implemented as part of the AutoML framework FEDOT. Two optimization objectives are used: solution quality and chain complexity.

Most of AutoML approaches are single-objective and to the best of our knowledge, existing multi-objective optimization approaches do not consider energy consumption. In this paper, we present the multi-objective proposal using GA, the specially designed crossover, mutation and selection operators.

3   Proposal of Multi-objective Optimization with GA

Among the optimization techniques used in AutoML formulation, including the hyperparameter and architecture optimization, Genetic Algorithms [11] have shown good results to optimize different ML models. GA are stochastic computational device that allows an effective search in very large search spaces. Mathematically GA mimic the mechanisms of natural evolution of species, comprising processes of genetic evolution of populations, survival and adaptation of individuals [4]. These algorithms have a lower probability of getting stuck in a local optima than most optimization algorithms because they perform a global search in the solution space. GA are considered one of the most efficient techniques for multi-objective optimization and, according to [25], one of the potential choices for AutoML creating stable pipelines and simpler to implement. But, in order to apply the GA we have to define specially designed genetic operators (crossover, mutation and selection) and a fitness functions. These special operators are important for processing the individuals described by a GA. They also offer the possibility to take into account multiple objective functions.

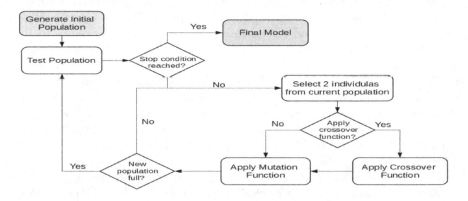

Fig. 1. Flowchart for the GA method workflow implemented in this work.

This section presents the GA design choices for a multi-objective approach centered in Green AI. The solution proposed in this work will be part of the development of AutoML framework, which aims to automate model construction process, focusing on HPO and AO/NAS optimization. The GA design has been developed for two ML models: XGBoost (XGB), for classification and regression (tabular data) and, CNN for image classification. The general workflow of our GA is presented in Fig. 1 and detailed next. The hyperparameter and architecture model tuning are treated as an optimization problem, where the multi-objective functions that we want to optimize are the predictive performance (accuracy or MSE), the energy and time to solution (Fitness function). The GA method (Fig. 1) and operators (in bold) we have designed are:

- The genetic evolution starts with the creation of an initial population consisting of randomly generated solutions, using the functions **generatePopulation** and **generateIndividual**. In our case, a solution is a combination of hyperparameters that defines the ML model's architecture and learning process, and are encoded in a binary format represented by a integer array of zeroes and ones, referred to as chromosome, and each element as allele. The **generatePopulation** initializes an array of pointers of length p that 'stores' the individuals created by the **generateIndividual** function that generates a random array of length l, where p is the number of individuals in the population and l is the minimum necessary number of "bits" to represent the highest value for each hyperparameter.
- After the population is created, each individual is tested and assigned a fitness score. This is the most time and energy-consuming part of the evolutive process. The fitness scores are given by the Eqs. 1 (for the XGB Classification) and 2 (XGB Regression):

$$f_i = \alpha * (a_i * 100) + \beta * \frac{1}{\frac{e_i}{\sum_{j=1}^{n} e_j}} + \gamma * \frac{1}{\frac{t_i}{\sum_{j=1}^{n} t_j}} \tag{1}$$

where a is the accuracy, e is the energy in Joules, t is the time in seconds, α, β, γ are the weights attributed to the relevance of the accuracy, energy and time respectively, and n is the total number of individuals in the population

$$f_i = \alpha * \frac{1}{\frac{m_i}{\sum_{j=1}^{n} m_j}} + \beta * \frac{1}{\frac{e_i}{\sum_{j=1}^{n} e_j}} + \gamma * \frac{1}{\frac{t_i}{\sum_{j=1}^{n} t_j}} \tag{2}$$

where m is the MSE, e is the energy in Joules, t is the time in seconds, α, β, γ are the weights attributed to the relevance of the MSE, energy and time respectively, and n is the total number of individuals in the population
- **Selection**: until the stop condition is not reached (for this work the number of generations), two individuals are selected from current population based on their fitness scores. The higher the score the more likely to be selected.
- **Crossover** function may be applied or not. The probability of the crossover parameter gives it: if set to 100 it is always applied; 0 it's never applied. If applied, a random allele is selected as the crossover point and all the alleles before it are copied directly to the offspring solutions. For example, individual "a" to offspring "a", and "b" to "b". The rest of the alleles, starting at the marked allele are copied from the other individual ("a" to "b" and "b" to "a"). If the crossover is not applied, the selected individuals' chromosomes are copied, as they are to the 2 new offspring solutions.

- The **Mutation** functions are applied to the offspring. The probability of mutations happening is given by the probability of mutation parameter, which is checked for every allele of offspring. If a mutation occurs, the value of the allele is inverted, i.e., 0 to 1 and 1 to 0; every time a mutation occurs, the probability is divided by 2, reducing the chance of another mutation happening in the same offspring.
- The two new offspring are then added to the new population pool along with the n best solutions of the current population, where n is given by the elitism parameter. This is due to preserving the best solution found up to that generation.
- The process of generating new individuals is repeated until the new population pool is full; once the new population is full, it is evaluated. Repeat this process until the stop condition is not reached.

This GA method was developed for XGB or CNN. But, it also could be used in general for other ML models-the same for the main operators like selection, mutation, and crossover functions. For example, at this moment, we implemented a simple selection function. In the future, other selection methods, such as the Tournament selection, could be implemented and evaluated. But, it will be applied to both ML models. However, the test population and fitness function presented are specific for XGB and, in future work, will be modified to include CNN HPO and NAS optimization. In addition, in our design, the strategy is a joint hyperparameter and architecture optimization. Most NAS methods fix the same setting of training-related hyperparameters during the whole search stage. After the search, the hyperparameters of the best-performing architecture are further optimized. However, this paradigm may result in sub-optimal results as different architectures tend to fit different hyperparameters [13].

4   Methodology and Experimental Setup

In order to evaluate the feasibility of the multi-objective optimization using our GA method and operators we evaluated our implementation for XGBoost algorithm using two datasets and seven experimental configurations.

XGBoost [3] is an implementation of gradient boosted tree algorithms. This technique, known as Boosting, is an ensemble learning technique that uses a set of base learners to improve the stability and effectiveness of an ML model. It's called gradient boosting because it uses a gradient descent algorithm to minimize the loss when adding new models. The central idea of boosting is the implementation of homogeneous ML algorithms in a sequential way, where each of these algorithms tries to improve the stability of the model by focusing on the errors made by the previous algorithm.

The dataset utilized for the classification task is Boson Higgs[3], whose objective is to classify if an event is Higgs bóson decay or not; and Seoul Bike Trip

---

[3] https://archive.ics.uci.edu/ml/datasets/HIGGS.

Duration[4] for regression, which predicts the travel duration of bicycles rented in Seoul, based on combining weather data. Boson Higgs has 11 million examples, 2 classes and 28 numerical features. Seoul has 9 million examples and 24 numerical features. None of the datasets have features with missing data.

A combination of four parameters of the XGBoost is tuned by GA, searching for the best optimization multi-objective. The HPO (N_jobs, ETA) and AO (N_estimators, *max_depth*) parameters are optimized jointly. The parameter N_jobs defines the number of cores in parallel that the processor will execute the algorithm. When N_jobs $= -1$, all cores are used, and for N_jobs $= 1$, only one core. The ETA parameter is the learning rate and defines the correction made at each boosting step. The N_estimators is the number of gradients boosted trees and the maximum tree depth for base learners is defined by *max_depth*. XGBoost splits up to the *max_depth* specified and then starts pruning the tree backward and removing splits beyond which there is no positive gain.

For the tests, we fixed the total length of the chromosome at 20 alleles, where the first 6 represent the ETA ranging from 0.01 to 0.64; the next 4 were used for the N_jobs, (1 to 16), followed by N_estimators, 9 alleles (9 to 520). The last 3 were used for the max_depth (3 to 10). The GA parameters were set as follows: population size $= 20$; number of generations $= 30$; elitism $= 4$; probability of crossover $= 80$; probability of mutation $= 10$.

Our scheme enables the multi-objectives of optimization to be defined according to the user's requirement. It's possible setting priorities for time, energy, and predictive performance. This defines the configuration of the fitness function of the GA (a combination of maximizing accuracy or minimizing MSE, minimizing time, and minimizing energy). These possibilities of configuration defined the methodology for experiments. We evaluated seven different experimental configurations, as described in Tables 1 and 2.

Table 1 shows the combinations of relevance, in percent, attributed to model accuracy, training time and energy consumed when training the model for the XGBoost for classification task (XGBC). Table 2 shows the combinations for the XGBoost for regression (XGBR). Note that the combination 1,1,1 indicates the 3 attributes have the same relevance, i.e., approximately 33.33%.

Table 1. Configurations of the relevance for the experimentes with XGBoost for classification

Configuration	Accuracy	Time	Energy
1	98	1	1
2	96	2	2
3	97	2	1
4	97	1	2
5	49	50	1

Table 2. Configurations of the relevance for the experimentes with XGBoost for regression

Configuration	MSE	Time	Energy
1	1	1	1
2	90	5	5

---

[4] https://www.kaggle.com/saurabhshahane/seoul-bike-trip-duration-prediction.

Test environment was a computer with an Intel Core i7-8700 CPU with 6 cores e 12 threads, frequency 3.20 GHz, memory 64 GB DDR4 (16 GB X4), 1 HD SATA Seagate de 2TB e 1 SSD Corsair MP510 de 240 GB of storage, and a GPU Nvidia GeForce RTX 2080Ti with a base clock of 1350 MHz e Boosted clock of 1545 MHz, and 11 GB of memory GDDR6 with ECC off, and OS Ubuntu 20.04.

XGBoost was implemented and executed with Python 3.8. The time and energy were measured with Perf Tool. The GA is implemented in C++. As a reproducibility compromise all the codes and configurations are available at: https://github.com/comcidis/GA_ML_XGB.git.

5   Results and Discussion

Figure 2 shows the fitness of each individual (bars) and the average of the population (line), for a sample test with a population of 10 individuals over 9 generations for XGBC. Generation 0 is the initial randomly generated population. For this test we used the elitism of 1, where only the best solution of the generation is passed on to the next, due to the small size of the population. The graph shows the evolution and that as new generations are created, it is possible to observe the improvement of the fitness function (the bigger, the better). This can be seen for the best individual (blue bar) and the general population (AVG line). This shows that you can improve the population, even with this small example with 10 individuals (for visualization purposes only).

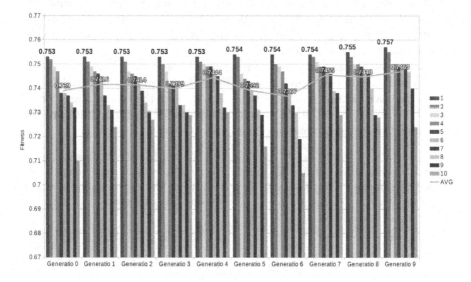

Fig. 2. Example of the evolution process with a population of 10 individuals.

Figures 3, 4, 5 and 6 show the Accuracy/MSE, time and energy, for the best solution for each configuration (Tables 1 and 2). Figure 3 compares the accuracy and time for the XGBC. Configuration 1 has a highest focus in the accuracy (Table 1). It had, as expected, the best accuracy at 76%, but the worst time at 266.99 s. For configuration 2, we slightly increased both time and energy relevance. Results had a significant drop in the time and a reduction of only one percentage point in the accuracy. Configurations 3 and 4 also show reductions in time, even though not as much as configuration 2, which had a smaller loss in accuracy. Configuration 5, which has high relevance to time, had the worse accuracy at 70.9%. It had the best time, only 26.28, less the 10% of the time of configuration 1. These results show the ability of GA to multi-objective optimization, even allowing for to definition of priorities. Furthermore, it is possible to see that the GA manages to optimize for the defined criteria.

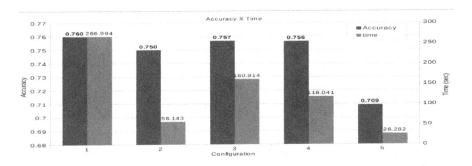

Fig. 3. Comparison of the Accuracy and Time for the XGBoost for classification.

In Fig. 4 it can be seem the the energy had the same behavior as the time.

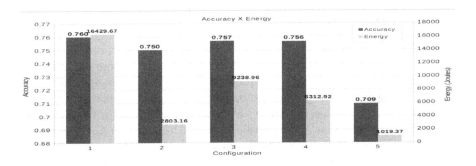

Fig. 4. Comparison of the accuracy and energy for the XGBoost for classification.

Figure 5 shows the comparison between the MSE and time for the XGBR, and Fig. 6 the MSE and Energy. The best time and energy were obtained in configuration 1, while the best MSE was obtained by configuration 2.

We compared these results with the work of [2] which trained XGBoost with the Higgs Boson dataset with 11 million examples searching by energy savings. We compared our results with the experiment of [2] with the default parameter of XGBoost, and also the configuration which obtained the highest accuracy. The configuration with the best accuracy achieved the same accuracy result as our configuration 1. However, their result was obtained in a shorter time (223.86 s in comparison with the 266.994 of ours and energy consumption of 13095.18J to 16429.67). For the default configuration, the accuracy of [2] was 74% with a time of 108.36 s and energy of 5647.08 J. Compared to our configuration 2 had the closest accuracy at 75%, but the total time was 56.143 s and energy 2803.16 J. It is about half of time and energy consumption.

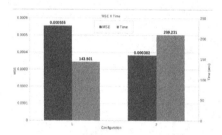 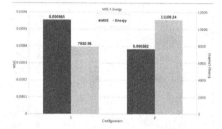

Fig. 5. Comparison of the accuracy and time for the XGBR.

Fig. 6. Comparison of the accuracy and energy for the XGBR.

Our results did not compare with other AutoML frameworks or GA approaches since it was not possible to find other works that implement multi-objective optimization evaluating energy and time in addition to precision.

6 Final Considerations

In this work, we designed, implemented, and evaluated a multi-objective optimization centered on Green AI, searching for an AI that is more environmentally friendly. It was implemented through GA using workflow and operators specially developed for this problem to be part of an AutoML solution. We performed the experimental set for HPO+AO joint optimization for the XGBoost algorithm for classification and regression tasks. Although the GA was evaluated using XGBoost, the workflow and some operators were designed to be used for ML models in general. The initial results showed our solution's effectiveness for multi-objective optimization using XGBoost. In addition, some experiments enable significant energy savings compared to manual hyperparameter tuning when searching for the same objectives. For future work, we are evolving this GA to include HPO+NAS for CNN model.

Acknowledgements. This work is funded by Faperj, CAPES and LNCC-MCTI. Projects GreenAI 21-CLIMAT-07 and SUSAIN Inria Associated Teams.

References

1. Bender, E.M., Gebru, T., McMillan-Major, A., Shmitchell, S.: On the Dangers of Stochastic Parrots: Can Language Models Be Too Big? Association for Computing Machinery, New York (2021)
2. Bernardo, F., Yokoyama, A., Schulze, B., Ferro, M.: Avaliação do consumo de energia para o treinamento de aprendizado de máquina utilizando single-board computers baseadas em arm. In: Anais do XXII Simpósio em Sistemas Computacionais de Alto Desempenho, pp. 60–71. SBC, Porto Alegre, RS, Brasil (2021). https://doi.org/10.5753/wscad.2021.18512
3. Chen, T., Guestrin, C.: XGBoost: a scalable tree boosting system. In: Proceedings of the 22nd ACM SIGKDD International Conference on Knowledge Discovery and Data Mining, pp. 785–794 (2016)
4. Colorni, A., Dorigo, M., Maniezzo, V.: Genetic algorithms and highly constrained problems: the time-table case. In: Schwefel, H.-P., Männer, R. (eds.) PPSN 1990. LNCS, vol. 496, pp. 55–59. Springer, Heidelberg (1991). https://doi.org/10.1007/BFb0029731
5. David, E., Greental, I.: Genetic algorithms for evolving deep neural networks. In: GECCO 2014 - Companion Publication of the 2014 Genetic and Evolutionary Computation Conference, July 2014. https://doi.org/10.1145/2598394.2602287
6. Doke, A., Gaikwad, M.: Survey on automated machine learning (AutoML) and meta learning. In: 2021 12th International Conference on Computing Communication and Networking Technologies (ICCCNT), pp. 1–5 (2021)
7. Ferreira, L., Pilastri, A., Martins, C.M., Pires, P.M., Cortez, P.: A comparison of AutoML tools for machine learning, deep learning and XGBoost. In: 2021 International Joint Conference on Neural Networks (IJCNN), pp. 1–8 (2021). https://doi.org/10.1109/IJCNN52387.2021.9534091
8. Ferro, M., Silva, G.D., de Paula, F.B., Vieira, V., Schulze, B.: Towards a sustainable artificial intelligence: a case study of energy efficiency in decision tree algorithms. Concurrency and Computation: Practice and Experience n/a(n/a), e6815, December 2021. https://doi.org/10.1002/cpe.6815
9. Feurer, M., Klein, A., Eggensperger, K., Springenberg, J.T., Blum, M., Hutter, F.: Auto-sklearn: efficient and robust automated machine learning. In: Hutter, F., Kotthoff, L., Vanschoren, J. (eds.) Automated Machine Learning. TSSCML, pp. 113–134. Springer, Cham (2019). https://doi.org/10.1007/978-3-030-05318-5_6
10. Ganapathy, K.: A study of genetic algorithms for hyperparameter optimization of neural networks in machine translation (2020)
11. Goldberg, D.E.: Genetic Algorithms in Search. 1st edn. Optimization and Machine Learning. Addison-Wesley Longman Publishing Co., Inc, USA (1989)
12. Hamdia, K.M., Zhuang, X., Rabczuk, T.: An efficient optimization approach for designing machine learning models based on genetic algorithm. Neural Comput. Appl. **33**(6), 1923–1933 (2021)
13. He, X., Zhao, K., Chu, X.: AutoML: a survey of the state-of-the-art. Knowl. Based Syst. **212**, 106622 (2021). https://doi.org/10.1016/j.knosys.2020.106622
14. Heffetz, Y., Vainshtein, R., Katz, G., Rokach, L.: DeepLine: AutoML tool for pipelines generation using deep reinforcement learning and hierarchical actions filtering. In: Proceedings of the 26th ACM SIGKDD International Conference on Knowledge Discovery & Data Mining, pp. 2103–2113. KDD 2020, Association for Computing Machinery, New York, NY, USA (2020). https://doi.org/10.1145/3394486.3403261

15. Holland, J.H.: Genetic algorithms. Scientific American, July 1992
16. Jian, W., Zhou, Y., Liu, H.: Densely connected convolutional network optimized by genetic algorithm for fingerprint liveness detection. IEEE Access **9**, 2229–2243 (2021). https://doi.org/10.1109/ACCESS.2020.3047723
17. Johnson, F., Valderrama, A., Valle, C., Crawford, B., Soto, R., Ñanculef, R.: Automating configuration of convolutional neural network hyperparameters using genetic algorithm. IEEE Access **8**, 156139–156152 (2020). https://doi.org/10.1109/ACCESS.2020.3019245
18. Kaggle: State of data science and machine learning 2021. Technical report (2021). https://www.kaggle.com/kaggle-survey-2021
19. LeDell, E., Poirier, S.: H2O AutoML: Scalable automatic machine learning. In: 7th ICML Workshop on Automated Machine Learning (AutoML), July 2020
20. Lee, S., Kim, J., Kang, H., Kang, D.Y., Park, J.: Genetic algorithm based deep learning neural network structure and hyperparameter optimization. Appl. Sci. (2021). https://doi.org/10.3390/app11020744
21. Nagarajah, T., Poravi, G.: A review on automated machine learning (AutoML) systems. In: 2019 IEEE 5th International Conference for Convergence in Technology (I2CT), pp. 1–6 (2019). https://doi.org/10.1109/I2CT45611.2019.9033810
22. Nikitin, N.O., et al.: Automated evolutionary approach for the design of composite machine learning pipelines. Future Gener. Comput. Syst. **127**, 109–125 (2022)
23. Olson, R.S., Bartley, N., Urbanowicz, R.J., Moore, J.H.: Evaluation of a tree-based pipeline optimization tool for automating data science. In: Proceedings of the Genetic and Evolutionary Computation Conference 2016, pp. 485–492. GECCO 2016, Association for Computing Machinery, New York, NY, USA (2016). https://doi.org/10.1145/2908812.2908918
24. Pfisterer, F., Coors, S., Thomas, J., Bischl, B.: Multi-objective automatic machine learning with AutoXGBoostMC (2019). https://doi.org/10.48550/ARXIV.1908.10796
25. Polonskaia, I.S., Nikitin, N.O., Revin, I., Vychuzhanin, P., Kalyuzhnaya, A.V.: Multi-objective evolutionary design of composite data-driven models. In: 2021 IEEE Congress on Evolutionary Computation (CEC), pp. 926–933 (2021). https://doi.org/10.1109/CEC45853.2021.9504773
26. Rani, R., Sharma, A.: An optimized framework for cancer classification using deep learning and genetic algorithm. J. Med. Imaging Health Inform. **7**, 1851–1856 (2017). https://doi.org/10.1166/jmihi.2017.2266
27. Schwartz, R., Dodge, J., Smith, N.A., Etzioni, O.: Green AI. Commun. ACM **63**(12), 54–63 (2020)
28. Strubell, E., Ganesh, A., McCallum, A.: Energy and policy considerations for deep learning in NLP. arXiv preprint arXiv:1906.02243 (2019)
29. Xiao, X., Yan, M., Basodi, S., Ji, C., Pan, Y.: Efficient hyperparameter optimization in deep learning using a variable length genetic algorithm (2020)
30. Young, S., Rose, D., Karnowski, T., Lim, S.H., Patton, R.: Optimizing deep learning hyper-parameters through an evolutionary algorithm. In: ACM Proceedings, pp. 1–5 (11 2015)
31. Yuan, Y., Wang, W., Coghill, G.M., Pang, W.: A novel genetic algorithm with hierarchical evaluation strategy for hyperparameter optimisation of graph neural networks. CoRR abs/2101.09300 (2021). https://arxiv.org/abs/2101.09300

Machine Learning

Evolving Node Embeddings for Dynamic Exploration of Network Topologies

Karen B. Enes[✉], Matheus Nunes, Fabricio Murai, and Gisele L. Pappa

Computer Science Department (DCC), Universidade Federal de Minas Gerais
(UFMG), Belo Horizonte, MG, Brazil
{karen,mhnnunes,fmurai,glpappa}@dcc.ufmg.br

Abstract. Static node embedding algorithms applied to snapshots of
real-world applications graphs are unable to capture their evolving process. As a result, the absence of information about the dynamics in these
node representations can harm the accuracy and increase processing
time of machine learning tasks related to these applications. We propose a biased random walk method named Evolving Node Embedding
(**EVNE**), which leverages the sequential relationship of graph snapshots
by incorporating historic information when generating embeddings for
the next snapshot. **EVNE** learns node representations through a neural network, but differs from existing methods as it: (i) incorporates
previously run walks at each step; (ii) starts the optimization of the current embedding from the parameters obtained in the previous iteration;
and (iii) uses two time-varying parameters to regulate the behavior of
the biased random walks over the process of graph exploration. Through
a wide set of experiments we show that our approach generates better
embeddings, outperforming baselines in a downstream node classification
task.

Keywords: Node embeddings · Evolving graphs · Representation
learning

1 Introduction

Real-world data and processes are often modeled as graphs, since these abstractions can express the relationships between entities of interest in an intuitive
and useful way. In many real networks, access to data associated to nodes and
edges is often difficult and/or costly. For this reason, it is not uncommon that,
at all times, a large fraction of the data we desire to model is unobserved. In
most cases, data is acquired through some sort of online search or exploration of
the graph, which can be seen as an evolving process that increases the available
knowledge about the network as the search progresses. At each step, information

---

Supplementary Information The online version contains supplementary material
available at https://doi.org/10.1007/978-3-031-22419-5_13.

about topology, nodes and edges is collected, but the full knowledge may never be attained [11].

In this work, we focus on addressing issues that arise with evolving graphs [7] whose network topology is partially unknown throughout the graph exploration. In this scenario, we start from a small set of nodes and edges and, at each step of the search (i.e. network snapshot), nodes and edges may be added or removed. As an example of evolving graphs, we have the process of spreading fake news on social media [17]. In this context, imagine that a few Twitter users post an initial tweet with fake news. Each of these users would then be modeled as a node in the graph, which is composed only by nodes in the beginning of the evolving process. After some time, other Twitter users start to post or retweet the same fake news. In this case, users posting the fake news are modeled as new nodes in the network and users retweeting the fake news are not only modeled as new nodes but also create an edge with the users from the original tweet. As the dissemination of the fake news goes on, the graph keeps growing with the addition of new nodes and edges.

Several machine learning methods have been developed for graph data in the past few years, mostly for tasks involving predictions over nodes and edges [3], by improving node and graph learning representations. Effective representations can improve both these types of predictions and the performance of downstream machine learning algorithms [1]. Also, graph embeddings were proven to be an excellent alternative for graph representation [1]. Graph embedding is a framework for building low dimensional representations of the entire graph, or parts of it, *e.g.* nodes, edges and sub-graphs, while preserving structural information and graph properties. In this work, we focus on *node embedding* techniques, which aim to represent each node as a d-dimensional feature vector (d is an input) that accurately captures its relationships to other nodes [1].

Although there is wide variety of node embedding approaches for static graphs [5,12,13] (see Sect. 2), all of the methods mentioned assume that the graph topology is fully observable and static (*i.e.*, does not change over time). In the scenario where the graph evolves, applying one of these methods to generate node embeddings consists of running it on snapshots taken from the network. In this case, each snapshot and respective embedding represents an independent network. This approach disregards the fact that each snapshot may contain relevant information about the next, and that incorporating past information into the learning procedure could improve the accuracy of downstream inference tasks. Moreover, the resources spent in computing previous embeddings are entirely wasted. As a result, this approach leads to decreased performance in machine learning tasks mainly due to the characteristic of adding or removing nodes and edges of growing graphs, which directly impacts on the stability of the generated embedding. Aiming to fill this gap, we provide the following contributions:

1: We propose a technique named **E**volving **N**ode **E**mbedding (**EVNE**), which instead of learning the node embeddings in each snapshot from scratch, incrementally builds the embedding at time t from the embedding at time $t - 1$.

2: We conduct an experimental study to evaluate the quality of the embeddings obtained for an evolving graph by **EVNE** in comparison to those obtained by state-of-the-art methods.

3: We evaluate the embeddings based on their performance on node classification tasks, using four different network datasets. We also conduct an ablation study to show the impact of each of the new mechanisms incorpotated into our model.

Our results show that **EVNE** generates better embeddings in a downstream node classification task when compared to node2vec and the other baselines in almost all scenarios, with gains up to 20%.

2   Related Work

Node embedding techniques have become the standard feature engineering paradigm for node representation in graphs [1], as they provide low dimensional representations of large adjacency matrices. The remainder of this section discusses the methods for two classes of node embedding methods: static and dynamic.

Static Node Embeddings: The first methods for static node embeddings were proposed in the past decade. We select 3 of these methods which will be used as baselines in our experiments: DeepWalk [12], LINE [13], and node2vec [5].

DeepWalk is an unsupervised feature learning model that learns latent representations of nodes in a network using information obtained from truncated random walks. It captures the structure of the graph regardless of node label distributions, allowing the same representation to be used across various classification problems. LINE, in turn, creates an embedding of large graphs while preserving their local and global structures, defined respectively in terms of first-order and second-order proximity. The model learns similar representations for nodes with high first and second order proximities.

Node2vec uses biased random walks to learn a low-dimensional embedding of a graph, preserving the neighborhoods of nodes, in a semi-supervised fashion. The model uses walks to relate nodes that belong to the same community (homophily) or nodes that play similar structural roles in the network (structural equivalence). These concepts define similarity in node2vec.

The methods described above have been developed for static graphic scenarios where the graph topology is completely observable upfront. Previous works have applied methods for generating static embeddings to evolving graph datasets. In this case, an embedding is generated for each snapshot of a network [2] that is treated by the algorithm as an independent network, not leveraging any information regarding the evolution of the network.

Dynamic Node Embeddings: Dynamic node embedding have been proposed as extensions to originally designed for static embeddings in order to capture the temporal evolution of real-world networks. The main idea behind these methods is to update the embeddings using information from the previous embeddings along with information from the current state of the graph [2].

Various methods proposed for generating embeddings in dynamic graphs are based on random walks and adapted to work with dynamic networks, including dynnode2vec [9] and HIN–DRL [10]. However, such approaches do not outperform the standard node2vec in well-know dataset (for more details, refer to [15]). The most of the recent methods are based on deep learning techniques, particularly, dyngraph2vec [2] and GraphSAGE [6]. Although they present competitive results in machine learning tasks when compared to static counterparts, the applicability of deep neural networks and deep autoencoders is restricted to graphs that are large enough to allow for learning a large number of parameters without overfitting.

Our Contributions. Unlike previous works, **EVNE** is a viable alternative for networks that evolve from very small graphs. As the method does not rely on deep learning architectures, it is possible to learn useful embeddings regardless of the graph size. **EVNE** updates the embeddings using information from the previous embeddings together with information from the current state of the graph without starting from scratch, as traditional dynamic node embeddings do. In sum, we explore the sequential relationship between graph snapshots by incorporating information from past snapshots into the generation of the embedding corresponding to the current snapshot. **EVNE** can handle small and large graphs with the addition and removal of nodes and edges. Table 1 highlights the most important differences among the main baselines for node embedding generation and **EVNE**.

Table 1. Comparison of baselines for node embedding generation. * refers to methods such as [2,4,6,8,14].

	DeepWalk	LINE	Dyn. DNN*	node2vec	**EVNE**
Fits to evolving graphs	-	-	✓	-	✓
Works with small graphs	✓	✓	-	✓	✓
Explor.-exploit. trade-off	-	-	-	✓	✓
Stable	-	✓	-	✓	✓
Scalable	✓	✓	✓	✓	✓

3   Evolving Node Embedding

Let $G = (V, E)$ be a graph where V is the set of nodes and E is the set of edges. We define a discrete evolving network as a sequence of graphs $\mathcal{G} = \{G_t\}_{t=1}^{T}$, in which $G_t = (V_t, E_t)$ is the network snapshot at exploration step $t = 1, \ldots, T$.

Given G, a node embedding is a mapping function f from nodes to feature vector representations, $f : V \rightarrow \mathbb{R}^d$, for some $d \ll |V|$. The function f is defined so that a similarity measure (e.g., cosine similarity) between $f(u)$ and $f(v)$ encodes some notion of proximity between $u, v \in G$. As a slight abuse of notation, we denote by $f_t(G_t) \in \mathbb{R}^{|V| \times d}$ the embedding matrix of all nodes in G_t. Hence, for evolving networks, we have $\mathcal{F} = \{f_1, f_2, \cdots, f_T\}$ as a series of $(|V_t| \times d)$-dimensional matrices, where f_t is the graph embedding matrix for G_t.

EVNE is an evolving node embedding algorithm that can be applied to both small and large evolving networks. We consider evolving graphs that start relatively small and evolve by the addition or removal of new nodes and edges, as in our fake news scenario described in Sect. 1. The purpose of **EVNE**'s embedding is to (i) preserve the aggregation of local and global structural characteristics from the exploration of the graph; (ii) account for information about the evolution of the network; and (iii) avoid the typical process of visiting and storing information about all network snapshots.

EVNE's challenge is to update the embedding for G_t from the embedding generated for G_{t-1}. Particularly, **EVNE** extends node2vec to handle evolving networks through three new mechanisms that allow for an evolving representation learning: (i) it concatenates the previous iteration's biased random walks to the current iteration's walks; (ii) it loads the previous snapshot's embeddings as the initial weights of the extended Skip-Gram model; and (iii) it uses a strategy for varying parameters p and q throughout the iteration over the snapshots.

In contrast to mechanism (i), in the original node2vec, random walks are independently generated for each time step, rendering it unable to capture temporal patterns across graph snapshots. To address this issue, we use both the walks sampled in step $t - 1$ and those sampled in step t when learning the embedding for G_t. This mechanism enables the inclusion of additional structural information identified in the previous step in the current embedding generation, improving structural knowledge and adding dynamic information to the embeddings generated by **EVNE**. Mechanism (ii) changes the way we start the training phase of Skip-Gram. The original Skip-Gram initializes the learning process from a vector of random weights. Instead, we initialize the weight vectors at time t using embeddings generated at time $t - 1$. This allows **EVNE** to capture structural changes in network nodes that were previously seen.

node2vec provides a flexible notion of neighborhood through the use of biased random walks that interpolate between a BFS and a DFS-like behavior. This interpolation is governed by two parameters: the return parameter p controls how far from the source node the walk will go, by defining the likelihood of returning to a node that was visited in the previous step; the in-out parameter q directs the walk towards nodes that are further away from the source. Random walks are in turn used to define the neighborhoods of the nodes by feeding the sequences of visited nodes to an extension of the Skip-Gram model [5].

We modify the original p and q parameters to change their values dynamically, allowing the balance between BFS and DFS biases to change over time. At the beginning, when the graph is small, we weaken the DFS behavior, reinforcing the BFS bias. Conversely, when the observed network becomes larger, **EVNE** reinforces the DFS bias in order to explore a larger region of the graph, while dimming the BFS behavior. This change can improve the quality of the embeddings by making the exploration more consistent over time, based on the size of the graph.

We define the values of p and q in two ordered lists of values ($\mathcal{P}$ and $\mathcal{Q}$), used as input to the algorithm. We start by creating the lists $\mathcal{P}$ and $\mathcal{Q}$ with all possible values p and q can assume during the network evolving process. Then, p and q will change over time according to the total number of iterations and size of the lists $\mathcal{P}$ and $\mathcal{Q}$ maintaining uniform intervals, so as to keep changes equally spaced in time. This means that, for example, if a network is observed for 200 timestamps and we have 4 different values for p in the input list $\mathcal{P}$, then the initial value will be set at iteration 0 and updated at iterations 50, 100 and 150. We opted for specifying the sequence of values for p and q to keep total control of the experimental process and ease of reproducibility.

In addition, in a scenario where nodes or edges are deleted, **EVNE** suppresses the representation of removed nodes and updates the representation of nodes kept in the graph. Moreover, the walks and valid weights from previous iterations are fed to the Skipgram model. Thus, structural changes in the graph associated with the removal of nodes/edges would be captured during **EVNE**'s representation update, a process analogous to the expansion of the graph.

EVNE's Algorithm: Algorithm 1 presents **EVNE**. It receives as inputs a set of snapshots $\mathcal{G} = \{G_t\}_{t=1}^T$, the embedding dimension d, the number of random walks r to be sampled per node, the walk length ℓ, the context window size for Skip-Gram k and the list of values for parameters p and q, $\mathcal{P}$ and $\mathcal{Q}$. It outputs a sequence of node embeddings $\mathcal{E} = \{G_t\}_{t=1}^T$, one for each graph G_t. When processing snapshot t, p_t and q_t are set as the $\lceil \frac{t}{T} L \rceil$ position on the lists $\mathcal{P}$ and $\mathcal{Q}$, where $L = |\mathcal{P}| = |\mathcal{Q}|$. At step t, the transition probability matrix $\mathbf{\Pi}_t$ – which governs the random walks – is initialized according to p_t, q_t and E_t. Next, for each node $u \in V_t$, we sample r walks of length ℓ per node using node2vec's biased random walk with transition probabilities given by $\mathbf{\Pi}_t$ (line 8). These walks are stored in $curWalks$ (line 9). Then, using stochastic gradient descent to minimize Skip-Gram's loss function, we obtain embedding f_t for snapshot G_t using both $prevWalks$ (from $t-1$) and $curWalks$ (from t), initializing the representations of all nodes observed in $t-1$ as f_{t-1} and those for new nodes, randomly (line 12). Last, we append f_t to the list $\mathcal{E}_t$ and update $prevWalks$ (line 13).

It is noteworthy that the mechanisms implemented to modify node2vec in order to build **EVNE** do not change the asymptotic complexity of node2vec. Thus, **EVNE** is as fast as node2vec. However, our task is different from the traditional task of node2vec, as our networks evolve over time. Thus, running **EVNE** is faster than running node2vec for each step of the evolving process.

4   Experimental Analysis

We analyze **EVNE** using 5 datasets representing undirected and unweighted networks. Although our experiments focus on undirected networks, **EVNE**'s execution generalizes and presents the same behaviour for both undirected and

---

Algorithm 1: Evolving **N**ode **E**mbedding Algorithm

Data: $\mathcal{G} = \{G_t\}_{t=1}^T$: list of graph snapshots, d: embedding dimension, r: walks per node, ℓ: walk length, k: context size, start values for p and q
Result: $\mathcal{F} = \{f_t\}_{t=1}^T$: list of embedding matrices for network snapshot
1   $prevWalks = \emptyset$; // stores walks from previous snapshot
2   **for** $G_t = (V_t, E_t) \in \mathcal{G}$ **do**
3      $p_t = \mathcal{P}[\texttt{ceiling}(|\mathcal{P}|t/T)]$; $q_t = \mathcal{Q}[\texttt{ceiling}(|Q|t/T)]$;
       /* initialize $\boldsymbol{\Pi}_t$ to cache RW transition probabilities   */
4      $\boldsymbol{\Pi}_t = initTransitionProbMatrix(p_t, q_t, E_t)$;
5      $curWalks = \emptyset$; // stores walks from current snapshot
6   **forall** $u \in V_t$ **do**
7   **forall** $i \in 1, \ldots, r$ **do**
8            $walks = sampleNode2vecWalk(E_t, \boldsymbol{\Pi}_t, u, \ell)$;
9   append $walks$ to $curWalks$
10   **end**
11   **end**
12     $f_t = \texttt{SGD}(k, d, prevWalks \cup curWalks, f_{t-1})$;
13   append f_t to $\mathcal{E}$; $prevWalks = curWalks$;
14 **end**

---

directed networks. We choose to apply our method to only undirected networks due to the high complexity associated with the growth in the number of edges in this scenario. The classification task is defined by choosing one node subpopulation of interest and defining them as targets and the remaning nodes as non-targets, yielding highly unbalanced classes. Table 2 summarizes the network characteristics of each dataset.

Table 2. Description and basic statistics of each network. "Targets" refers to the subpopulation of interest, $|\mathcal{G}|$ is the number of snapshots, $|\mathcal{V}|$ $|\mathcal{E}|$ are the number of nodes and edges in the last snapshot, respectively, and $|\mathcal{V}_+|/|\mathcal{V}|\%$ is the percentage of target nodes in the last snapshot of the network.

ID	CS	DBP	DC	KS	WK
Dataset	CiteSeer	DBpedia	DonorsChoose	Kiskstarter	Wikipedia
Nodes	Papers	Places	Donors	Donors	Wikipages
Edges	Citations	Hyperlinks	Co-donors	Co-donors	links
Targets	Top venue	adm.regions	P donors	DFA donors	OOP pages
$|\mathcal{G}|$	1482	677	133	680	377
$|\mathcal{V}|$	3825	4931	677	18644	4536
$|\mathcal{E}|$	31984	44204	7406	439970	39302
$|\mathcal{V}_+|/|\mathcal{V}|\%$	41.38	14.72	8.27	7.89	4.47
#SH	1482	677	133	680	379
#BFS	832	39	10	107	133

Graph Snapshots: The evolving networks were modeled as a sequence of graph snapshots representing consecutive observation steps. We consider snapshots generated by two different network search algorithms: Selective Harvesting (SH) [11] and a modified version o Breadth-First Search (BFS).

SH performs a type of online network search. The goal of SH is to maximize the number of nodes found, belonging to a certain target subpopulation, given a partial view of the graph and assuming the cost to query node labels is high. It covers scenarios where good temporal embeddings for nodes could boost performance in a given machine learning task. SH resembles a DFS, with the addition of a guiding procedure, which decides which node to query next. Figure 1(left) illustrates two consecutives steps of SH execution. On the left snapshot, three nodes (black) have already been queried for labels. Four other nodes (gray) are known but unqueried and the remaning nodes (white) are still unknown at this point. Solid edges represent the known graph structure in this step, dashed edges are still unseen. In the next snapshot, on the right, a node is queried for its label, also revealing its outgoing edges and neighboring nodes. Nodes above the traced line are included in the current snapshot.

The second algorithm modifies the BFS original search procedure in order to explore one node of the same level at each query. Figure 1(right) illustrates an example of two consecutives steps of BFS procedure execution. On the left, the gray nodes compose the current exploration queue. Each query removes one node from the queue, turning it to black and revealing its connections. Figure 2 shows the growth in number of observed nodes and edges, respectively, across snapshots for SH's and BFS.

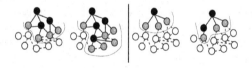

Fig. 1. Example of two consecutives steps of Selective Harvesting (left) and BFS (right). Black, gray and white colors represent queried, unqueried and unknown nodes respectively. Solid and dashed lines represent known and unknown edges.

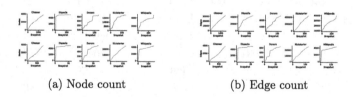

(a) Node count   (b) Edge count

Fig. 2. Node and Edge counts for each dataset through the snapshots for SH's (top) and BFS (bottom) search procedure.

The number of snapshots varies for each search procedure. As the search begins from a random node, the neighborhood exploration differs for each node. The last two rows in Table 2 presents the number of generated snapshots for both SH's and BFS. For reproducibility, all graph snapshots used in this work are publicly available[1].

Experimental Setup for Node Embedding Generation: We obtain node embeddings for each of the snapshots of the 5 networks using 4 static node embedding models as baselines, namely DeepWalk, LINE1st, LINE2nd (1st and 2nd order, respectively), and node2vec. The embeddings were trained with 128 dimensions. DeepWalk and node2vec have three common parameters: size of the context window d and number r and length ℓ of the random walks. These were set to their default values: $(d = 80, r = 10, \ell = 80)$. Node2vec's in-out parameter p and the return parameter q were both set to 1.

We tested all variants of our method using the same parameters as node2vec's. The only difference is the value of the parameters p and q, which are static in node2vec and adaptive in our methods. We designed a strategy for changing these parameters as a function of the knowledge about the network. We define the start values for $\mathcal{Q} = 4$ and $\mathcal{P} = 0.25$. We increase $\mathcal{P}$ and decrease Q by a factor of 2 at every $\frac{1}{5}$ of the total number of snapshots.

4.1   Node Classification Task

Our experimental study focuses on the node classification task. We use the node embeddings generated for each snapshot and the corresponding node labels to train a classifier along the graph evolution process. We consider 4 standard classifiers, namely Adaboost (AD), Logistic Regression (LG), Naive Bayes (NB) and Random Forest (RF). Classifiers were run with their default parameters, since the goal is to compare the embedding techniques. The performance of each combination *dataset* $\times$ *embedding technique* $\times$ *classifier* was computed over 5 executions of a 10-fold cross-validation procedure, each corresponding to a sequence of snapshots obtained from a different initial node. Macro-F1 was used as the evaluation metric.

Also, to provide a holistic assessment of the embeddings, we compute two evalutation metrics that account for all executions over all datasets used in the experimental study: Mean Penalty (MP) and Mean Rank (MR). These metrics were proposed in [16] as a simple and intuitive way of comparing ensemble methods. Lower values for these metrics imply better performance. We computed a value of MR and MP for each classifier, and present the sum of all values.

Let $R_{e,d}$ be the rank of embedding method e on dataset $d \in D$, where D is the set of all datasets we run a classifier. The Mean Rank of embedding method MR_e is given by $MR_e = \frac{\sum_{d \in D} R_{e,d}}{|D|}$. Let E be the set of embedding methods and $S_{e,d}$ be the score achieved by embedding method e on same dataset d for a classifier. The Mean Penalty MP_e is given by $MP_e = \frac{\sum_{d \in D} \max(S_{e',d}) - S_{e,d}}{|D|}; e' \in E$.

Mean Rank determines which embedding yields the best results more often. Mean Penalty, in turn, identifies embeddings that are robust and consistent to changes in datasets and classifiers through the evolution process. By robustness,

---

[1] **EVNE** implementation and all the datasets, snapshots, generated embedding and results are available at: Here.

we mean the method should have small variations of Mean Penalty values across different datasets and classifiers. By consistency, a robust method should also be consistent over time regardless of changes in the network's evolution scenario.

Best Static Embedding Method: We evaluate whether node2vec overperforms standard baselines DeepWalk, LINE1[st] and LINE2[nd]. Also, we provide results to identify the method with best robustness and consistency, in the node classification task. Figure 3 presents the results of Mean Rank and Mean Penalty for all 5 datasets and each classifier. The results are presented for different iterations steps of the evolution process, namely 25%, 50%, 75% and 100% of the total number of steps. Figure 3b shows that, for BFS, node2vec outperforms all other methods except at the first search stage. Also, node2vec's performance is superior to LINE1 and LINE2 in all cases. For SH, node2vec outperforms LINE2 in all cases, LINE1 in 2 cases and Deepwalk in 1 out of 4 cases. Thus, we conclude that node2vec works best for BFS. For SH, Deepwalk yields competitive results when compared to node2vec. Figure 3b presents the Mean Penalty results for each baseline and search procedure. Note that the methods' performance is consistent across classifiers and iteration steps. Node2vec achieves lower values of Mean Penalty in 5 out of 8 cases. From the results discussed above, we find that node2vec displays great robusteness across datasets and classifiers, and great consistency over time.

EVNE's Performance: EVNE introduces three new mechanisms to node2vec. In order to understand the impact of each, we analyze all seven combinations of these mechanisms. Combinations are referred to by the following acronyms: *dpq* for dynamic *p* and *q* parameters, *rw* for the evolving random walks, *wgt* for initializing weights using those of previous snapshots. Combinations that include 2 mechanisms are denoted by their acronyms separated by a "+", whereas those including the 3 mechanisms are denoted by *all*.

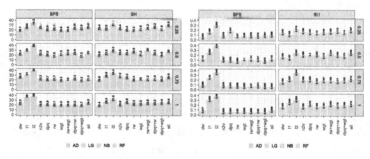

(a) Mean Rank for BFS and SH.   (b) Mean Penalty of node2vec and
                                     the 7 combinations of **EVNE**.

Fig. 3. Results of Mean Rank (left) and Mean Penalty (right) for all datasets on all classifiers, regarding 4 iteration steps and 2 search methods (BFS and SH). Dw, l1, l2 and n2v represent, respectively, DeepWalk, LINE1[st], LINE2[nd] and node2vec. Lower values represent better performance.

In Fig. 3a, the results of the BFS indicate that in each iteration step, at least one variation of **EVNE** embeddings presents better performance compared to the baselines. For the SH search, we find the same behavior, except at stage 100%, where node2vec outperforms all seven combinations of **EVNE** . In short, **EVNE** substantially outperforms all baselines in 7 out of 8 cases. Also, we observe that **EVNE** variations *dpq+wgt* and *rw+wgt* perform better than the other variations for BFS. In addition, *dpq+wgt* outperforms all baselines in all 4 scenarios (BFS and SH for both mean rank and mean penalty metrics). Regarding the SH search, **EVNE** variations *dpq+wgt*, *rw* and *wgt* perform better than the other variations. Additionaly, variations *dpq+wgt* and *wgt* outperform node2vec in 3 out of 4 scenarios.

From these analysis, we highlight 3 key facts: (i) If we choose the same classifier for both **EVNE** and node2vec, regardless of which, at least one of **EVNE**'s variations outperforms node2vec. This result suggests that **EVNE**'s performance results are not biased by the choice of the classifier, (ii) the mechanisms of **EVNE** are responsible for improving node classification performance in evolving scenarios, and (iii) **EVNE** *wgt* is the variation with the best results considering all iteration steps. In addition, in a general representation scenario, **EVNE** *wgt* is the best variation of the proposed model.

EVNE's Robustness: We analyze the Mean Penalty results to assess the consistency and robusteness of all **EVNE** variations and node2vec over time. Figure 3b presents the results of Mean Penalty for all **EVNE** variations and all baselines. Two interesting facts: (i) the mean penalty values are very low for both BFS and SH, which indicates great robustness of the proposed method variants; (ii) **EVNE** is more robust than the baselines evaluated. The similar values of Mean Penalty observed at different iteration steps indicate that changes during the evolution of the network do not interfere with the consistency of the method over time. In particular, we observe that our method and its variations are more consistent than the other baselines.

The Effect of the Modifications on node2vec Over Time: From Fig. 3, we observe that leveraging information on the network evolution process through each of **EVNE**'s mechanisms can have a positive impact on the node classification performance for all classifiers. Our analysis shows two main results: (i) most of **EVNE**'s variants outperform all baselines w.r.t. Macro-F1 based on the Mean Rank evaluation; and (ii) **EVNE** provides results that are more robust and methods more consistents throughout the network evolution process.

For a more in-depth anaylisis of modifications over time, we analyze how the aggregation of evolving information affects the node classification performance during the exploitation process when compared to the baselines, we quantify this effect at different points of the evolving process. To provide an overview of **EVNE** 's performance, we show, in Fig. 4 the Macro-F1 values for all classifiers though all stages of the evolving process, for both BFS (upper) and SH (bottom) search method. The results of Macro-F1 for the BFS search are generally lower than the values of Macro-F1 for the SH search process. In addition, the values of Macro-F1 tend to increase over time for both BFS and SH search. Also from

Fig. 4, for both SH and BFS, the variations *dpq*, *dpq+wgt* and *wgt* tend to provide the best results, while the variation *all* tends to provide lower values of Macro-F1 for both BFS and SH search process. **EVNE** is a skip-gram model, based on biased random walks. There are some random features inherent in the model. These characteristics may be responsible for causing variations in the results of **EVNE** variants. In a general scenario, **EVNE** *wgt* is the best variation of the proposed model, showing better results than node2vec and the other **EVNE** variations in most cases, for both BFS and SH search process.

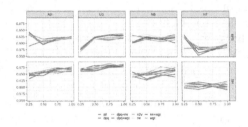

Fig. 4. Mean absolute Macro-F1 results of node2vec and the seven variations of **EVNE**, by classifier and search method, at all iteration steps.

5   Final Remarks

This work proposed **EVNE**, a new node embedding technique that learns representation vectors from evolving networks. Our proposals to **EVNE** are generic enough to be applied to all SGNE and Random Walks based methods in a dynamic setting. **EVNE** yields embeddings capable of adapting to structural changes that take place in evolving graphs. Experimental analysis showed that it generated better embeddings in a downstream node classification task, where our embeddings achieved better performance than the other static methods in most cases. There are several directions of future work. We plan to improve the generation of the time-varying parameters with a dynamic adaptation strategy. We also plan to use **EVNE** representations in a hybrid scenario where an evolving graph becomes large enough that a deep learning method can be applied.

References

1. Chen, F., Wang, Y.C., Wang, B., Kuo, C.C.J.: Graph representation learning: a survey. Trans. Signal Inf. Process. **9**, 1–21 (2020)
2. Goyal, P., Chhetri, S.R., Canedo, A.: Dyngraph2vec: capturing network dynamics using dynamic graph representation learning. Knowl. Based Syst. **187**, 104816 (2020)
3. Goyal, P., Ferrara, E.: Graph embedding techniques, applications, and performance: a survey. Knowl. Based Syst. **151**, 78–94 (2018)

4. Goyal, P., Kamra, N., He, X., Liu, Y.: Dyngem: Deep embedding method for dynamic graphs. arXiv. cs.LG 1805.11273 (2018)
5. Grover, A., Leskovec, J.: node2vec: Scalable feature learning for networks. In: ACM SIGKDD (2016)
6. Hamilton, W., Ying, Z., Leskovec, J.: Inductive representation learning on large graphs. In: NIPS (2017)
7. Kazemi, S.M., et al.: Representation learning for dynamic graphs: a survey. JMLR. **21**, 1–73 (2020)
8. Ma, Y., Guo, Z., Ren, Z., Zhao, E., Tang, J., Yin, D.: Streaming graph neural networks. arXiv:1810.10627 (2018)
9. Mahdavi, S., Khoshraftar, S., An, A.: dynnode2vec: Scalable dynamic network embedding. In: IEEE International Conference on Big Data (2018)
10. Meilian, L., Danna, Y.: HIN_DRL: a random walk based dynamic network representation learning method for heterogeneous information networks. Expert Syst. App. **158**, 113427 (2020)
11. Murai, F., Rennó, D., Ribeiro, B., Pappa, G.L., Towsley, D., Gile, K.: Selective harvesting over networks. DMKD **32**, 187–217 (2018)
12. Perozzi, B., Al-Rfou, R., Skiena, S.: DeepWalk: Online learning of social representations. In: ACM SIGKDD (2014)
13. Tang, J., Qu, M., Wang, M., Zhang, M., Yan, J., Mei, Q.: Line: Large-scale information network embedding. In: WWW (2015)
14. Trivedi, R., Farajtbar, M., Biswal, P., Zha, H.: Representation learning over dynamic graphs. arXiv:1803.04051 (2018)
15. Vázquez, C.O., Mitrović, S., De Weerdt, J., Broucke, S.: A comparative study of representation learning techniques for dynamic networks. In: WorldCIST (2020)
16. Vijayan, P., Chandak, Y., Khapra, M.M., Ravindran, B.: Fusion graph convolutional networks. In: ACM SIGKDD, Mining and Learning with Graphs (MLG) (2018)
17. Zhang, X., Ghorbani, A.A.: An overview of online fake news: characterization, detection, and discussion. Inf. Process. Manage. **57**, 102025 (2020)

Fast Kernel Density Estimation with Density Matrices and Random Fourier Features

Joseph A. Gallego$^{(\boxtimes)}$, Juan F. Osorio , and Fabio A. Gonzalez

Universidad Nacional de Colombia, Bogota, Colombia
{jagallegom,josorior,fagonzalezo}@unal.edu.co

Abstract. Kernel density estimation (KDE) is one of the most widely used nonparametric density estimation methods. The fact that it is a memory-based method, i.e., it uses the entire training data set for prediction, makes it unsuitable for most current big data applications. Several strategies, such as tree-based or hashing-based estimators, have been proposed to improve the efficiency of the kernel density estimation method. The novel density kernel density estimation method (DMKDE) uses density matrices, a quantum mechanical formalism, and random Fourier features, an explicit kernel approximation, to produce density estimates. This method has its roots in the KDE and can be considered as an approximation method, without its memory-based restriction. In this paper, we systematically evaluate the novel DMKDE algorithm and compare it with other state-of-the-art fast procedures for approximating the kernel density estimation method on different synthetic data sets. Our experimental results show that DMKDE is on par with its competitors for computing density estimates and advantages are shown when performed on high-dimensional data. We have made all the code available as an open source software repository.

Keywords: Density matrix · Random fourier features · Kernel density estimation · Approximations of kernel density estimation · Quantum machine learning

1 Introduction

In many applications we have a finite set of data and we would like to know what probability distribution has generated the data. From the point of view of statistical inference, this problem has played a central role in research and has inspired many methods that are based on the use of the density function. Also in machine learning there are many methods base on density estimation, such as anomaly detection methods [16], generative models [15], agglomerative clustering [18], spatial analysis [4], sequence-to-sequence models [22], among others.

Making certain assumptions on the probability model that generated the data leads to parametric estimation. Another common approach is non-parametric

© The Author(s), under exclusive license to Springer Nature Switzerland AG 2022
A. C. Bicharra Garcia et al. (Eds.): IBERAMIA 2022, LNAI 13788, pp. 160–172, 2022.
https://doi.org/10.1007/978-3-031-22419-5_14

estimation. The most representative non-parametric method is called Kernel Density Estimation (KDE) [21,25] and can be understood as a weighted sum of density contributions that are centered on each data point. In this method, one has to choose a function called kernel and a smoothing parameter that controls the dispersion of the estimate. Given n as the number of training data points, direct evaluation of the KDE on m test data points requires $O(mn)$ kernel evaluations and $O(mn)$ additions and multiplications, i.e., if the number of training n is sufficiently large, similar to the number of m test points, its complexity is quadratic. This makes it a very expensive process, especially for large data sets and higher dimensions as stated in [10]. One approach to the problem of scalability of KDE focuses on finding a fast approximate kernel evaluation. According to [27], we can identify three main lines of work: space partitioning methods, random sampling, and hashing-based estimators. In the Subsect. 3 we elaborate on each type of approximation. In [9], a novel KDE approach using random Fourier features and density matrices was proposed, which appears to be a promising way to scale up the KDE. This method uses an explicit Gaussian kernel approximation and a density matrix to produce density estimates. A more detailed explanation can be found in the Subsect. 2.3.

The goal of this paper is to compare, within a statistical experimental setup, several fast KDE implementations that are based on famous theoretical approaches for kernel density estimation approximation, including DMKDE, all the code is available as an open source software repository [8]. The paper is organized as follows: Sect. 2 covers the background of random features, kernel density estimation; Sect. 3 presents some methods for kernel density estimation approximation; Sect. 4 presents the experimental setup and systematic evaluation, where seven synthetic data sets are used to compare DMKDE against other KDE approximation methods. This section shows that DMKDE outperformed other KDE approximation methods in terms of time consumed to make new predictions; finally, Sect. 5 discusses the conclusions of the paper and future research directions.

2   Background and Related Work

2.1   Kernel Density Estimation

The multivariate kernel density estimator at a query point $\boldsymbol{x} \in \mathbb{R}^d$ for a given random sample $X = \{\boldsymbol{x}_i\}_{i=1}^n$, where n is the number of samples drawn from an unknown density f, is given by

$$\hat{f}(\boldsymbol{x}) = \frac{1}{n} \sum_{i=1}^n |h|^{-1/2} k \left(h^{-1/2} (\boldsymbol{x} - \boldsymbol{x}_i) \right) \tag{1}$$

where H is the $d \times d$ bandwidth matrix, which is positive definite and symmetric and $k : \mathbb{R}^d \to \mathbb{R}_{\geq 0}$ is the kernel function. Defining $k_h(\boldsymbol{u}) = |h|^{-1/2} k \left(h^{-1/2} \boldsymbol{u} \right)$ we can write 1 as

$$\hat{f}(\boldsymbol{x}) = \frac{1}{n} \sum_{i=1}^n k_h (\boldsymbol{x} - \boldsymbol{x}_i) \tag{2}$$

Throughout the paper we will work with the *rescaled* multivariate KDE version written in the Eq. 2. We will also assume that $h = \sigma^2 \mathbf{I}_d$ where σ^2 represents the bandwidth in each dimension, $\mathbf{I}_d$ is the $d \times d$ identity matrix, and the kernel will be Gaussian expressed by

$$K_\gamma(\boldsymbol{x}, \boldsymbol{y}) = (2\pi)^{-d/2} \exp\left(-\gamma \|\boldsymbol{x} - \boldsymbol{y}\|^2\right). \tag{3}$$

where $\gamma = 1/(2\sigma^2)$.

All of these assumptions lead to the functional form

$$\hat{f}_{\gamma, X}(\boldsymbol{x}) = \frac{1}{n(\pi/\gamma)^{\frac{d}{2}}} \sum_{i=1}^{n} e^{-\gamma \|\boldsymbol{x} - \boldsymbol{x}_i\|^2} \tag{4}$$

of the KDE estimator for a query point $x \in \mathbb{R}^d$ where we define $\gamma = \frac{1}{2\sigma}$.

Kernel density estimation has multiple applications some examples include: to visualize clusters of crime areas using it as a spatio-temporal modification [18]; to assess injury-related traffic accidents in London, UK [1], to generate the intensity surface in a spatial environment for moving objects couple with time geography [6]; to make a spatial analysis of city noise with information collected from a French mobile application installed on citizens' smartphones using their GPS location data [11]; to generate new samples from a given dataset by dealing with unbalanced datasets [12]; to process the image by subtracting the background in a pixel-based method [14]; to propose an end-to-end pipeline for classification [13]; to propose a classification method using density estimation and random Fourier features [9].

2.2 Random Fourier Features

Using random Fourier features (RFF) to approximate kernels was initially presented in [23]. In this method, the authors approximate a shift invariant kernel by an inner product in an explicit Hilbert spaces. Formally, given a shift invariant kernel $k : \mathbb{R}^d \times \mathbb{R}^d \to \mathbb{R}$ they build a map $\phi_{\text{rff}} : \mathbb{R}^d \to \mathbb{R}^D$ such that

$$k(x, y) \approx \phi_{\text{rff}}(\boldsymbol{x})^* \phi_{\text{rff}}(\boldsymbol{y}) \tag{5}$$

for all $\{\boldsymbol{x}, \boldsymbol{y}\} \subseteq \mathbb{R}^d$, and in the case of the Gaussian kernel the mapping is defined as follows

$$\phi_{\text{rff}}(\boldsymbol{x}) = \sqrt{2}\cos(\boldsymbol{\omega}^* \boldsymbol{x} + b) \tag{6}$$

where $\boldsymbol{w}$ is sampled from $\mathcal{N}(\boldsymbol{0}, \boldsymbol{I}_D)$ and b is samples from Uniform$(0, 2\pi)$. The main theoretical result supporting 5 comes from an instance of Bochner's theorem [26] which states that a shift invariant continuous kernel is the Fourier transform of a nonnegative probability measure. This methodology suggests that we can compute shift-invariant kernel approximations via a sampling strategy.

2.3 Quantum Kernel Density Estimation Approximation

The central idea of the method density matrix kernel density estimation (DMKDE), introduced by [9] and systematically evaluated in [7], is to use density matrices along with random Fourier features to represent arbitrary probability distributions by addressing the important question of how to encode probability density functions in $\mathbb{R}^n$ into density matrices. The overall process is divided into a training and a testing phases. The training phase is defined as follows:

1. Input: a set of n d-dimensional samples $x_1, \cdots, x_n$, D number of random Fourier features and a bandwidth parameter $\gamma \in \mathbb{R}$
2. Sampling of the vector w and b as explained in Subsect. 2.2
3. Mapping: compute the random Fourier feature vector for each data point as explained in Eq. 6.
4. x_i as $z_i = \phi_{\mathrm{rff}}(x_i)$
5. Compute a density matrix as $\rho = \frac{1}{n} \sum_{i=1}^n z_i z_i^T$

The testing phase is as follows:

1. Apply step 3 to each testing data point.
2. The density estimation of a testing point x is calculate using the Born's rule

$$\hat{f}_\rho(x) = \frac{\phi_{\mathrm{rff}}(x)^T \rho \phi_{\mathrm{rff}}(x)}{\mathcal{Z}}$$

where the normalizing constant is: $\mathcal{Z} = (\pi/(2\gamma))^{d/2}$

Note that this algorithm does not need to store each training data point, but a square matrix whose dimensions are equal to $D \times D$ suffices computed in step 5. Moreover, the time consumed in estimating a new data point does not depend on the training size. The ρ-density matrix is one of the main building blocks used in quantum physics to capture classical and quantum probability in a given physical system. This formalism was conceived by [28] as the foundation of quantum statistical mechanics. The density matrix describes the states of the quantum system and explains the relationship between the pure state and the mixed states of the system. Define the quantum state of a system in pure state as ψ. Then the ρ-density matrix for the pure state ψ is defined as $\rho := |\psi\rangle\langle\psi|$. Let us now consider a quantum ensemble system of n systems (objects) that are not in the same state. Let $p_i := n_i/n$ where n_i is the number of systems that are in state $|\psi_i\rangle$ and $\sum n_i = n$. Therefore, the mixed density matrix is defined as $\rho_{mix} = \sum_i p_i \rho_i^{pure} = \sum_i |\psi_i\rangle\langle\psi_i|$.

A matrix factorization of the ρ-matrix simplifies the computation of the matrix as $\rho = V^* \Lambda V$, where $V \in \mathbb{R}^{r \times D}, \Lambda \in \mathbb{R}^{r \times r}$ is a diagonal matrix and $r < D$ is the reduced rank of the factorization. Thus, using this matrix factorization, the method called DMKDE-SGD is expressed as: $\hat{f}_\rho = \frac{1}{\mathcal{Z}} ||\Lambda^{1/2} V \phi_{\mathrm{rff}}(x)||^2$. This method reduces the time required for a new density to $O(Dr)$.

3   KDE Approximation Methods

Fast density estimation methods are of paramount importance in several applications, as shown above in Subsect. 2.1. There are four main approaches for fast kernel density estimation approximation. First, space partitioning methods do not work well in high dimensions and make use of geometric data structures to partition the space, and achieve speedup by limiting the contribution of data points to the kernel density using partitions. Second, random sampling focuses on randomly sampling the kernel density and gives good results in high dimensions. Third, the most recent methods and the most promising state-of-the-art tools are hash-based estimators for kernel density estimation, in which a hash structure is used to construct close distance boxes that allow the calculation of few distances in the prediction step [2]. Here, a sampling scheme is used, where points are sorted into buckets thanks to a hash function whose main objective is to send similar objects to the same hash value. Finally, the novel DMKDE as explained above in Sect. 2.3 can approximate kernel density estimation. The following algorithms including DMKDE are used in the experimental setup.

- *Tree kernel density estimation (TREEKDE)*: the density approximation is obtained using the partitioning of the space through recursive segmentations of the space into smaller sections. By obtaining the partition, we can approximate the kernel of a specific point by traversing the tree, without having to explicitly evaluate the kernel function [17].
- *Kernel Density Estimation using k-dimensional Tree (KDEKDT)*: this method addresses the k-nearest neighbor problem using a kd-tree structure that generalizes two-dimensional Quad-trees and three-dimensional Oct-trees to an arbitrary number of dimensions. Internally, it is a binary search tree that partitions the data into nested orthotopic regions that are used to approximate the kernel at a specific point [3].
- *Kernel Density using Ball Tree (KDEBT)*: in this method the Ball Tree structure is used to avoid the inefficiencies of KD trees in high dimensionality spaces. In these spaces, the Ball Tree divides the space into nested hyperspheres [19].

4   Experimental Evaluation

In this section, we systematically assess the performance of DMKDE on various synthetic data sets and compare it with kernel density estimation approximation methods.

Data Sets and Experimental Setup. We used seven synthetic data sets to evaluate DMKDE against kernel density approximation methods. The data sets are characterized as follows:

- The data set *Arc* corresponds to a two-dimensional random sample drawn from a random vector $X = (X_1, X_2)$ with probability density function given by

$$f(x_1, x_2) = \mathcal{N}(x_2|0, 4)\mathcal{N}(x_1|0.25x_2^2, 1)$$

where $\mathcal{N}(u|\mu, \sigma^2)$ denotes the density function of a normal distribution with mean μ and variance σ^2. [20] used this data set to evaluate his neural density estimation methods.

- The data set *Potential 1* corresponds to a two-dimensional random sample drawn from a random vector $X = (X_1, X_2)$ with probability density function given by

$$f(x_1, x_2) = \frac{1}{2}\left(\frac{||x|| - 2}{0.4}\right)^2 - \ln\left(\exp\left\{-\frac{1}{2}\left[\frac{x_1 - 2}{0.6}\right]^2\right\} + \exp\left\{-\frac{1}{2}\left[\frac{x_1 + 2}{0.6}\right]^2\right\}\right)$$

with a normalizing constant of approximately 6.52 calculated by Monte Carlo integration.

- The data set *Potential 2* corresponds to a two-dimensional random sample drawn from a random vector $X = (X_1, X_2)$ with probability density function given by

$$f(x_1, x_2) = \frac{1}{2}\left[\frac{x_2 - w_1(x)}{0.4}\right]^2$$

where $w_1(x) = \sin\left(\frac{2\pi x_1}{4}\right)$ with a normalizing constant of approximately 8 calculated by Monte Carlo integration.

- The data set *Potential 3* corresponds to a two-dimensional random sample drawn from a random vector $X = (X_1, X_2)$ with probability density function given by

$$f(x_1, x_2) = -\ln\left(\exp\left\{-\frac{1}{2}\left[\frac{x_2 - w_1(x)}{0.35}\right]^2\right\} + \exp\left\{-\frac{1}{2}\left[\frac{x_2 - w_1(x) + w_2(x)}{0.35}\right]^2\right\}\right)$$

where $w_1(x) = \sin\left(\frac{2\pi x_1}{4}\right)$ and $w_2(x) = 3\exp\left\{-\frac{1}{2}\left[\frac{x_1-1}{0.6}\right]^2\right\}$ with a normalizing constant of approximately 13.9 calculated by Monte Carlo integration.

- The data set *Potential 4* corresponds to a two-dimensional random sample drawn from a random vector $X = (X_1, X_2)$ with probability density function given by

$$f(x_1, x_2) = -\ln\left(\exp\left\{-\frac{1}{2}\left[\frac{x_2 - w_1(x)}{0.4}\right]^2\right\} + \exp\left\{-\frac{1}{2}\left[\frac{x_2 - w_1(x) + w_3(x)}{0.35}\right]^2\right\}\right)$$

where $w_1(x) = \sin\left(\frac{2\pi x_1}{4}\right)$, $w_3(x) = 3\sigma\left(\left[\frac{x_1-1}{0.3}\right]^2\right)$, and $\sigma(x) = \frac{1}{1+\exp(x)}$ with a normalizing constant of approximately 13.9 calculated by Monte Carlo integration.

- The data set *2D mixture* corresponds to a two-dimensional random sample drawn from the random vector $X = (X_1, X_2)$ with a probability density function given by

$$f(x) = \frac{1}{2}\mathcal{N}(x|\mu_1, \Sigma_1) + \frac{1}{2}\mathcal{N}(x|\mu_2, \Sigma_2)$$

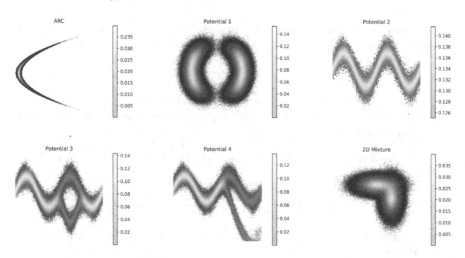

Fig. 1. True density of each data set: arc, potential 1 to 4, and 2D mixture. High density points are colored as yellow and low-density points are colored as white. (Color figure online)

with means and covariance matrices $\mu_1 = [1, -1]^T$, $\mu_2 = [-2, 2]^T$, $\Sigma_1 = \begin{bmatrix} 1 & 0 \\ 0 & 2 \end{bmatrix}$, and $\Sigma_1 = \begin{bmatrix} 2 & 0 \\ 0 & 1 \end{bmatrix}$

- The data set *10D-mixture* corresponds to a 10-dimensional random sample drawn from the random vector $X = (X_1, \cdots, X_{10})$ with a mixture of four diagonal normal probability density functions $\mathcal{N}(X_i | \mu_i, \sigma_i)$, where each μ_i is drawn uniformly in the interval $[-0.5, 0.5]$, and the σ_i is drawn uniformly in the interval $[-0.01, 0.5]$. Each diagonal normal probability density has the same probability of being drawn $1/4$.

The functions from *Potential 1* to *4* were presented in [24] to test their normalizing flow algorithms. The *ARC* data set was presented in [20] to test his autoregressive models.

For this experiment, we compared DMKDE with different methods of approximate kernel density estimation. We used for this comparison: (1) a raw implementation of kernel density estimation using Numpy (RAWDKE, (2) a naive implementation of kernel density estimation using the KDE.py library (NAIVEKDE)[1], (3) tree-based kernel density estimation (TREEKDE), (4) kernel density estimation using a k-dimensional tree (KDEKDT), and (5) kernel density using a ball tree (KDEBT). Details of NAIVEKDE, KDEKDT, and KDEBT can be found above in Sect. 3. In this experimental setup, we did not compare with the hashing-based approach due to the restriction of the algorithm implementations. For all experiments, the Gaussian kernel was used. Two different types of experiments were performed. In the first, we evaluated the

---

[1] Implementation of KDE.py: https://github.com/Daniel-B-Smith/KDE-for-SciPy/blob/master/kde.py.

accuracy of each of the algorithms on each data set. In the second, we evaluated the time it takes to make a new density prediction on the training set. Each run was performed with different training set sizes, using a scale of 10^i where $i \in \{1, 2, 3, 4, 5\}$, and we set the test set to 10^4 test examples. The spread parameter was found using a 5-fold cross validation for each data set and each training size in a logarithmic scale $\gamma \in \{2^{-20}, \cdots, 2^{20}\}$. The optimal number of random Fourier features used by DMKDE-SGD was searched in the set $\{50, 100, 500, 1000\}$. After finding the best-hyperparameter, several attempts were performed for each algorithm in each data set.

We measure the efficacy of each algorithm on each data set using the L_1-error also known as *average error*. Some advantages over L_2-error are outlined in [5]. The L_1-error is defined over n samples by the following equation: $MAE = \frac{1}{n} \sum_{i=1}^{n} |\hat{f}(x) - f(x)|$. In [5], the author shows that the loss L_1 loss $\int |\hat{f} - f|$ is invariant under monotone transformations of the coordinate axes and points out that it is related to the maximum error made if we were estimating the probabilities of all Borel sets of $\hat{f}$ and f respectively. Efficiency was evaluated using CPU time in milliseconds (ms), which defines the amount of time it takes the central processing unit (CPU) to execute its processing instructions to compute the evaluation query. We used the built-in **time** package in the Python programming language to measure the elapsed time in prediction time for each algorithm.

Table 1. Efficacy test results measured in MAE for training size 1×10^5

DATA SET	RAW	NAIVE	TREE	KDBTREE	KDKDTREE	DMKDE-SGD
ARC	$0.0012 \pm 2E-4$	$0.0012 \pm 1E-4$	$0.0012 \pm 2E-4$	$0.0012 \pm 1E-4$	$0.0012 \pm 1E-4$	$0.0080 \pm 1E-4$
2D MIXTURE	$0.0010 \pm 1E-4$	$0.0010 \pm 1E-4$	$0.0010 \pm 3E-4$	$0.0010 \pm 5E-4$	$0.0010 \pm 5E-4$	$0.0016 \pm 2E-4$
10D MIXTURE	$2.5282 \pm 3E-4$	$2.5282 \pm 3E-4$	583000 ± 0.0004	$2.6216 \pm 2E-4$	$2.5282 \pm 3E-4$	$1.7420 \pm 1E-4$
POTENTIAL 1	$0.0046 \pm 1E-4$	$0.0046 \pm 1E-4$	$0.0046 \pm 3E-4$	$0.0046 \pm 2E-4$	$0.0046 \pm 6E-4$	$0.00334 \pm 3E-4$
POTENTIAL 2	$0.0456 \pm 2E-4$	$0.0456 \pm 2E-4$	0.0456 ± 0.0004	0.0456 ± 0.0007	0.0456 ± 0.0007	$0.0332 \pm 2E-4$
POTENTIAL 3	$0.0182 \pm 1E-4$	$0.0182 \pm 1E-4$	$0.0182 \pm 5E-4$	$0.0182 \pm 6E-4$	$0.0182 \pm 6E-4$	$0.0299 \pm 2E-4$
POTENTIAL 4	$0.0190 \pm 2E-4$	$0.0190 \pm 1E-4$	$0.0190 \pm 2E-4$	$0.0190 \pm 5E-4$	$0.0190 \pm 8E-4$	$0.0235 \pm 3E-4$

Table 2. Efficiency test results in millisecond (ms) for training size 1×10^5

DATA SET	RAW	NAIVE	TREE	KDBTREE	KDKDTREE	DMKDE-SGD
ARC	52400 ± 500	103000 ± 447	24700 ± 171	49200 ± 700	56000 ± 141	$\mathbf{4330 \pm 145}$
2D MIXTURE	42400 ± 282	68000 ± 707	35900 ± 282	54800 ± 505	62000 ± 577	$\mathbf{3500 \pm 190}$
10D MIXTURE	88000 ± 115	123000 ± 1527	583000 ± 435	190000 ± 436	193000 ± 577	$\mathbf{504 \pm 110}$
POTENTIAL 1	46100 ± 378	76000 ± 547	22100 ± 141	39901 ± 212	52400 ± 141	$\mathbf{3640 \pm 311}$
POTENTIAL 2	28800 ± 212	70000 ± 0.000	4920 ± 210	29400 ± 302	36700 ± 700	$\mathbf{3630 \pm 106}$
POTENTIAL 3	30200 ± 0.000	72000 ± 0.000	19900 ± 410	41100 ± 520	48900 ± 0.000	$\mathbf{3280 \pm 176}$
POTENTIAL 4	30700 ± 0.000	68000 ± 707	54700 ± 424	54800 ± 0.000	64000 ± 0.000	$\mathbf{3700 \pm 158}$

Results and Discussion. Table 1 shows the comparison of the mean error of each algorithm on each data set against the true density. The results obtained by each approximation method are better than those of DMKDE-SGD, except

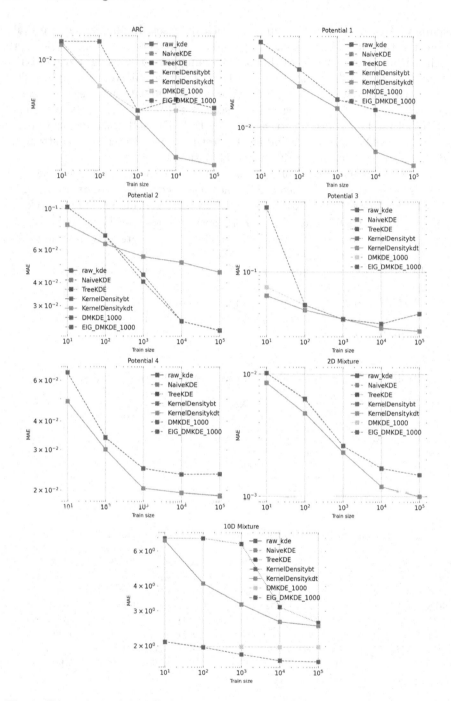

Fig. 2. Comparison of the efficacy of each algorithm on six synthetic data sets. The x-axis is a logarithmic scale of 10^i where $i \in \{1, \cdots, 5\}$. The y-axis represents the mean average error between the prediction of the algorithm and the true density.

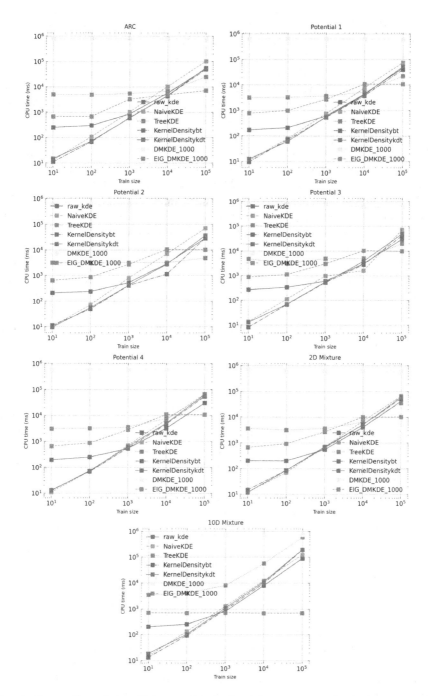

Fig. 3. Comparison of the efficiency of each algorithm on six synthetic data sets. The x-axis is a logarithmic scale of 10^i where $i \in \{1, \cdots, 5\}$. The y-axis represents the time consumed by each algorithm in milliseconds used of the central processing unit (CPU).

for the 10-dimensional mixture data set. Table 2 shows the time consumed by each algorithm on each data set. The DMKDE-SGD method is at least 6 times faster than the TREE algorithm, 10 times faster than RAW, KDBTREE and KDKDTREE. And 20 times faster than NAIVE. It is worth noting that if we increase the number of training sizes, all algorithms except DMKDE-SGD will consume more time to produce a prediction. Figure 2 shows the comparison of the efficacy measure with the mean average error (MAE) of each algorithm on six synthetic data sets. The MAE of DMKDE and DMKDE-SGD is close to other KDE approximation methods in ARC, Potential 1, Potential 3, Potential 4, and 2d mixture. In Arc, however, their performance does not improve after 10^3 training data points. In Potential 2, both DMKDE and DMKDE-SGD are better than the KDE approximation methods. On 10D Mixture, DMKDE and DMKDE-SGD outperform other KDE approximation methods. Figure 3 shows the comparison of the efficiency measure in time taken of the central processing unit (CPU) of approximation methods of KDE. All approximation methods, except DMKDE and DMKDE-SGD, increase their prediction time when the number of points increases. However, it is observed that DMKDE does not increase linearly like the other methods. DMKDE-SGD has a larger initial footprint, but it remains constant as the number of data points increases. If we evaluate it with more than 10^5 points, we would expect all methods to exceed the time consumed by DMKDE-SGD.

5   Conclusion

In this paper we systematically evaluate the performance of the method called Density Matrix Kernel Density Estimation (DMKDE). This method uses the kernel approximation given by random Fourier features and density matrices which are a fundamental tool in quantum mechanics. The efficiency and efficacy of the model was compared with three kernel density approximation methods: tree kernel density estimation method, kernel density estimation using a k-dimensional tree and kernel density using a ball tree. Systematic comparison shows that the new method is close in terms of mean error to these kernel density estimation approaches, but uses ten times less computational resources. The method can be used in domains where the size of the training data set is large ($> 10^4$), where kernel density estimation will suffer given its memory-based behavior.

References

1. Anderson, T.K.: Kernel density estimation and K-means clustering to profile road accident hotspots. Acc. Anal. Prev. **41**(3), 359–364 (2009)
2. Backurs, A., Indyk, P., Wagner, T.: Space and time efficient kernel density estimation in high dimensions. In: 33rd Conference on Neural Information Processing Systems (NeurIPS 2019), Vancouver, Canada, vol. 32 (2019)
3. Bentley, J.L.: Multidimensional binary search trees used for associative searching. Commun. ACM **18**(9), 509–517 (1975)

4. Borruso, G.: Network density estimation: a GIS approach for analysing point patterns in a network space. Trans. GIS **12**(3), 377–402 (2008)
5. Devroye, L.: Nonparametric density estimation. The L₋1 View (1985)
6. Downs, J.A.: Time-geographic density estimation for moving point objects. In: Fabrikant, S.I., Reichenbacher, T., van Kreveld, M., Schlieder, C. (eds.) GIScience 2010. LNCS, vol. 6292, pp. 16–26. Springer, Heidelberg (2010). https://doi.org/10.1007/978-3-642-15300-6_2
7. Gallego M., J.A., González, F.A.: Quantum adaptive Fourier features for neural density estimation (2022). https://doi.org/10.48550/ARXIV.2208.00564, https://arxiv.org/abs/2208.00564
8. Gallego M., J.A., Osorio, J.F., Gonzalez, F.A.: Fast kernel density estimation with density matrices and random Fourier Features Software, July 2022. https://doi.org/10.5281/zenodo.6941020
9. González, F.A., Gallego, A., Toledo-Cortés, S., Vargas-Calderón, V.: Learning with density matrices and random features. Quantum Mach. Intell. **4**, 23 (2021)
10. Gramacki, A.: Nonparametric Kernel Density Estimation and Its Computational Aspects. Springer, Cham (2018). https://doi.org/10.1007/978-3-319-71688-6
11. Guardnaccia, C., Grimaldi, M., Graziuso, G., Mancini, S.: Crowdsourcing noise maps analysis by means of kernel density estimatio. J. Phys. Conf. Ser. **1603**(1):, 1691–1697 (2021)
12. Kamalov, F.: Kernel density estimation based sampling for imbalanced class distribution. Inf. Sci. **512**, 1192–1201 (2020)
13. Kristan, M., Leonardis, A., Skočaj, D.: Multivariate online kernel density estimation with gaussian kernels. Pattern Recogn **44**, 2630–2642 (2011)
14. Lee, J., Park, M.: An adaptive background subtraction method based on kernel density estimation. Sensors (Switzerland) **12**(9), 12279–12300 (2012)
15. Liu, D., Yao, Z., Zhang, Q.: Quantum-Classical Machine learning by Hybrid Tensor Networks. Tech. rep. (2020)
16. Lv, P., Yu, Y., Fan, Y., Tang, X., Tong, X.: Layer-constrained variational autoencoding kernel density estimation model for anomaly detection. Knowl. Based Syst. **196** (2020)
17. Maneewongvatana, S., Mount, D.M.: It's okay to be skinny, if your friends are fat. In: Center for Geometric Computing 4th Annual Workshop on Computational Geometry. vol. 2, pp. 1–8 (1999)
18. Nakaya, T., Yano, K.: Visualising crime clusters in a space-time cube: an exploratory data-analysis approach using space-time kernel density estimation and scan statistics. Trans. GIS **14**, 223–239 (2010)
19. Omohundro, S.M.: Five Balltree Construction Algorithms. International Computer Science Institute Berkeley (1989)
20. Papamakarios, G., Pavlakou, T., Murray, I.: Masked autoregressive flow for density estimation. In: 31st Conference on Neural Information Processing Systems (NIPS 2017), Long Beach, CA, USA, May 2017. http://arxiv.org/abs/1705.07057
21. Parzen, E.: On estimation of a probability density function and mode. Ann. Math. Stat. **33**(3), 1065–1076 (1962)
22. Peng, K., Ping, W., Song, Z., Zhao, K.: Non-autoregressive neural text-to-speech. In: Proceedings of the 37th International Conference on Machine Learning, (PMLR), pp. 119:7586-7598, 2020, May 2019. http://arxiv.org/abs/1905.08459
23. Rahimi, A., Recht, B.: Random features for large-scale kernel machines. In: Proceedings of the 20th International Conference on Neural Information Processing Systems, pp. 1177–1184. NIPS'07, Curran Associates Inc. (2007)

24. Rezende, D.J., Mohamed, S.: Variational inference with normalizing flows (2015)
25. Rosenblatt, M.: Remarks on some nonparametric estimates of a density function. Ann. Math. Statist. **27**(3), 832–837 (1956)
26. Rudin, W.: Fourier Analysis on Groups, vol. 121967. Wiley Online Library (1962)
27. Siminelakis, P., Rong, K., Bailis, P., Charikar, M., Levis, P.: Rehashing kernel evaluation in high dimensions. I: Proceedings of the 36th International Conference on Machine Learning (PMLR), vol. 97, pp. 10153–10173 (2019)
28. Von Neumann, J.: Wahrscheinlichkeitstheoretischer aufbau der quantenmechanik. Nachrichten von der Gesellschaft der Wissenschaften zu Göttingen, Mathematisch-Physikalische Klasse **1927**, 245–272 (1927)

A Novel Methodology for Engine Diagnosis Based on Multiscale Permutation Entropy and Machine Learning Using Non-intrusive Data

Juan Camilo Mejía Hernández$^{(\boxtimes)}$ (ID), Federico Gutiérrez Madrid(ID),
Héctor Fabio Quintero Riaza(ID), Carlos Alberto Romero Piedrahita(ID),
and Juan David Ramírez Alzate(ID)

Universidad Tecnológica de Pereira,, Risaralda Pereira 630004, Colombia
{contactenos,j.mejia1}@utp.edu.co
https://www.researchgate.net/profile/Juan-Hernandez-188

Abstract. Internal Combustion Engines (ICE) are widely used in everyday life regardless of its contaminant emissions production, in order to reduce the impact of these emissions it is necessary to maximize the engines' efficiency and diagnosing them it is the first step. Usually, these diagnoses are based on the analysis of signals (pressure, rotational speed, emissions, temperature, air flux, fuel consumption, etc.), taken through multiple means, some intrusive and expensive. In order to reduce the complexity, cost and avoid intrusive measurements, in the present work it is presented a robust and efficient ICE diagnosis methodology based on vibrations, by means of Multiscale Permutation Entropy (MPE) measurements, combined with the feature selection technique Variance Relevance Analysis (VRA) and the supervised classifier of the K Nearest Neighbors (KNN). This methodology based on non-intrusive signals manages to perform a diagnosis of performance and identification of ICE parameters (rotational speed, pressure and emissions) with over 91.6% of accuracy.

Keywords: Multiescale · Combustion · Machine · Diagnosis · Non-intrusive · Vibrations

1 Introduction

In order to improve competitiveness in the engine industry, it is necessary to research and develop ways to reduce fuel consumption, emissions and increase the specific power. Maintaining engines in an optimal state is an alternative and effective way to mitigate this problem in the industry [16]. Steady state performance studies can give a broad understanding of the real state of the engine under different regimes. There are multiple methodologies to diagnose the real state of an ICE [17], of those methodologies, signal-based use measured signals rather than input-output models for diagnosis. Therefore, the device states are inferred from the measured data, usually based on the study and analysis of signals as the pressure, temperature, fuel-air flow ratio, acoustic emissions, exhaust emissions, temperature of gases, soot, etc. However, most of these variables need the use of a variety of different sensors, controlled environments and modifications to the engine system for being able to capture the signals. One of the problems of measurement procedures it is the uncertainty and energy provided by the measurement system to the object measured, which modifies the purity of the physical phenomenon. Therefore, it is ideal for a measurement system not to interfere in any way in the morphology of the object to measure; nonetheless the impossibility of this feat leaves the only option of reducing the interference of the sensor over the objective system. In the case of the ICE measurement system, multiple technologies try to reduce the number and volume of sensors or to avoid intrusion of the instrumentation system over the ICE, such as the use of vibrations and acoustic signals [5,39]. Some of them search to evaluate the state of the combustion based on the pressure in the combustion chamber which is reconstructed based on vibrations [14,18,30,32]. Moreover patter recognition approaches are used commonly based on representation of the signals in time, frequency or other special domains [2,8]. The most common methods are based on frequency representations employing Fast Fourier Transform (FFT) yet inadequate for identifying non-stationary data and limited by the sampling rate [34]. Time-frequency usually uses Wavelet Transform (WT) [27] and shows to be promising for diagnosis; yet ignore information due to the limited length of the wavelet. Typical approaches in time domain have focused on statistical measures intended for stationary processes [13], overlooking early faults; nonetheless, those can find information invisible for frequency domains when used the proper feature extraction technique. Feature extraction techniques are mathematical approaches to characterize signals and describe them using features. The most common techniques for these analyses, commonly include Principal Component Analysis (PCA) [23], Kernel Principal Component Analysis (KPCA) [33], entropy based approaches as Simple Entropy (ApEn) [33], Approximate Entropy (SampEn) [31], Multiscale Entropy (MSE) [9], Permutation Entropy (PE) [4] and Multiscale Permutation Entropy (MPE) [3,19,24,25,28]. The features determine the quantity of the information to provide in the learning phase, therefore, unimportant features increase both time and complexity of the learning process also driving to an overfitted model [10]. Then, in order to determine the quality of the information, feature selection techniques are often employed for dimensionality reduction. Techniques such as Laplacian Score (LS) [12], Relief (REL) [37] and Variance based on Relevance Analysis (VRA) [21] are widely used to achieve this reduction. Finally, the techniques to perform classification step are commonly found as Neural Networks [40], Hidden Markov Models [42], Genetic Algorithms [15], Support Vector Machines [36] proves to be accurate even though they are complex and require a long computation time, different from Decision trees (TREE) [6], Naive Bayes (BAYES) [35] and K-Nearest Neighbors (KNN) [20,22], which are also accurate but do not require as long computation time.

Complex classification algorithms let the system achieve accurate results, however not always the most complex it is the most accurate. It is possible for the combination of suitable characterization method, selection technique and classification algorithm to also provide high accurate results. In this work it is described a methodology for ICE diagnosis based on non-intrusive data, specifically vibrations signals. These signals are processed based on the robust MPE feature extraction method, and VRA feature selection algorithm, to use as base information to classify different variables of the current state of an ICE, with the help of the efficient KNN supervised classification machine learning algorithm. In this paper, it is described the methodology, and each of its components for the vibration-based diagnosis of the ICE. Then, it is described the experimental setup and discussion of the Results obtained. Finally, it is outlined the main ideas in the Conclusions.

2   Materials and Methods

In this section is described the mathematical models and base knowledge of the MPE, VRA and KNN in Sects. 2.1, 2.2, 2.3, respectively. Then, in Sect. 2.4 it is described the ICE, signals and engine parameters used for the diagnosis.

2.1   Multiscale Permutation Entropy

Entropy is defined as a physical quantity capable of interpreting the disorder, positive information, lack of information or ignorance, among others [7]. Specifically, when it comes to information content, Shannon's entropy is often considered the fundamental, the most natural [43] and has been used in the characterization of a wide variety of systems. However, Shannon's entropy is very sensitive to conditions such as stationarity, time series length, parameter variation, noise pollution level, among others [43]. To overcome this problem, multiple variations of Shannon's entropy have been developed, such as Multiscale Entropy (MSE), Permutation Entropy (PE) and Multiscale Permutation Entropy (MPE). The MPE is positioned as a particularly useful and robust tool in the presence of dynamic noise and its definition comes from the Shannon Entropy [45], as is shown in Eq. 1.

$$H(X) = -\sum_i p(x_i) \ln p(x_i) \tag{1}$$

Where $x_i \in R$ and $p(x_i)$ is the marginal probability of value x_i over the entire time series. Entropy is a measure of information by which it is seen as the sum of the marginal probability of each point multiplied with the natural logarithm of each probability. The natural logarithm is implemented with the objective of giving greater weight to the low probabilities, because the values that are less repeated are the ones that carry more information. If considered a time series $[x_t]_{t=1}^T$, where T is the length of it, the time series can be represented with a time and dimension delay, as shown in Eq. 2.

$$X_i^{m,\tau} = (x_i, x_{i+\tau}...x_{i+(m-1)\tau}) \tag{2}$$

where m is the dimension, τ the delay and $j = 1, 2...T - (m - 1)\tau$ is the new length in terms of the delay and scale, this means, it goes from a space of characteristics to a space of permutations. The process begins by truncating the time series in $N = T - (m - 1)\tau$ different vectors. To each vector the calculation of the Shannon's Entropy is performed, but represented in a permutations space of $m!$ different symbols $[\pi_i^{m,\tau}]_{i=1}^{m!}$ denoted as $\prod$, as show in Eq. 3.

$$H(m, \tau) = - \sum_{i:\pi_i^{m,\tau} \in \prod} p(\pi_i^{m,\tau}) \ln p(\pi_i^{m,\tau}) \tag{3}$$

where probability $p(\pi_i^{m,\tau})$ is as show in Eq. 4.

$$p(\pi_i^{m,\tau}) = \frac{\sum_{j \leq N} 1_{u:type(u)=\pi_i}(X_j^{m,\tau})}{\sum_{j \leq N} 1_{u:type(u)\in\prod}(X_j^{m,\tau})} \tag{4}$$

where the judgment $type$ denotes the map from pattern space to symbol space. Also, $1_A(u) = 1$ if $u \in A$ and $1_A(u) = 0$ if $u \notin A$. The MPE can take values between the ranges $[0, ln(m!)]$ and it is invariant under nonlinear monotonic transformations (signal rearrangement).

2.2   Variance Relevance Analysis

The VRA is based on the Principal Component Analysis (PCA) and is a technique used to describe a data set in terms of new uncorrelated variables (components). The components are sorted by the amount of original variance they describe, so the technique is useful for reducing the dimensionality of a data set [26]. Suppose we have a dataset $x^{(1)}, x^{(2)}, ..., x^{(m)}$ with n dimension inputs. To reduce the data from n dimension to k dimension ($k << n$) using PCA. The procedure of PCA starts standardizing the raw data making it to have zero as mean and unit variance, like is show in Eq. 5.

$$x_j^{(i)} = \frac{x_j^{(i)} - \bar{x}_j}{\sigma_j} \forall j \tag{5}$$

For a symmetric matrix, such as the covariance matrix, it is possible to calculate an orthogonal base given by its eigenvalues λ_j and eigenvectors. The new orthogonal base is created with the first eigenvector that points to the direction of the greatest variance of the data. In this way, the first eigenvectors are ($m < p$) can be selected and the transformation matrix can be constructed as shown in Eq. 6.

$$u^T \Sigma = \lambda u \tag{6}$$

where y is the vector of projected characteristics in a new sub-space with a lower dimensionality.

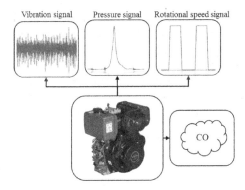

Fig. 1. Signals measured from the ICE.

2.3   K Nearest Neighbors

The K Nearest Neighbors (KNN) method is a supervised and non-parametric classification method that estimates the subsequent probability that an element x belongs to the class C_j from a set of information provided [11]. A point in space is assigned to the class C_j, this is the most frequent class among the K nearest samples of training. Assuming, p as the dimension of the feature set and the vector $X_i = (x_{1i}, x_{2i}, \ldots, x_{pi})X$, the calculation of the probabilities is commonly made using the Euclidean distance. Said procedure is shown Eq. 7.

$$d(X_i, X_j) = \sqrt{\sum_{r=1}^{p}(X_{ri} - X_{ij})^2} \tag{7}$$

It should be noted that the choice of K is based fundamentally on the nature of the data, since at large values of K it reduces noise effects but creates boundaries between similar classes.

2.4   Internal Combustion Engine

In this work a stationary 406 cm3 four-stroke cycle vertical OHV air-cooled single cylinder CHANGFA 186F engine with a compression ratio of 14:1 is used. The mechanical vibrations, pressure in the combustion chamber, the crankshaft speed, and gas emissions (CO) signals were measured. The schematic of the experimental set-up is illustrated in Fig. 1.

With the signals and configurations shown in Fig. 1, an extensive diagnosis of the motor can be made since in those signals a large part of the physical phenomena that occur in its operation is reflected. The problem lies in the instrumentation needed to capture the signals, since (except for vibrations and emissions) the signals are intrusive. The intrusive signals are those in which it is necessary to physically affect the ICE to make its acquisition, specifically in this case, the speed and pressure. On the one hand, ICE's have strict requirements for emissions to be accepted commercialy, on the other hand, the adquisition of this signal requires expesive emission analyzers. To meet this need, in this article an unconventional methodology for the diagnosis of ICE is exposed, which can be seen in Fig. 2. The classical methodology needs multiple signals which are complex and expensive to implement, so a new methodology is proposed that only needs signal to diagnose an ICE. The methodology only needs the ICE vibration signals, which contains a lot of information about the operation.

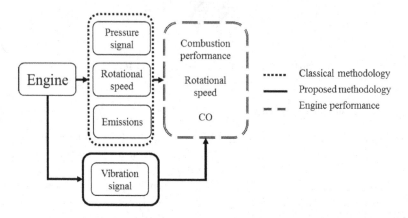

Fig. 2. Classical methodology and proposal for the diagnosis of ICE.

2.5   Data Base

The database was created by an instrumentation system that allowed the synchronous capture of signals from the engine. Pressure signals were captured through KISTLER pressure sensor model 7061B with range until 250 bar. Vibration signals were captured for 5 s 51200 Hz sampling rate. Each of the 24 signals consists of 256.000 samples, which are divided into 62 sub signals of 4096 samples with the objective of having more samples per class and reducing the computation time. In addition, dividing the original signals allows to follow the experimental framework of works such as [44], which look for more stationary signal segments. Classification tags for the CI engine diagnosis are as follows. It is used the phase angle between the TDC (Top Dead Center) and the peak pressure which should vary from 10.5 to 12.5 °C [40]; this gives three classes, under the interval, in the interval and over the interval. Moreover, the different fuels were classified to demonstrate the relationship between vibrations and general information about the engine working state, which here are three different blends. In addition, three different rotational speeds were tested for each fuel. Last, it was classified the state of the CO emissions, been permissible or non-permissible under the standard. A second test was done for the comparison of the methodology, in which the Case Western Reserve database was used [1]. It was taken the vibration signals of the normal bearings, bearings with faults in the internal train, external train, and bearings with ball failure for the classification, which makes a total of four classes for the test. To guarantee the independence of the model, a cross validation for both tests of k = 5 was performed (20% for testing, 70% for training and 10% for validation).

3   Results

Before testing the proposed methodology in the ICE database, its effectiveness is established by experimenting with fault diagnosis in bearing systems (Case Western Reserve database) and comparing its accuracy with results from the literature. The result of this experiment can be seen in the Table 1.

Table 1. Accuracy comparison for bearings diagnosis.

Author	Classes	Feature extraction	Classifier	Features	Acc.[%]
Zhang et al. [36]	3	PE+EMD	SVM	12	97.8
Yuwono et al. [41]	3	WPT	HMM	12	95.8
William et al. [38]	4	ZC	ANN	10	97.1
Ocak et al. [29]	3	LPM	HMM	30	99.6
Wei et al. [37]	6	FR+WPT	AP	18	96.0
Zheng et al. [45]	6	GCMPE	SVM+PSO	2	98.8
Liang et al. [22]	4	TP+FR	KNN	3	92.9
Muru et al. [28]	4	SSA	ANN	10	95.1
This work	4	MPE	KNN	9	99.7

Table 2 positions the proposed methodology as one of the best results in the state of the art. It should be noted that the high number of features is compensated by the low-time computation due to the KNN classifier. This can be seen in Table 3.

Table 2. Accuracy comparison for bearings diagnosis.

Classifier	Computation time
KNN	3.297 ± 0.035
SVM	108.822 ± 3.503
ANN	45.232 ± 0.578

The computation time was computed based on 10 classification repetitions in a computer with an i7 processor and 16 Gb of ram (dedicated only to the classification process). KNN has a computation time 36 and 15 times lower than SVM and ANN, respectively. Now, the proposed methodology seeks to diagnose ICE based on vibration signals, classifying different configurations and characteristics through machine learning. The characteristics and configurations are divided into classes to be classified by the algorithms and thus be able to generate a correct diagnosis. The division into classes can be seen in Fig. 3. The defined classes allow the creation of the database, which is composed of vibration signals that have the six labels to perform the automatic classification. The results of the classification can be seen in Fig. 4. All classification results exceed 91.6% accuracy, demonstrating the effectiveness of the proposed methodology to generate a diagnosis of the ICE. In addition, it is verified that the vibration signals have relevant information on the operating status of the ICE, making acquisition of other signals unnecessary. The highest precision is presented by the variable CO with 96.6% ± 2.1%. The high accuracy of CO classification verifies the high correlation this variable has with mechanical vibrations. On the other hand, speed tag has the lowest with an accuracy of 91.6% ± 0.63%,

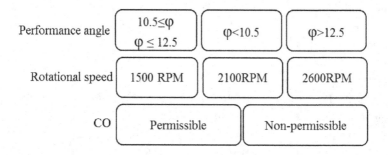

Fig. 3. Configuration in classes and characteristics.

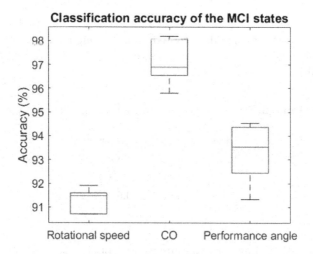

Fig. 4. Classification results for each label.

although without significant differences with the other tags. The Performance angle is a very important feature in the operation of the ICE and the proposed methodology can identify it with $93.2\% \pm 1.6\%$ accuracy.

4 Conclusion

In the present work, a robust and efficient ICE diagnosis methodology based on vibrations, a non-intrusive signal, is presented in detail by means of Multiscale Permutation Entropy measurements, processing them with the feature selection technique Variance Relevance Analysis in the combination with the supervised classifier of the K Nearest Neighbors. The proposed methodology generates a model that classifies characteristics as rotational speed, performance angle, and CO for an ICE diagnosis. The model obtained can be considered independent after cross validating the data with 20% for test data. The non-linear classification method KNN used to analyze the data within the vibration signal with the MPE characterization method derives an innovative alternative in place of classical analyzes in frequency domain. Improving its performance hierarchizing and selecting beforehand, by variance with the VRA technique, the features, which greatly improves the efficiency and effectiveness of the model avoiding overfitting. The characteristics and configurations were classified with an accuracy of over 91.6% with low standard deviation for all variables predicted, allowing to confirm the relevance of the information presented in the vibration signals of the ICEs. The best characteristic classified was CO emissions, which achieved 96.6% accuracy, which would help to make fast diagnose for maintenance in order to have a clean environment. Another important contribution of this research is the application of a methodology for computing the ICE combustion performance angle, which together with the proposed classification methodology would help to take measures to maintain the ICE with the best possible combustion, extending components' useful life and reducing emissions by taking advantage of a maximum power, since knowing the Performance angle even if ICE configurations and characteristics are altered, allows corrective actions (advance the spark in the combustion chamber or modify the fuel).

Acknowledgement. The authors thank the Master's in Mechanical Engineering program and the Faculty of Mechanical Engineering of the Universidad Tecnológica de Pereira UTP (Technological University of Pereira) for their support throughout the research.

References

1. Bearing Data Center, Case Western Reserve University. http://csegroups.case.edu/bearingdatacenter/pages/download-data-file
2. Al-Badour, F., Sunar, M., Cheded, L.: Vibration analysis of rotating machinery using time-frequency analysis and wavelet techniques. Mech. Syst. Signal Process. **25**(6), 2083–2101 (2011)

3. Aziz, W., Arif, M.: Multiscale permutation entropy of physiological time series. In: 2005 Pakistan Section Multitopic Conference, pp. 1–6 (2006)
4. Bandt, C., Pompe, B.: Permutation entropy: a natural complexity measure for time series. Phys. Rev. Lett. **88**, 174102 2002)
5. Barelli, L., Bidini, G., Buratti, C., Mariani, R.: Diagnosis of internal combustion engine through vibration and acoustic pressure non-intrusive measurements. Appl. Themolr Eng. **29**, 1707–1713 (2009)
6. Bilton, P., Jones, G., Ganesh, S., Haslett, S.: Classification trees for poverty mapping. Comput. Stat. Data Anal. **115** (2017)
7. Brissaud, J.B.: The meanings of entropy. Entropy **7**(1), 68–96 (2005)
8. Chen, J., Randall, R.: Vibration signal processing of piston slap and bearing knock in IC engines. In: 6th Conference on Surveillance, January 2011
9. Costa, M., Goldberger, A., Peng, C.K.: Multiscale entropy analysis of complex physiologic time series. Phys. Rev. Lett. **89**, 068102 (2002)
10. Das Gupta, M., Xiao, J.: Non-negative matrix factorization as a feature selection tool for maximum margin classifiers. In: CVPR 2011, pp. 2841–2848, June 2011
11. Daza-Santacoloma, G., Arias-Londono, J.D., Godino-Llorente, J.I., Sáenz-Lechón, N., Osma-Ruíz, V., Castellanos-Dominguez, G.: Dynamic feature extraction: an application to voice pathology detection. Intelll. Autom. Soft Comput. **15**(4), 667–682 (2009)
12. Daza-Santacoloma, G., Arias-Londoño, J.D., godino llorente, J., Saenz-Lechon, N., Osma-Rutz, V., Castellanos-Dominguez, G.: Dynamic feature extraction: An application to voice pathology detection. Intell. Autom. Soft Comput. **15**, 667–682 (2009)
13. Ericsson, S., Grip, N., Johansson, E., Persson, L.E., Sjöberg, R., Strömberg, J.O.: Towards automatic detection of local bearing defects in rotating machines. Mech. Syst. Signal Process. **19**, 509–535 (2005)
14. Fernández, J., Alvarez, A., Quintero, H., Echeverry-Correa, J., Orozco, A.: Multilayer-based HMM training to support bearing fault diagnosis. In: Proceedings of 6th International Workshop, IWAIPR 2018, Havana, Cuba, 24–26 September 2018, pp. 43–50, September 2018
15. Filippetti, F., Franceschini, G., Tassoni, C., Vas, P.: Recent developments of induction motor drives fault diagnosis using AI techniques. Ind. Eloctr., IEEE Tians. **47**, 994–1004 (2000). https://doi.org/10.1109/41.873207
16. Flett, J., Bone, G.: Fault detection and diagnosis of diesel engine valve trains. Mech. Syst. Signal Proces. **72–73**, 316–327 (2015)
17. Gao, Z., Cecati, C., Ding, S.: A survey of fault diagnosis and fault-tolerant techniques-part i: Fault diagnosis with model-based and signal-based approaches. IEEE Trans. Ind. Electr. **62**, 3757–3767 (2015)
18. Grajales, J., Quintero, H., Lopez Lopez, J., Romero, C., Henao, E., Cardona, O.: Engine diagnosis based on vibration analysis using different fuel blends. Diagnostyka **18**, 27–36 (2017)
19. Hernández, J.M., Echeverry, J.D., Riaza, H.F.Q.: Use of multiscale permutation entropy feature selection and supervised classifiers for bearing failures diagnosis. Sci. Tech. **26**(4), 448–449 (2021)
20. Hernández, J.C.M., Madrid, F.G., Quintero, H.F., Alzate, J.D.R.: Diesel engine diagnosis based on entropy of vibration signals and machine learning techniques. Electr. Lett. **58** (2022)
21. Jinde, Z., Junsheng, C., Yang, Y.: A rolling bearing fault diagnosis approach based on LCD and fuzzy entropy. Mech. Mach. Theory **70**, 441–453 (2013)

22. Liang, L., Liu, F., Li, M., He, K., Xu, G.: Feature selection for machine fault diagnosis using clustering of non-negation matrix factorization. Measurement **94**, 295–305 (2016)
23. Malhi, A., Gao, R.: Pca-based feature selection scheme for machine defect classification. Instrum. Measur. IEEE Trans. **53**, 1517–1525 (2005)
24. Malhi, A., Gao, R.: Pca-based feature selection scheme for machine defect classification. Instrumentation and Measurement, IEEE Trans. **53**, 1517–1525 (2005)
25. Mejía, J.C., Quintero, H.F., Echeverry-Correa, J.D., Romero, C.A.: Detection of ice states from mechanical vibrations using entropy measurements and machine learning algorithms. Diagnostyka **21** (2020)
26. Müller, P.,et al.: Scent classification by k nearest neighbors using ion-mobility spectrometry measurements. Exp. Syst. Appl. **115** (2018)
27. Moosavian, A., Najafi, G., Ghobadian, B., Mirsalim, S., Jafari, S., Sharghi, P.: Piston scuffing fault and its identification in an IC engine by vibration analysis. Appl. Acous. **102**, 40–48 (2016)
28. Muruganatham, B., Krishnakumar, S., Murty, S.: Roller element bearing fault diagnosis using singular spectrum analysis. Mech. Syst. Signal Process. **35**(1), 150–166 (2013)
29. Ocak, H., Loparo, K.: Hmm-based fault detection and diagnosis scheme for rolling element bearings. J. Vib. Acoust. **127**, 2–15 (2005)
30. Payri, F., Luján, J., Martín, J., Abbad, A.: Digital signal processing of in-cylinder pressure for combustion diagnosis of internal combustion engines. Mech. Syst. Signal Proces. **24**, 1767–1784 (2010)
31. Pincus, S.: Approximate entropy as a measure of system complexity. In: Proceedings of the National Academy of Sciences of the United States of America, vol. 88, pp. 2297–301 (1991)
32. Saraswati, S.: Reconstruction of cylinder pressure for SI engine using recurrent neural network. Neural Comput. Appl. **19**, 935–944 (2010)
33. Shao, R., Hu, W., Wang, Y., Qi, X.: The fault feature extraction and classification of gear using principal component analysis and kernel principal component analysis based on the wavelet packet transform. Measurement **54**, 118–132 (2014)
34. Taghizadeh-Alisaraei, A., Ghobadian, B., Tavakoli-Hashjin, T., Mohtasebi, S.S., Rezaei-asl, A., Azadbakht, M.: Characterization of engine's combustion-vibration using diesel and biodiesel fuel blends by time-frequency methods: A case study. Renew. Energy **95**, 422–432 (2016)
35. Vencalek, O., Pokotylo, O.: Depth-weighted Bayes classification. Comput. Stat. Data Anal. **123**, 1–12 (2018)
36. Wang, Y., Ma, Q., Zhu, Q., Liu, X., Zhao, L.: An intelligent approach for engine fault diagnosis based on Hilbert-hyang transform and support vector machine. Appl. Acoust. **75**, 1–9 (2014)
37. Wei, Z., Wang, Y., He, S., Bao, J.: A novel intelligent method for bearing fault diagnosis based on affinity propagation clustering and adaptive feature selection. Knowl.-Based Syst. **116**, 1–12 (2017)
38. William, P., Hoffman, M.: Identification of bearing faults using time domain zero-crossings. Mech. Syst. Signal Process. **25**(8), 3078–3088 (2011)
39. Wu, J.D., Chuang, C.Q.: Fault diagnosis of internal combustion engines using visual dot patterns of acoustic and vibration signals. NDT e Int. **38**(8), 605–614 (2005)
40. Wu, J.D., Liu, C.H.: An expert system for fault diagnosis in internal combustion engines using wavelet packet transform and neural network. Expert Syst. Appl. **36**, 4278–4286 (2009)

41. Yuwono, M., Qin, Y., Zhou, J.: Automatic bearing fault diagnosis using particle swarm clustering and hidden Markov model. Eng. Appl. Artif. Intell. **47**, 88–100 (2016),

42. Zaidi, S.S., Aviyente, S., Salman, M., Shin, K.K., Strangas, E.: Prognosis of gear failures in dc starter motors using hidden markov models. Industrial Electronics, IEEE Trans. Ind. Electr.**58**, 1695–1706 (2011)

43. Zanin, M., Zunino, L., Rosso, O.A., Papo, D.: Permutation entropy and its main biomedical and econophysics applications: a review. Entropy **14**(8), 1553–1577 (2012)

44. Zeng, K., Ouyang, G., Cheng, H., Gu, Y., Liu, X., Li, X.: Characterizing dynamics of absence seizure eeg with spatial-temporal permutation entropy. Neurocomputing **275** (2017)

45. Zheng, J., Pan, H., Yang, S., Cheng, J.: Generalized composite multiscale permutation entropy and Laplacian score based rolling bearing fault diagnosis. Mech. Syst. Signal Process. **99**, 229–243 (2018)

Markers of Exposure to the Colombian Armed Conflict: A Machine Learning Approach

María Isabel Cano[1]([✉]), Claudia Isaza[1], Angela Sucerquia[2], Natalia Trujillo[3], and José David López[1]

[1] SISTEMIC, Facultad de Ingeniería, Universidad de Antioquia UdeA, calle 70 No. 52-21, Medellín, Colombia
mariai.cano@udea.edu.co
[2] Instituto Tecnológico Metropolitano ITM, Medellín, Colombia
[3] GISAME, Facultad Nacional de Salud Pública, Universidad de Antioquia UdeA, calle 62 No. 52 - 59, Medellín, Colombia

Abstract. The Colombian armed conflict has affected in some degree its entire population. Health authorities require markers to determine this exposure and provide proper mental-health interventions. Unsupervised learning techniques allow clustering subjects with similar features. Here, we propose a novel methodology to automatically finds the features that best relate to levels of exposure to the armed conflict and associated risks (drug dependency, alcoholism, etc.) through cluster centers. Unlike previous studies on the armed conflict field, we do not use key predefined labels to cluster the data. We test this methodology with a mixed-response type characterization database of 528 features obtained from 346 volunteers with different estimated levels of exposure to extreme experiences in the frame of the Colombian armed conflict. As a result, using the proposed approach we identified 62 features related to exposure. In order to confirm the selected features as violence exposure markers, we created a model based on artificial neural networks (ANN). The ANN model uses the 62 features as input and it was able to estimate the subjects' level of exposure to conflict with 100 % accuracy in training and over 76% in validation.

Keywords: Armed conflict · Mental health · Feature selection · Unsupervised learning · Clustering

1 Introduction

The internal armed conflict is the main cause of violence in Colombia. This has brought economic, humanitarian, and social consequences, as well as emotional impacts on the population for more than six decades. The RUV (from Spanish: Unique Registry of Victims) in its latest report shows that more than 8.2 million people have been internally displaced [15]. Previous studies have evidenced that emotional affectation is extended to the civilian population and not only

© The Author(s), under exclusive license to Springer Nature Switzerland AG 2022
A. C. Bicharra Garcia et al. (Eds.): IBERAMIA 2022, LNAI 13788, pp. 185–195, 2022.
https://doi.org/10.1007/978-3-031-22419-5_16

ex-combatants and victims [4,11,14]. In these studies, we have observed higher levels of aggression and incidence of mental-health disorders in populations who show high exposure to conflict-related extreme experiences, even in subjects who do not perceive themselves as victims. This exposure could lead to long term outcomes in mental health if not timely and properly accounted. Therefore, a better characterization of the population is required to help the health authorities, for example by identifying affectation markers with computational intelligence tools. The markers are features included in psychological questionnaires. The features can be analyzed with artificial intelligent algorithms in order to identify the most relevant to emotional affectation.

Machine learning techniques allow finding relationships among features. These techniques have their foundations in mathematics and statistics, but they are more focused on prediction than traditional statistical inference methods. The main advantage of unsupervised learning techniques is their model independency and their aim to maximize precision and minimize replicability issues [10]. Unsupervised techniques naturally find associations among data with similar features, e.g., clustering techniques establish their operation on finding natural groups according to their data structure. The goal of clustering is to find a characteristic pattern per group, which is represented by a vector located in the center of the cluster [1]. Here, we propose to use cluster centers to identify features that could be used as health markers, using a database with several different aspects: demographic, economic, psychiatric, among others; and several different types of response (dichotomous, Likert, categorical, etc.) To our knowledge, there are not previous works using cluster centers analysis to find key features in this kind of problem.

Other authors who have studied violence and armed conflict using machine learning techniques [5,13] have used regression methods, random forests, and deep neural networks to make dimensionality reductions looking for the most relevant features in the classification of violence. But these studies use supervised techniques to select the variables (commonly veteran and victim), leaving aside the civilian population who has been affected by the conflict but misclassified as controls, as stated by [11]. In addition to the fact that most dimensionality reduction techniques are supervised, they cannot be applied to mixed data types, such as PCA or LDA [7,9]. Besides, the number of initial variables they use is significantly lower than ours.

In this work, we propose a methodology to provide health markers of exposure to violence using unsupervised learning over a database with 528 variables surveyed from 346 subjects, i.e., an ill-posed problem. We use the cluster centers to find relevant features related to the armed conflict, always keeping traceability and interpretability. Once they are determined, we test them with an artificial neural network (ANN) as a supervised technique. The ANN is used to find a model to identify levels of exposition in civilians, according to the Extreme Experiences Scale (EX^2) [4]. Our aim with the ANN is to analyze the estimation capability of the relevant features to relate the subjects with the armed conflict, so they are reliable health markers useful for decisions takers.

2   Materials and Methods

2.1   Materials

Antioquia is one of the Colombian regions with more historical armed conflict events, especially in rural areas. We analyze data from a sample of 346 university students from different municipalities of Antioquia, who participated in two different surveys: A general characterization (virtual survey) which contains 401 questions on the following aspects: basic information, health and social safety, academic, psychological, socioeconomic, socio-family, work, sexuality and affectivity, nutrition, sport, recreation, and culture [16]. Within the questions that evaluate psychological aspects, 22 of these partially correspond to the self-report questionnaire (SRQ), a psychometric scale that evaluates five aspects: depression, anxiety, alcoholism, psychosis, and epilepsy. This scale results in an alert of possible clinical indicators of the above aspects. [3]. The database is composed of different types of responses: open responses, Likert-based, numerical, categorical, and dichotomous.

This characterization is performed by the Wellness Unit of the Universidad de Antioquia www.udea.edu.co, public university with presence in several urban and rural municipalities of Antioquia, and its objective is to identify risks. Based on the answers given by the subjects, 88 risks are computed and classified as high, medium, or low (for example: low risk for drug use, high risk for academic desertion, etc.) In total, the database contains 507 variables (including 18 default items: dates, undergraduate program, etc.)

This database is complemented with the Scale of Extreme Experiences (EX^2) [4], which is an 18-items questionnaire adapted to the context of the Colombian armed conflict. The items are classified in two dimensions: direct extreme experiences (dEX^2) and indirect extreme experiences (iEX^2). The first one focused on personal physical situations (e.g. been kidnapped or beaten) and the second one on third parties with whom there is an emotional bond, such as family or friends (e.g. murder or kidnapping of a close person). The responses are dichotomous (yes/no), and the final score is the sum of the affirmative responses per item. In [4], authors found that more than two affirmative questions (EX^2 scores over 2.5) were indicative of high exposure to extreme experiences in the frame of the armed conflict.

For data protection, all open responses (name, address, etc.) were removed from the study. Then, all categorical features were quantified expanding each category to a new feature. The new feature denotes the presence or absence numerically with 0 or 1 (e.g. a categorical feature with 3 categories was replaced by 3 binary features). Finally, all variables were normalized to avoid biases. The EX^2 results were not used in the clustering stage (as they could work as labels). In the same way, items "victim of the armed conflict", "internally displaced person", and risk variables given by the characterization survey were excluded. These variables were later used for further testing the automatically generated clusters. In total, 303 variables were used to form the clusters and the rest of them for testing.

This research was approved by the ethics committee of the National School of Public Health at the Universidad de Antioquia, adopting the ethical considerations of resolution 8430 of 1993 of the Colombian Ministry of Health and the declaration of Helsinki [6]. Privacy, confidentiality, the right to not participate, data anonymity, and information custody and management conditions were guaranteed. The risk of the investigation was minimal. However, due to the sensitivity that some items of the EX^2 can cause, support networks were established to attend to eventual crises, but no incidents were reported.

2.2   Methods

The proposed methodology consists of the flowchart presented in Fig. 1. The process is iterative and begins by automatically grouping the subjects in nine partitions (from two to ten clusters) via k-means. Then, the adequate number of clusters is selected by applying an internal cluster index to each partition. With this index, the partition with the most (within) compacted and (between) separated clusters is selected. The next step consists of finding the cluster center values (for each feature). Based on the measured center distances, the most relevant features are selected, and a new iteration begins with this new reduced set.

Starting from the second iteration and once the adequate number of clusters is determined, their correlation with conflict related and risks variables is obtained. If this correlation improves compared to the previous iteration, the algorithm continues by pruning non-relevant features and starting a new iteration with the reduced set. Otherwise, the process stops and the set of features from the previous iteration is kept. These steps are explained below.

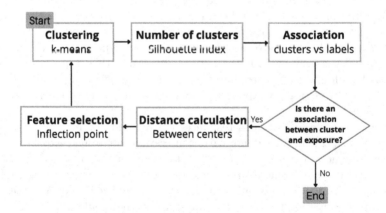

Fig. 1. Proposed methodology to find markers relating a general characterization of civilian population with their level of exposure to the armed conflict.

Clustering: The first step consists of grouping the N_c subjects in clusters using the k-means algorithm [8]. Let define the objects $X = \{x_1, x_2, \ldots, x_{N_c}\} \in \mathbb{R}^{N_f}$, where each object is formed with the N_f features (or attributes) of a subject; and the clusters $C = \{c_1, c_2, \ldots, c_{N_k}\}$, where the number of clusters N_k is provided by the user. Our aim is to minimize the squared error $\epsilon_{k-means}$ between the empirical mean of each k-th cluster (represented by its center $\mu_k \in \mathbb{R}^{N_f}$) and the dimensions of its own objects $X^{(k)}$:

$$\epsilon_{k-means} = \sum_{k=1}^{N_k} \sum_{i=1}^{N_c^{(k)}} \left\| x_i^{(k)} - \mu_k \right\|^2 \tag{1}$$

where $x_i^{(k)}$ is the i-th object belonging to the k-th cluster, which contains $N_c^{(k)}$ objects. The algorithm consists of the following steps:

1. Select the parameter N_k corresponding to the number of clusters.
2. Randomly create a starting position for the centers of each cluster.
3. Compute the Euclidean distance between each object and the centers.
4. Assign each object to the closest center.
5. To minimize the error of (1), μ_k is relocated from the calculation of the average of the objects that belong to the cluster c_k.
6. Repeat the steps 3 to 5 until it stabilizes or until reaching a maximum number of iterations.

To avoid local minima, several runs with different seeds should be made (we ran 500). The algorithm records the final value of $\epsilon_{k-means}$ per run, and returns the minimum one.

We repeated this procedure for nine partitions, varying the number of clusters in step 1 from $N_k = \{2, \ldots, 10\}$.

Finding the Adequate Number of Clusters: Once all partitions are created, the one with the number of clusters that better groups the subjects (in terms of intra-group cohesion and inter-group separation) is validated through the silhouette index [12]. For each partition, the average dissimilarity $a \in \mathbb{R}^{N_c}$ among all objects from cluster c_k is calculated:

$$a(i) = \frac{1}{N_c^{(k)} - 1} \sum_{j, j \neq i}^{N_c^{(k)} - 1} d\left(x_i^{(k)}, x_j^{(k)}\right) \tag{2}$$

where $d\left(x_i^{(k)}, x_j^{(k)}\right)$ is the Euclidean distance between $x_i^{(k)}$ and $x_j^{(k)}$. Then, the average dissimilarity $b \in \mathbb{R}^{N_c}$ between the objects of c_k and the ones from the nearest cluster $c_{\widehat{k}}$ ($\widehat{k} \neq k$) is calculated:

$$b(i) = \frac{1}{N_c^{(\widehat{k})}} \sum_{j=1}^{N_c^{(\widehat{k})}} d\left(x_i^{(k)}, x_j^{(\widehat{k})}\right) \tag{3}$$

With a and b, silhouette values for each subject $s \in \mathbb{R}^{N_c}$ are computed:

$$s(i) = \frac{b(i) - a(i)}{\max\left(a(i), b(i)\right)} \tag{4}$$

each giving a score between -1 and 1. Finally, an average silhouette width per partition is computed: $S(j) = \mathrm{mean}(s)$, with $N_k^{(j)} = 2, \ldots, 10$; and the chosen partition corresponds to $N_k = \arg\max(S)$.

Distance Calculation: With the proper number of clusters N_k found, the maximum distances among all cluster centers per feature $\delta \in \mathbb{R}^{N_f}$ are computed:

$$\delta(f) = \max\left(\mu_f^{(k)} - \mu_f^{(\widehat{k})}\right); \ \forall k, \widehat{k} = 1, \ldots, N_k; \ k \neq \widehat{k} \tag{5}$$

with $f = 1, \ldots, N_f$. A graphic example is presented in Fig. 2, where the largest distance among centers for Feature X is 0.7 (between $\mu_x^{(1)}$ and $\mu_x^{(3)}$), while $\delta(Z) = 0.15$. All distances are computed from the largest center position, so all values are positive.

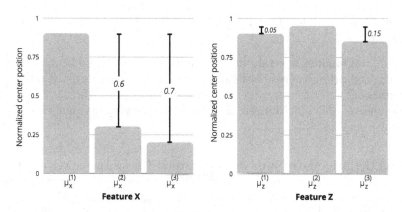

Fig. 2. Examples of distances among cluster centers for each feature. The y-axis represents the position of the centroid in each feature dimension. Distances for Feature X are significantly larger than for Feature Z, which will be accounted in the following stage.

Finding the Most Relevant Features: In this step, the distances δ are arranged from largest to smallest and plotted as in Fig. 3 (left). Then, the slope between the minimum and maximum values is calculated (see Fig. 3 (mid)) and the graph is rotated until the slope is zero (Fig. 3 (right)). Finally, the new minimum value is the inflection point (red dot in Fig. 3 (right)), features found before that value are considered as the most relevant for the clusters (toy features X, V, and Q in this example).

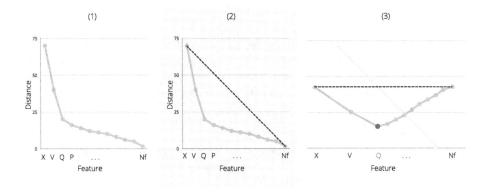

Fig. 3. Procedure to select the variables that contribute the most to the clustering. The y-axis represents the largest distance δ found for each feature. Toy names X, V, Q, and Nf represent the features ordered in descending order. Left: Distances calculated with (5). Middle: The slope between the maximum and minimum distances is calculated. Right: The plot is rotated, and a new minimum is found. In this example, the inflection point occurs in feature Q; therefore, features X, V, and Q are the more relevant for the clustering.

Association of Clusters with Level of Exposure to Conflict: Starting in the second iteration and once the partition with the adequate number of clusters is found, the correlation between the set of clusters and the variables related to conflict and risks (labels) is computed. If the correlation is higher than in the previous iteration (i.e., if the reduced set of features behaves better), there is a chance that further reducing the number of features would improve the results. Otherwise, the set of features from the previous iteration is kept as the one that better differentiates the clusters in terms of levels of exposure to conflict and risks.

Testing the Final Set: Although the EX^2 test is a direct measure of exposure to extreme experiences in the frame of the conflict, it is not part of the general characterization. Therefore, it would be desirable to determine the capabilities of the selected set of features to estimate high exposure to the conflict without including the EX^2 test.

To this aim. We created an artificial neural network (ANN) fed with the selected features and the EX^2 score. The ANN consisted of a feedforward neural network with four hidden layers, each with the same number of neurons (corresponding to the final number of features). An 80/20 partition was made for training and testing. The activation function for the hidden and output layers was a symmetric sigmoid transfer function. A gradient descent with adaptive learning rate was used for training.

3   Results

Applying the proposed methodology led to find 62 features as relevant markers (a reduction of 79% from the 303 initial ones), grouping the subjects in three clusters after just three iterations. They are listed in Table 1. It shows which features were more relevant per aspect, numbering them in relevance order (1 being the most discriminant).

Table 1. Relevant features defined as markers with the proposed methodology. They are separated in the aspects to which each characteristic belongs, and each feature includes its relevance within the set.

	Description		Description
	psychological		*academic*
1	SRQ: Do you feel sad?	43	Study habits: Do you have a quiet space to study?
2	SRQ: Do you feel bored?	46	Why did you choose the academic program? possibility of exchange to another program
3	SRQ: Do you feel tired all the time?	47	Study habits: Do you have a desk to study?
4	SRQ: Have you lost interest in things?	59	Why did you choose the academic program? professional vocation
5	SRQ: Do you feel nervous or tense?	61	Why did you choose the academic program? low cost
7	SRQ: Do you sleep poorly?	62	Study habits: Do you have a solitary space to study?
8	SRQ: Is it difficult for you to do your job? / Has your job been affected?		*socio-economic*
11	SRQ: Is it difficult for you to enjoy your daily activities?	20	Type of home you live in: Room
12	SRQ: Do you have difficulty making decisions?	29	Type of housing: Own
14	SRQ: Have you had the idea of ending your life?	31	Type of housing: Leased
16	SRQ: Do you have frequent headaches?	44	Who do you depend on financially? Parents
17	SRQ: Have you noticed interference or something strange in your thinking?	45	Who do you depend on financially? Yourself
18	SRQ: Do you have a bad appetite?	54	What type of employment relationship do you have? independent
19	Did you use this substance in the last year? Marijuana	55	Type of housing you live in: House
22	SRQ: Are you unable to think clearly?		*socio-family*
25	SRQ: Do you cry very often?	13	Family dynamics: Are you satisfied with the time you and your family spend together?
30	Do you consider that working has affected any aspect of your life?	24	Family dynamics: Do you discuss with each other the problems you have at home?
32	SRQ: Do you suffer from tremor in your hands?	27	Family dynamics: Are you satisfied with the help you receive from your family?
33	SRQ: Do you get scared easily?	28	Family dynamics: Are important decisions made together at home?
34	SRQ: Are you unable to play a useful role in your life?	50	Do you feel that your family loves you?
35	SRQ: Do you suffer from poor digestion?	51	Have you suffered pressure from your family in making decisions?
39	Do you consider that the Internet has affected any aspect of your life?		*nutrition*
40	Do you consider that chatting has affected any aspect of your life?	6	On most days of the week, do you have a snack between lunch and dinner?
48	Do you consider that online games have affected any aspect of your life?	15	On most days of the week, do you have set times for your meals?
49	SRQ: Do you feel that someone has tried to hurt you?	21	On most days of the week, do you eat snacks between breakfast and lunch?
52	Do you consider that social networks have affected any aspect of your life?	23	Are you currently using any strategy to keep from gaining or losing weight?
58	Do you consider that sex has affected any aspect of your life?	36	On most days of the week, how much fruit do you consume?
	academic	38	In a typical week, how many days do you eat fish?
9	Study habits: Do you have a chair to study?	57	In a typical week, how many days of the week do you consume legumes?
10	Study habits: Do you have a desk to study?		*others**
26	Study habits: is the place where you study illuminated?	53	In general terms, how do you think your health is?
37	Have you thought about quitting school?	56	do you have social security?
41	Study habits: Do you have time to study?	60	sex: female or male
42	Study habits: Do you have a smartphone to study?		

*Others refers to health, social safety, and basic information

The features that contributed the most to the separation of the groups were: psychological aspects (specially the SRQ scale), followed by academic, and socio-economic variables. From the 22 SRQ features, the 20 selected (see Table 1) are focused on depression and anxiety disorders. To a lesser extent, relevant aspects related to socio-family, nutrition, health and social safety, and basic information. In past studies, demographic factors, age, sex, and education have been used for traditional analyses in populations related to the armed conflict [2,11,14]. However, our methodology uses the data to find natural clusters that in turn may be related to exposure to conflict, so that other useful aspects not considered before can be included.

Figure 4 (left panel) shows the distances of the final set of features. This information was used to define the relevance list of Table 1 and allows the users to quantify this relevance, i.e., these distances provide traceability in post-hoc studies, where psychologist might orient their intervention by focus in cognitive and conduct trainings in specific aspects.

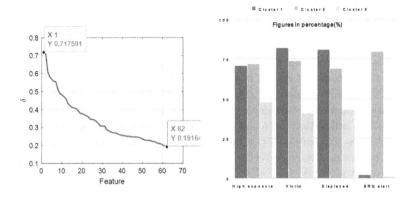

Fig. 4. Left: Maximum distance δ of each of the final 62 selected features. This chart provides the relative relevance among features. Right: Percentage of subjects in each cluster that present high exposure, consider themselves as victims of the conflict, are displaced, or presented high SQR alert. These charts provides the profiles later used by psychologists to define new alerts and provide cognitive and conduct trainings.

The clustering identified representative groups associated with exposure to conflict and risks. Figure 4 (right panel) shows that Clusters 1 and 2 included a higher percentage of subjects with those variables more directly related to the conflict (e.g. "high exposure" from the EX^2) and mental health risks (e.g. SRQ results). Note how the SQR based alert was highly associated to Cluster 2. This coincides with the top features shown in Table 1.

Estimation of Exposure to Conflict: An ANN was implemented to analyze the capabilities of the final 62 features to estimate individual levels of exposure (using the EX^2 score as a label). We first needed to balance the classes, leaving 120 samples per class and a total population of 240 subjects. A 5-fold cross-validation was performed, reaching a mean accuracy of 99.58% (SD=0.83) for the training stage and 75.36% (SD = 0.98) in the testing stage. This supervised learning method confirms that the variables selected from the previous procedure are discriminatory. Therefore, in absence of the EX^2 test (which is the usual situation), health authorities (the Wellness Unit of the University in this case) can automatically create a high-precision alert on students who were more exposed to the conflict and may present undesired outcomes if an intervention is not made. Additionally, the correlation with other risks could help on the design of trainings for this population.

4   Conclusions

We proposed a methodology for finding relevant features that better relate with a set of conflict related and risk variables, extracted from a characterization database with mixed personal information (demographic, socioeconomic, psychological, etc.) The selected features were chosen with a novel pipeline based on distance into the cluster centers; which resulted in a set of 62 markers from the initial set of 303 features, with a high relevance of the mental health SRQ scale. This latter relation has been suggested in the literature but not tested before in the Colombian armed conflict context.

An ANN was implemented to test the effectiveness of the set of 62 markers to estimate exposure to extreme experiences in the frame of the armed conflict. As a result, a validation accuracy close to 80% was obtained. The low number of subjects compared to the number of variables could be a limitation to achieve a higher accuracy; nevertheless, the results show consistency and a good approach to study this population. With this approach, mental health professionals can prepare intervention programs with features that could be neglected by traditional statistical or supervised learning analyses.

Acknowledgement. This work was supported by MinCiencias (Colombia) grant 111584467273.

References

1. Aggarwal, C.C.: Data Mining. Springer, Cham (2015). https://doi.org/10.1007/978-3-319-14142-8
2. Bell, V., Méndez, F., Martínez, C., Palma, P.P., Bosch, M.: Characteristics of the Colombian armed conflict and the mental health of civilians living in active conflict zones. Confl. Heal. **6**(1), 1–8 (2012)
3. Beusenberg, M., Orley, J.H.: A user's guide to the self reporting questionnaire SRQ. World Health Organization, Technical report (1994)
4. Giraldo, L.S., Aguirre-Acevedo, D., Trujillo, S., Ugarriza, J.E., Trujillo, N.: Validation of the extreme experiences scale (ex2) for armed conflict contexts. Psychiatr. Q. **91**(2), 495–520 (2020)
5. Goin, D.E., Rudolph, K.E., Ahern, J.: Predictors of firearm violence in urban communities: a machine-learning approach. Health Place **51**, 61–67 (2018)
6. Goodyear, M.D.E., Krleza-Jeric, K., Lemmens, T.: The declaration of Helsinki. BMJ. **335**(7621), 624–625 (2007). https://doi.org/10.1136/bmj.39339.610000.BE, https://www.bmj.com/content/335/7621/624
7. Guyon, I., Gunn, S., Nikravesh, M., Zadeh, L.A.: Feature Extraction: Foundations and Applications, vol. 207. STUDFUZZ. Springer, Cham (2008). https://doi.org/10.1007/978-3-540-35488-8
8. Jain, A.K.: Data clustering: 50 years beyond k-means. Pattern Recogn. Lett. **31**(8), 651–666 (2010)
9. Liu, H., Motoda, H.: Computational Methods of Feature Selection. CRC Press, Boca Raton (2007)
10. Orrù, G., Monaro, M., Conversano, C., Gemignani, A., Sartori, G.: Machine learning in psychometrics and psychological research. Front. Psychol. **10**, 2970 (2020)

11. Quintero-Zea, A., et al.: Characterization framework for Ex-combatants based on EEG and behavioral features. In: Torres, I., Bustamante, J., Sierra, D. (eds.) VII Latin American Congress on Biomedical Engineering CLAIB 2016, Bucaramanga, Santander, Colombia, October 26th -28th, 2016. IP, vol. 60, pp. 205–208. Springer, Singapore (2017). https://doi.org/10.1007/978-981-10-4086-3_52
12. Rousseeuw, P.J.: Silhouettes: a graphical aid to the interpretation and validation of cluster analysis. J. Comput. Appl. Math. **20**, 53–65 (1987)
13. Santamaría-García, H., et al.: Uncovering social-contextual and individual mental health factors associated with violence via computational inference. Patterns **2**(2), 100176 (2021)
14. Trujillo, S., Giraldo, L.S., López, J.D., Acosta, A., Trujillo, N.: Mental health outcomes in communities exposed to armed conflict experiences. BMC Psychol. **9**(1), 1–9 (2021)
15. Unidad de Víctimas: Registro único de víctimas. https://www.unidadvictimas.gov.co/es/registro-unico-de-victimas-ruv/37394
16. Universidad de Antioquia: Caracterización de estudiantes. https://bit.ly/caracterizacionEstudiantes

Model Compression for Deep Reinforcement Learning Through Mutual Information

Jesús García-Ramírez$^{(\boxtimes)}$ ⏺, Eduardo F. Morales ⏺, and Hugo Jair Escalante ⏺

Instituto Nacional de Astrofísica Óptica y Electrónica (INAOE),
Sta. Maria Tonantzintla, 72840 Puebla, CP, Mexico
{gr_jesus,emorales,hugojair}@inaoep.mx

Abstract. One of the most important limitation of deep learning and deep reinforcement learning, is the number of parameters in their models (dozens to hundreds of millions). Different model compression techniques, such as policy distillation, have been proposed to alleviate this limitation. However, they need a high number of instances to obtain acceptable performance and the use of the source model. In this work, we propose a model compression method based on the comparison of mutual information between the distribution layers of the network. This method automatically determines how much the model should be reduced, and the number of instances required to obtain acceptable performance is considerably lower than the state-of-the-art solutions (19M). It also requires lower resources because only the last two layers of the network are fine-tuned.

Keywords: Model compression · Deep reinforcement learning · Mutual information

1 Introduction

Recently, deep Learning (DL) techniques have shown important advances in different tasks, such as natural language processing [2], object recognition [15], and reinforcement learning [12] (RL), among others. Nevertheless, these algorithms have some limitations such as the long training times (even with specialized hardware), the large number of instances required to obtain acceptable performance, and the number of parameters that form such deep models. Among these limitations, we approach in this paper the latter one. To illustrate its impact, consider some *classical* deep learning models: VGG19 [19] has 143.6M parameters, ResNet [8] has 60M, BERT [2] has 345M, DRL models such as Deep Q-Network(DQN) [12] has 4M parameters. Such a number of parameters has a direct impact into the memory required to load the model and the number of operations required to evaluate the model. Also, model complexity challenges the machine learning field that has to deal with vanishing gradients, overfitting, and related issues.

To alleviate the previous limitation, some researchers have proposed techniques to reduce the number of parameters in deep models [4]. Such techniques

A. C. Bicharra Garcia et al. (Eds.): IBERAMIA 2022, LNAI 13788, pp. 196–207, 2022.
https://doi.org/10.1007/978-3-031-22419-5_17

can be categorized into three groups: quantization, which reduces the number of bits to represent the weights in the network; pruning, which removes the irrelevant elements of a trained model to reduce the number of parameters in the network; and knowledge distillation, which trains a new model with fewer parameters using the outputs of the source model.

In the context of Deep Reinforcement Learning (DRL), Rusu et al. [16] propose policy distillation using the outputs of a pre-trained DQN (teacher) to train a new Convolutional Neural Network (CNN) with fewer parameters (student). They propose two schemes, using a single teacher to train a new model or using multiple teachers to obtain a useful model for different tasks. Some extensions use one [20] or more than one [21] teacher in a specific task to train all of them at the same time. Nevertheless, these methods require more than one model to train the new network because the teacher needs to provide the examples to train the student, and they need a large number of instances to obtain an acceptable performance (270 and 80 million, respectively, for one and two teachers).

In this work, we propose a method to compress a DQN network trained in Atari games [1]. This method is based on the mutual information between the distribution of the source model and a compressed model, and we focus on the fully connected layer, that has a large number of parameters (using the DQN architecture each unit of the fully connected layer contains 7.5K parameters). Our method uses a fine-tuning process to train the new model and it requires, in the worst case, 19 million instances to obtain an acceptable performance; also, the models for the games are compressed to 47.22% of the parameters.

The contributions of this work are the following: (i) a method that reduces a DQN network in the fully connected layer; (ii) the proposed method determines how much to reduce from the model using the mutual information between the source model and the compressed model; (iii) compared with the state-of-the-art methods, our proposed method needs a considerably lower number of instances to obtain the highest scores (19M instances).

The remainder of this paper is structured as follows: In the next section, we present some background material for the proposed method; in Sect. 2 the related work of model compression in DRL is presented; Sect. 4 introduces the proposed method for model compression; then, in Sect. 5 we present the experimental results; finally, we present conclusions and future work in Sect. 6.

2   Background

2.1   Deep Reinforcement Learning

In RL, an agent interacts with the representation of the environment (state, s) and selects an available action a. Then, the agent receives a positive or negative reward depending on the next state and which action was taken. This process is repeated until the agent reaches a final state. The function that decides which action to take (policy) is updated based on the rewards obtained by the agent to maximize the accumulated reward [22].

An RL problem can be modeled as a Markov Decision Process, which is described by a 4-tuple=(S, A, P, R): S is a set of states that are a representation of the environment; A is a set of actions available to the agent during training; P is a probabilistic state transition function that gives the probability of reaching a state s' when taking action a in state s; finally, a reward function R returns a positive or negative reward depending on the action selected by the agent.

With a large state-action space, we can approximate the policy [17], the Q value [12], or both [11] with parameterized functions. In this work we use DQN [12] that approximates the Q value function with a deep neural network. DQN used the same hyperparameters and CNN architecture to train agents for 49 different Atari games. This work uses as base the DQN algorithm, as other related work methods do in model compression for DRL.

2.2 Entropy and Mutual Information

In this section, we describe two important concepts of information theory that are used in the proposed method: entropy and mutual information. In this work, entropy is used to rank the importance of the units of a fully connected layer, under the assumption that units that produce outputs with a high entropy are more important than those that produce lower values. While mutual information is used to compare two probability distributions to determine how much to compress a model. In the next paragraphs we provide an introduction of these concepts.

The entropy represents the number of bits necessary to represent a random variable [18] and gives a measure of the uncertainty of a random variable. Considering a probability mass function $p(x)$ the entropy $H(x)$ of a discrete random variable is defined by $H(x) = -\sum_{x \in X} p(x) \log p(x)$, where the log is to the base 2 and the entropy is expressed in bits [5] in a discrete probability distribution with X dims. The value of entropy is higher when the distribution $p(x)$ (in this case the distribution corresponds to outputs of a fully connected layer in a pre-trained model) has diverse values, and when uniform outputs are obtained it will have a lower entropy. In this work, we want to remove those units that produce a uniform output.

Mutual information is a measure of the amount of information that a random variable contains about another random variable. Consider the joint probability mass of two random variables $p(x, y)$, the mutual information is defined by $I(x, y) = \sum_{x \in X} \sum_{y \in Y} p(x, y) \log \frac{p(x,y)}{p(x)p(y)}$. In this work, we use mutual information to determine the units to remove from a fully connected layer [5].

3 Related Work

Policy distillation [16] is a method based on knowledge distillation [9] and model compression [14], also, some of them use a similar process to train a model for different tasks (multi-task DRL). These methods consist of a teacher-student scheme, where a pre-trained model (teacher) is used to guide the training of a new model (student) with fewer parameters than the teacher. Nevertheless, it require high computer resources to train the student model because the teacher (one or more) and the student models are used during this process.

Policy Distillation [16] compresses a model for a certain task and a similar approach is used to obtain a model for the same task. Unlike the reference DQN architecture [12], this model is pre-trained in the source task and it is used to obtain the experience replay to train a model with fewer units in its layers. However, this method requires one to evaluate the instances in the source model to update the compressed one, which requires high computational resources. In contrast to them, we propose to compress the model and train this model as a new DQN, using fewer computational resources.

Parisotto et al. [13], propose an actor-mimic approach that consists of two parts, first they find a representation with a deep model for different source tasks (Atari games) and then they train different output modules for each source task. Also, they propose to transfer the learned representation to new games; however, those games with different dynamics obtain poor performance in the fine-tuning process. The last two methods focus on a multitask scheme, and are related to policy distillation.

Some other extensions of policy distillation have been proposed. For example, Green et al. [7] propose an extension called proximal policy optimization [17]. Sun et al. propose to train the teacher and the student at the same time; they use two variants: using only one teacher [20] and using more than one teacher [21]. These methods have similar limitations to policy distillation, as they require high computational resources to load different source models and evaluate them to extract instances for training the new target model. In the next section, we will describe the proposed method for model compression in DRL.

Also, policy distillation have been used for applications in economics [6], where a teacher-student scheme to learn an optimal policy from imperfect information, however they do not use this scheme to reduce the model size. Other approach combine a pruning strategy with policy distillation [24], they propose to prune the less important units according to explainability metrics and then mimic the behavior of the teacher model with the pruned model. Explainability techniques are combined with policy distillation to reduce the size of experience replay [10] and to obtain saliency maps more interpretable for humans [23].

4   Model Compression for Deep Reinforcement Learning

The proposed method can be seen graphically in Fig. 1. Our method is based on ranking the units of fully connected layers and then comparing the mutual information between the distribution of the output with all the units and without those with lowest ranked units. In the next subsection, we introduce the steps of our proposed method. First, we present how to rank the units based on the entropy of their outputs. Then, we present a method to reduce the number of units in the fully connected layer.

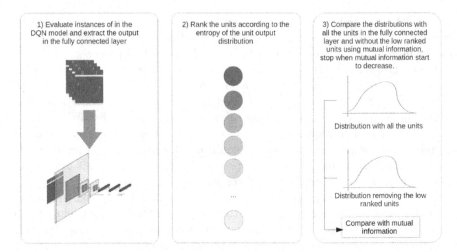

Fig. 1. Proposed method for model compression: 1) evaluate a sample in the pre-trained model; 2) extract the outputs in the hidden layer, and find the distribution of each unit, then rank the units according to the entropy; 3) finally, compare the distributions with the units and without the lower ranked ones with mutual information, the stop criterion is when the mutual information starts to decrease.

4.1   Ranking the Units

In this subsection, we introduce a method for ranking fully connected layers using a sample of the task used to train a DQN. In Algorithm 1 the procedure for ranking the units is shown, we use the entropy value of the output in the fully connected layers, we hypothesize that the units that produce diverse values are more useful that those that produce uniform values.

As input of the algorithm we need a pre-trained model in an Atari game using the DQN algorithm (DQN_{source}) and a set of instances ($Data$) of the game. Then, we obtain the output in the fully connected layer using the instances of the game. If the architecture proposed by Mnhi [12] is used, we obtain 512 outputs per instance. In the next step, we obtain the discrete probability distribution of each unit in the fully connected layer using 100 dims in order to represent in a good way the distributions. Finally, we find the entropy of each distribution that was obtained in the previous step. Finally, we rank the units according to their entropy values; the units that obtain low values can be removed from the model, and then the model can be fine-tuned to obtain similar performance with all the units.

In the next subsection, we will define how many units we will preserve in the fully connected layer using mutual information of the distributions with the entire layer and without the units that obtain the lowest entropy values. We hypothesize, and experimentally show, that the units with lowest entropy do not contribute as much to the performance of the model, as those with highest values.

Algorithm 1. Algorithm for obtaining the ranking of the units in a fully connected layer.

Input: A pre-trained model with the DQN architecture DQN_{source}, and a set of instances of the task $Data$
Output: A list with the ranking of the units $entropies$

for instance in $Data$ **do**
    Evaluate the **instance** in DQN_{source} until the fully connected layer
    Save the outputs in a list $Outs$
end for
Obtain the discrete distribution in each unit using $Outs$
Obtain the entropy of each distribution
return the rank of the entropies of the units $entropies$

4.2 Model Compression

In this subsection, we will present the proposed method to build the new model without the units that have the lowest entropy values, this process can be seen in Algorithm 2. To apply this algorithm we need a pre-trained model, a set of instances (that we produce evaluating the model for some episodes), and the ranked entropy values obtained from Algorithm 1.

Algorithm 2. Method to compress a pre-trained model of a DQN.

Input: A pre-trained DQN DQN_{source}, a set of instances $Data$, and the ranked units $entropy$ values
Output: A new compressed model DQN_{target}

Obtain the rank of units in DQN_{source} using $Data$ with Algorithm 1
Obtain the distribution of $outs$ ($entire\_distribution$)
$mutual\_information\_entire \leftarrow mutual(entire\_distribution, entire\_distribution)$
$current\_mutual\_info \leftarrow mutual\_information\_entire$
while $mutual\_information\_entire = current\_mutual\_info$ **do**
    Remove the unit with lowest entropy
    Find the new distribution now with the unit removed
    Compute $current\_mutual\_info$ with the new distribution
end while
return The compressed model DQN_{target}

The next step is to obtain the outputs of the fully connected layer in the pre-trained model and the outputs are stored in a list, then we obtain the distribution of the entire outputs. After that, we find the mutual information with the same output distribution, this value will be used to determine the stop criterion of the algorithm.

Then, the unit with lowest entropy is removed, and we find its distribution without this unit and compare it with the distribution of the entire layer through mutual information. The process continues removing the units with lowest entropy and comparing their distributions with mutual information, until the mutual information starts to decrease. Finally, we obtain a compressed model in the fully connected layer, and fine-tune the last two layers to improve the performance of the compressed model.

The main advantages of the proposed method are presented below. The method can obtain the number of units to compress the layer without adjusting more parameters. Nevertheless, there can be biases for the number of bins used to obtain the discrete probability distribution. Also, the experimental results show that it can obtain similar performance compared to the source model. The number of parameters of the target model is lower than the source model and we need few number of instances to fine-tune the model (a DQN requires 200M of instances).

5   Experimental Results

In this section, we present an experimental evaluation of our proposed method. The aim of the experiments is to show that our method is able to automatically obtain the number of units to retain in the fully connected layer of a pre-trained DQN without significantly decreasing its performance.

To evaluate the proposed method, we use the Atari Learning Environment with 32 games. The implementation of the DQN algorithm of Castro et al. [3] in the dopamine-rl library is used for our experiments. First, we use the ten games that are used in the policy distillation paper [16]. Also, we use the games that overcome human level performance according to the DQN paper. The baseline algorithm used in this work is DQN [12]. We compare the proposed method with policy distillation [16], and the ensemble policy distillation that are the methods that use model compression in their experiments.

Table 1 shows a comparison between the state-of-the-art and a comparison of percentages obtained by the three methods, those experiments that are not performed by the authors are marked by "X". We can see that in different teacher models the method proposed by Sun [21] obtain poor performance (i.e., Beam Rider or Breakout). Furthermore, they avoid using the Freeway and Space Invaders games in their comparison with policy distillation [16]. On the other hand, policy distillation obtains good results in all the games used with models with fewer parameters. Nevertheless, the methods of the state-of-the-art require more than one model to train the student one; consequently, they require computational resources higher than ours. The proposed method requires in the worst case 19M instances, and only one model is fine-tuned in the last two fully connected layers, and the experiments for all the games are fast (in the worst case 3 days in Google Colab notebooks).

Table 1. Comparison with the state-of-the-art, for each algorithm we report: the source model score (**S**), the compressed model score (**C**) and the percentage score of the compressed model respect to the source model (**%S**). The proposed method obtains higher performance in some games, but in some others obtains poor scores. In the games marked with X the experiment is not reported.

Game	Policy distillation			Ensemble PD			Ours		
	S	C	%S	S	C	%S	S	C	%S
Beam Rider	8672.4	**7552.8**	87.09	147.0	698.0	474.82	5517.6	3881.4	70.34
Breakout	303.9	**321.0**	105.62	13.0	16.0	123.09	142.9	177.2	124.00
Enduro	475.6	677.9	142.53	157.9	317.0	200.75	792.3	**915.6**	115.62
Freeway	25.8	26.7	103.42	X	X	X	32.8	**31.0**	94.51
MsPacman	763.5	782.5	102.48	959.4	1041.0	108.50	3721.0	**2886.7**	77.57
Pong	16.2	**16.8**	103.72	12.7	14.1	111.02	17.1	**16.8**	98.24
Qbert	4589.8	**5994.0**	130.59	2990.1	3295.0	110.19	12403.2	646.0	5.20
Riverraid	4065.3	**4442.7**	109.28	3608.0	4142.0	114.80	10829.0	3627.0	33.49
Seaquest	2793.3	**4567.1**	163.50	1241.3	1682.0	135.50	1225.0	2480.3	202.47
Space Invades	1449.7	1140.0	78.63	X	X	X	1841.47	**1878.6**	102.01

To provide further evidence of the performance of the proposed approach, we used twenty-two additional games. Figure 2 shows a comparison of the performance and parameters that are reduced with the proposed method. The two extreme cases as Krull, where the percentage of parameters of the target model is 7.29% with respect to the source one, and the performance is increased in 85%, and Qbert, where the parameters are only slightly reduced (85.25%, with respect to the source model) and obtains a very poor performance (5% when compared with the entire model).

According to all experiments, the proposed method obtains a competitive performance to the source model in 22/32 experiments (higher than 90% with respect to the source model); in 7/22 cases, the proposed method obtains performances between 50%–90% with respect to the source model; finally, in 3 games, the proposed method obtains poor performance (lower than 35%). Using the mean of the different metrics that are shown in Table 2: the performance mean is a little higher than the source models; the average of the number of frames that are needed in the training of the agents is 7.35 million, which is considerably lower than the state-of-the-art (80 and 120 million, respectively); finally, the mean number of parameters in the compressed models is roughly half when compared to the source models.

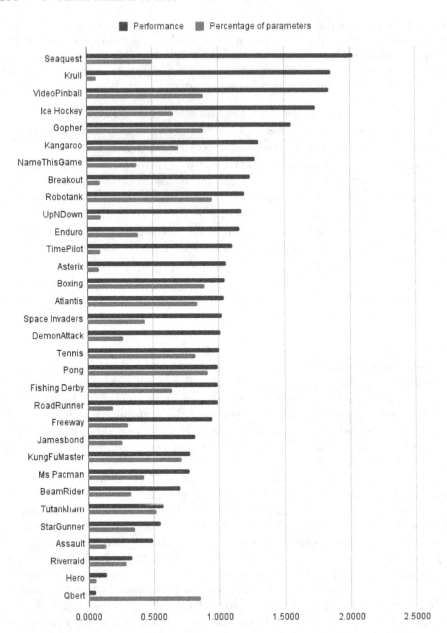

Fig. 2. Comparison of the experiments with 32 different games. In blue we can see the performance of the compressed model compared with the source model, and in red the percentage of the parameters reduction. (Color figure online)

Table 2. Results of the proposed method in 32 Atari games, we show the normalized performance respect to the source model, the necessary frames to obtain the best performance (in millions) and the percentage of parameters respect to the source model.

Game	Performance	No. Frames	Percentage of parameters
Seaquest	2.0248	0.5 M	0.5039
Krull	1.8589	6.0 M	0.0729
VideoPinball	1.8380	4.0 M	0.8889
Ice Hockey	1.7358	3.5 M	0.6590
Gopher	1.5529	5.0 M	0.8851
Kangaroo	1.3045	5.0 M	0.6973
NameThisGame	1.2761	4.5 M	0.3775
Breakout	1.2402	5.0 M	0.0997
Robotank	1.1972	6.5 M	0.9521
UpNDown	1.1758	12.5 M	0.1093
Enduro	1.1556	8.0 M	0.3909
TimePilot	1.1016	1.0 M	0.1035
Asterix	1.0563	6.0 M	0.0940
Boxing	1.0454	8.0 M	0.8946
Atlantis	1.0399	3.0 M	0.8372
Space Invaders	1.0202	7.0 M	0.4368
DemonAttack	1.0112	12.5 M	0.2740
Tennis	1.0000	4.0 M	0.8199
Pong	0.9927	5.0 M	0.9119
Fishing Derby	0.9925	1.0 M	0.6456
RoadRunner	0.9884	14.5 M	0.1936
Freeway	0.9446	7.5 M	0.3085
Jamesbond	0.8183	13.0 M	0.2625
KungFuMaster	0.7778	3.5 M	0.7146
Ms Pacman	0.7758	13.0 M	0.4292
BeamRider	0.7035	13.5 M	0.3315
Tutankham	0.5728	1.0 M	0.5211
StarGunner	0.5538	19.0 M	0.3564
Assault	0.4928	19.0 M	0.1361
Riverraid	0.3349	12.0 M	0.2913
Hero	0.1415	3.0 M	0.0576
Qbert	0.0521	8.5 M	0.8525
Mean	1.0242	7.35 M	0.4722

6   Conclusions and Future Work

In this work, we propose a method for compressing a DQN in the fully connected layer. The proposed method consists on the comparison of the distribution of the outputs in a fully connected layer with all the units and without those that have low entropy values. We observe that with the proposed method we can automatically obtain the number of units to preserve in the layer. Also, the proposed method obtains competitive results when compared with the state-of-the-art, while using less computational resources (we only fine-tune the last two layers) and uses fewer instances (19 million in the worst case) to obtain an acceptable performance. The proposed method is also evaluated in 22 more games.

As future work we will propose a method to compress the convolutional layers and with this the number of operations will be reduced, instead of only reducing the number of parameters, also we want to apply this method in other domains such as classification.

Acknowledgements. The authors thank the computer resources, technical advice, and support provided by the Laboratorio Nacional de Supercómputo del Sureste de México (LNS), a member of the CONACYT national laboratories with projects No. 201901047C and 202002030c. Jesús García-Ramírez acknowledges CONACYT for the scholarship that supports his PhD studies associated with CVU number 701191. This work was supported by CONACyT under grant CB-S-26314.

References

1. Bellemare, M.G., Naddaf, Y., Veness, J., Bowling, M.: The arcade learning environment: an evaluation platform for general agents. J. Artif. Intell. Res. **47**, 253–279 (2013)
2. Brown, T.B., et al.: Language models are few-shot learners. In: Neural Information Processing Systems, pp. 1877–1901 (2020)
3. Castro, P.S., Moitra, S., Gelada, C., Kumar, S., Bellemare, M.G.: Dopamine: A research framework for deep reinforcement learning. arXiv preprint arXiv:1812.06110 (2018)
4. Cheng, Y., Wang, D., Zhou, P., Zhang, T.: A survey of model compression and acceleration for deep neural networks. arXiv preprint arXiv:1710.09282 (2017)
5. Cover, T.M.: Elements of Information Theory. John Wiley & Sons (2006)
6. Fang, Y., et al.: Universal trading for order execution with oracle policy distillation. In: Proceedings of the AAAI Conference on Artificial Intelligence, vol. 35, pp. 107–115 (2021)
7. Green, S., Vineyard, C.M., Koç, C.K.: Distillation strategies for proximal policy optimization. arXiv preprint arXiv:1901.08128 (2019)
8. He, K., Zhang, X., Ren, S., Sun, J.: Deep residual learning for image recognition. In: Proceedings of the IEEE Conference on Computer Vision and Pattern Recognition, pp. 770–778 (2016)
9. Hinton, G., Vinyals, O., Dean, J.: Distilling the knowledge in a neural network. In: Deep Learning and Representation Learning Workshop, pp. 1–9 (2014)

10. Liu, X., Liu, S., Li, W., Yang, S., Gao, Y.: Keeping minimal experience to achieve efficient interpretable policy distillation. arXiv preprint arXiv:2203.00822 (2022)
11. Mnih, V., et al.: Asynchronous methods for deep reinforcement learning. In: International Conference on Machine Learning, pp. 1928–1937 (2016)
12. Mnih, V., et al.: Human-level control through deep reinforcement learning. Nature **518**(7540), 529 (2015)
13. Parisotto, E., Ba, J.L., Salakhutdinov, R.: Actor-mimic: Deep multitask and transfer reinforcement learning. In: International Conference on Learning Representations, pp. 1–16 (2016)
14. Polino, A., Pascanu, R., Alistarh, D.: Model compression via distillation and quantization. In: International Conference on Learning Representations, pp. 1–21 (2018)
15. Redmon, J., Divvala, S., Girshick, R., Farhadi, A.: You only look once: Unified, real-time object detection. In: Proceedings of the IEEE Conference on Computer Vision and Pattern Recognition, pp. 779–788 (2016)
16. Rusu, A.A., et al.: Policy distillation. In: International Conference on Learning Representation,. pp. 1–13 (2015)
17. Schulman, J., Wolski, F., Dhariwal, P., Radford, A., Klimov, O.: Proximal policy optimization algorithms. arXiv preprint arXiv:1707.06347 pp. 1–12 (2017)
18. Shannon, C.E.: A mathematical theory of communication. Bell Syst. Tech. J. **27**(3), 379–423 (1948)
19. Simonyan, K., Zisserman, A.: Very deep convolutional networks for large-scale image recognition. In: International Conference on Learning Representations (2014)
20. Sun, Y., Fazli, P.: Real-time policy distillation in deep reinforcement learning, pp. 1–6 (2019)
21. Sun, Y., Fazli, P.: Ensemble policy distillation in deep reinforcement learning. In: Workshop on Reinforcement Learning in Games, pp. 1–9 (2020)
22. Sutton, R.S., Barto, A.G.: Reinforcement Learning: An Introduction. MIT Press, London (2018)
23. Xing, J., Nagata, T., Zou, X., Neftci, E., Krichmar, J.L.: Policy distillation with selective input gradient regularization for efficient interpretability. arXiv preprint arXiv:2205.08685 (2022)
24. Xu, R., Luan, S., Gu, Z., Zhao, Q., Chen, G.: LRP-based policy pruning and distillation of reinforcement learning agents for embedded systems. In: 2022 IEEE 25th International Symposium On Real-Time Distributed Computing (ISORC), pp. 1–8. IEEE (2022)

Early Detection of Abandonment Signs in Interactive Novels with a Randomized Forest Classifier

Javier Navarro[1]([✉]) [iD], Iván García-Magariño[1] [iD], Jorge J. Gómez Sanz[1] [iD],
Raquel Lacuesta[2], Rubén Fuentes Fernández[1] [iD], and Juan Pavón[1] [iD]

[1] Research Group on Agent-Based, Social and Interdisciplinary Applications (GRASIA),
Complutense University of Madrid, 28008 Madrid, Spain
{jnavar05,igarciam,jjgomez,jpavon}@ucm.es, ruben@fdi.ucm.es
[2] Department of Computer Science and Engineering of Systems, University of Zaragoza,
50009 Zaragoza, Spain
lacuesta@unizar.es

Abstract. Interactive applications are becoming increasingly popular to gather feedback from users in different fields (e.g., Urbanism, Design, Economy, or Sociology). However, it is difficult to keep users engaged with an application to provide high-quality answers, as there are plenty of competitors for their attention (e.g. other applications) and their attention time is short. In this context, the interactive adaptation of applications to the actual interaction with users is a key element to improve users' engagement. It allows modifying the interface and story of the application to make it more attractive to the users while they play with it. This paper addresses this issue with early detection of potential signs of fatigue or abandonment by users in interactive visual novels. It applies Randomized Forests over a variety of events common in this type of application and analyzes which of them are best predictors of those signs. The results with a variety of novels and the selected features show promising results (a minimum accuracy of 81%).

Keywords: User engagement · Adaptation of applications · Visual novels · Abandonment predictor · Randomized Forest

1 Introduction

Interactive applications (and contents in general) have been gaining increasing presence in our lives [1, 2]. They can be used to gather people feedback on multiple domains [3, 4], as they offer multiple interaction possibilities and users are already familiar with their common interaction mechanisms. However, this use is also very challenging [5], since users spend less time on each application, switching among them frequently, so their level of attention can be low. For these reasons, there is a risk to get answers that not always have the quality and confidence that is required.

Nowadays, the success of interactive applications to attract and retain users largely depends on their ability to adapt to their users, contexts and specific interactions [6, 7].

A. C. Bicharra Garcia et al. (Eds.): IBERAMIA 2022, LNAI 13788, pp. 208–217, 2022.
https://doi.org/10.1007/978-3-031-22419-5_18

In turn, this adaptation needs techniques to analyze user engagement and retention. For instance, the identification of early signs of fatigue or potential abandonment will make it possible to design some changes on applications trying to improve users' attention and interest.

Though there are multiple works in the area of the assessment of user engagement from the interaction in applications, the problem is still open. This is largely due to the variety of types of interactive applications and the information that they can gather.

Our work focuses on visual novels as a tool to provide a context for interactive (and sometimes geolocated) questions to gather users' feedback. Visual novels [8] are a hybrid between written stories and video-games that involve the use visual assess to convey characters, backgrounds, and actions. They also frequently request users' interactions to choose paths of action or answers to specific questions.

This paper presents a model that can predict early abandonment in visual novels with interactive and geolocated questions. There is still no information on predictions in this type of applications as opposed to traditional systems. For this purpose, the paper also reports a study of the temporal variables that can be obtained from the use of a visual novel to identify those that can contribute to better predictors.

The rest of the paper is structured as follows. Section 2 reviews the related work on user retention and abandonment predictions. Section 3 describes the proposed methodology for the design of the method to predict when the user wants to abandon the application, while Sect. 4 presents the results and Sect. 5 discusses them. Finally, Sect. 6 discusses the conclusions and future work.

2   Related Work

The problem of user engagement in applications and interactive content has been well studied. Acquiring new users is a difficult process, so applications must retain those users who already use them as a first step to increasing their number of users. In recent times, this problem has gained increased attention due to its relevance for the monetization of, for instance, websites [9] and mobile apps [10].

The study of user engagement from application data has been characterized as a type of data mining problem that includes five steps [11]: data-preprocessing, feature selection, sampling of data, training the classifier, testing for prediction, and output of prediction. There are multiple approaches to this perspective, like [12] that studies the acquisition and retention of clients in the area of marketing with an algorithm for prediction based on clustering, [11], whose goal was to know when a customer would leave the services using several techniques (e.g. Boruta algorithm, Support Vector Machines, and Random Forests), or [13], which analyzed the performance of eight classifiers for this purpose. In a previous paper working on designing an emotion-based prediction system [14], a comparison was made with other predictor techniques, such as support vector machines (SVC), k-nearest neighbors (KNN) and decision trees. It was concluded that decision trees allowed better predictions.

Given the results in the literature, our work uses Randomized Forests (RFs). Though the literature [15] shows that Decision Trees (DTs) facilitate explaining to users the result of classification, they tend to overfit. This means that they usually learn the training data very well, but their generalization to other data is not straightforward. One way to improve the generalization of decision trees is to use regularization, which involves limiting the model's capabilities in some way to obtain a better generalizing machine learning model, and beyond this, combining several trees. In general, it improves classification but harms explainability. This combination is used, for instance, in the proposed RFs.

The nonparametric RFs [16] use supervised learning to extend DTs. The RF algorithm arises as to the grouping of several classification trees. Basically, it randomly selects a number of variables with which each of the individual trees is built, and predictions are made with these variables that will later be weighted through the calculation of the most voted class of the trees that were generated, to finally make the prediction.

The problem of classification is not only about the algorithms, but also about the chosen data. A poor choice causes poor prediction results. In the case of abandonment, some articles (e.g. [17]) show a relationship between it and the time users spend playing.

Our study uses variables related to the time users expend in the application because they are frequently collected in visual novels. In our application, some questions in the novel also have geolocation data.

3 Methodology

Our research has designed five visual novels using a framework based on Monogatari.io [18] (https://monogatari.io). Their application domains are different: three of them are based on awareness and sustainability; the other two are intended for data collection regarding citizen opinions about potential changes in the Urbanism of a city. These visual novels have immersive questions and can also have geolocation. The application asks these to users while visiting the place in the application (i.e. immersive) or also the real world (i.e. geolocated).

The tests of the novels collected various data, like the section/chapter of the novel and the time spent in each section. For the novels with geolocation, the tests also collected the time spent going from one place to another. Finally, the experiments recorded the time at which the user finishes the novel, which serves as an arrival point in our ranking system.

The implementation of the prediction algorithm and the data analysis is made with Python. It uses the libraries: Scikit, which provides an implementation of the RF classifier algorithm; Numpy to works with vectors, matrices, and arrays, and implements high-level mathematical functions; Pandas which provides flexible data structures and allows working with them very efficiently.

With these elements, the work with the visual novel is as follows. After analyzing the data and executing the prediction algorithm, possibilities of users' abandonment can be detected. Then, the application can change the story to try to avoid that abandonment. This adaptation process has to meet two requirements: (1) That abandonment indicators can be detected quickly in the users' interactions, so they can be made when changing the user's mind is still possible; (2) That the accuracy of the prediction algorithm is good to detect the possibility of abandonment, so there are not too many false positives that could lead to changes when the users are actually engaged with the application.

4   Results

Our datasets contain measurements from tests performed with several visual novels. For analysis, we split them into two datasets, to test how the fact that some visual novels have geolocation affects them. The samples include data on the elapsed time between different sections of the novel. For these datasets, we differentiate between dataset1, which includes whether the visual novel has geolocation or not (e.g., 1: True, 0: False), and dataset2, which only has data from geolocated novels, the elapsed time it takes to move between section locations.

4.1   Sample

There are 60 data samples obtained in total, each from different users and distributed in 5 visual novels. The novels have been made in the cities of Madrid and Teruel, both in Spain. 21 out of those 60 samples correspond to novels with geolocation.

4.2   Data Description

The independent variables to be considered for the analysis of the RF model are the following:

- **geo:** Parameter that indicates if the novel has geolocation or not (true or false).
- **time1, time2, time3, time4, time5:** The different times spend in each section of the novel (in seconds).
- **time1geo, time2geo:** The different times spend between geolocations in the novel (in seconds), They indicate the time users use to travel a section between two locations.
- **last_time:** Last entry of time taken by the user to complete the last section of the novel (in seconds).
- **last_time_geo:** Last entry of the time it takes to traverse the location of the last section of the novel (in seconds), It corresponds to the time users need to traverse the section between the penultimate and the last location.
- **total_time:** Completion time (in seconds).
- **finished:** This variable is only saved at the end of the novel if the user has completed the whole experience. (true or false). It allows identifying users' abandonment of the novel.

If abandonment occurs, the times corresponding to the sections that the user do not play will be coded as zero. Zero time is also used when novels have a different number of sections. In this way, results are unified to allow comparison, so they do not affect the prediction.

The different datasets corresponding to the different visual novels that we are going to use in our random forest are the following:

- **dataset1:** geo, time1, time2, time3, time4, time5, last_time, total_time and finished.
- **dataset2:** time1, time1geo, time2, time2geo, last_time, last_time_geo, total_time and finished.

4.3 Model Construction

The construction of the RF model considers the following setup: (1) the number of estimators (DTs) is 100, (2) for the construction of each estimator, the minimum number of observations to split a node is four random variables, and (3) different sample sizes are used for training, so the complete training dataset is never used.

To determine the most important variables for the classification of finished novels, the Gini variable importance method [19] was used, v_importance variable is measured in Gini Importance or Mean Decrease in Impurity (MDI) calculates each feature importance as the sum over the number of splits (across all tress) that include the feature, proportionally to the number of samples it splits. Table 1 and Table 2 show the results.

Table 1. Relevant variables for dataset1.

	feature	v_importance (MDI)
7	last_time	0.286018
1	total_time	0.203590
2	time1	0.152313
3	time2	0.135189
6	time5	0.085493
5	time4	0.071685
4	time3	0.056398
0	geo	0.009315

Table 2. Relevant variables for dataset2.

	feature	v_importance (MDI)
6	last_time_geo	0.200849
4	time1geo	0.199045
1	time1	0.168858
3	last_time	0.147107
5	time2geo	0.142674
2	time2	0.079342
0	total_time	0.062124

The first four variables in importance are the following. For dataset1, the time it takes to perform the last section of the novel, with the highest variable importance value, of the two datasets (0.29), followed by the total time (0.20), the time of the first sector (0.15) and the time of the second sector (0.14). For dataset2, the time it takes to go through the last section of the novel, with the highest variable importance value (0.20), followed by the time it takes to go through the first sector (0.199), the time of the first sector (0.17) and the time of the last sector (0.15).

4.4   Model Validation

OOB Error
The OOB error [20] rate estimate is calculated from the out-of-bag observations. The error estimate suggests that when the model is applied to new observations, the responses will have an OOB error of 18.8% for dataset1 and 11.8% for dataset2. It can also be said that the model is 81.2% accurate for dataset1 and 88.2% accurate for dataset2.

Confusion Matrix
To proceed with the validation of the model, the analysis considers the "Test" dataset to calculate the confusion matrix. It shows small values in the diagonal of the misclassified, that is, there is an error of 1.2% (Fig. 1) for dataset1 and 0.42% (Fig. 2) for dataset2. Thus, the proposed RF model classifies with a very low number of errors.

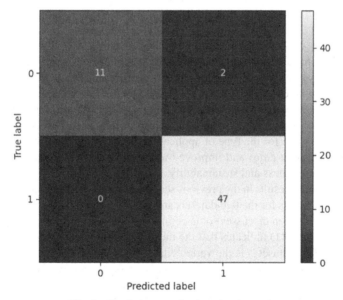

Fig. 1. Confusion matrix chart for dataset1.

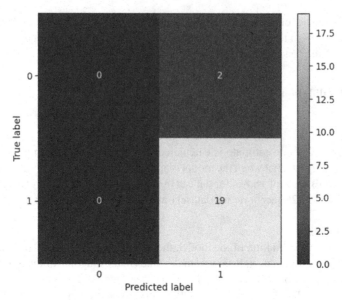

Fig. 2. Confusion matrix chart for dataset2.

5 Discussion

Early abandonment detection is novel in immersive applications based on visual novels. In the literature, we see how it has been studied in other areas such as gaming or gamification [17] or with enterprise clients [11] (more focused on not losing monetization [9, 10]). Our case focuses on obtaining data more accurately by preventing users from abandoning the app and completing all the fields until the end. This is something we consider of great value for the type of applications we are going to use, to collect data for the design of smart cities and improve the attention of users when they are being explained about awareness and sustainability.

Going back to the results in the previous section, some additional conclusions can be drawn. Since the results for the two datasets are slightly different, this section considers them separately and then discusses some common observations.

Table 1 (for dataset1) indicates that the most important variable is last_time; in the case of Table 2 (for dataset2) is the variable last_time_geo. If these times are low, the users will not finish the application. We can also observe how in dataset2, the most important variable in dataset1 (last_time) is in the fourth position below time1. This change may be because in dataset2, there are fewer dropouts and therefore, the variable corresponding to the last time is less important since most of them arrive in that section.

The previous variables are close to the end of the novel, so they have a limited impact to enable changes in the novel. Looking for other relevant variables, the first two sections of time are also important for the two datasets. So, our system is able to detect cases of abandonment in an early way, as it can also use the first sections (time1 and time2) for prediction.

Regarding the data with geolocation, the values time1 and time1geo, (Table 2), are of similar relevance. Being two variables related to the same section, the first one allows to detect the potential abandonment with less requirements (i.e. geolocation).

Using the previous values, the application allows anticipating if users are going to abandon, meaning that they will complete the novel but not do the last survey of the application. This survey is the key elements to ask users about their opinion of the novel. If the application could anticipate the abandonment, it could change some parts of the story to keep users' attention as soon as the novel starts, so there would be an increase in the number of users completing the survey. This will allow design applications to consider these values in order to increase engagement and reduce the number of abandonments.

The results of dataset2 (Table 2) are slightly worse than those of dataset1 (Table 1). Probably, once there are more geolocated records, these differences will be reduced.

The RF appears as a simple and quite fast and precise model. Although it lacks the interpretability of DTs, it reduces overfitting and is a more robust classifier. The RF model gave good results in terms of classification (82.2% and 88.2% of accuracy) and, above all, in determining the importance of the variables, all of it with small datasets. This is very useful for subsequent variable debugging and more precise analysis.

Our study makes it reasonable to use time as an important variable for the detection of abandonment. We have identified the most important variables when using visual novels seeing that with the first-time intervals we can detect them with good accuracy in the results. Just as in [16] they see a relationship with game time, we see it between the time it takes users to perform different activities within the application with abandonment. However, the study has some limitations such as the small size of the data sample.

6   Conclusions

This paper shows an algorithm based on RFs that analyzes time variables to predict application abandonment. This is applied to an application that tries to collect data for the design of smart cities by using visual novels. The statistical analysis of the data collected in the tests with the applications shows a significant relationship between the variables that are used for the prediction (of time and geolocation) and abandonment. These variables are relevant for this kind of application (i.e., playing scenarios in a smart city), and time should always have an impact, but others should be considered for other cases.

Among the most important variables, those in the first sections have a high degree of importance, so early detection is possible. This would give enough time to make changes in the application and keep the attention of the user on its content.

With accuracy in the two datasets above 80%, the detection of future abandonments offers enough confidence for adaptation. Changes in the story have a high probability of improving the user experience. The changes will be classified in two ways, if we want to do it for specific histories (i.e. If the user is walking around the city, we will add some kind of gamification to make the user interact in an immersive way with the monuments) or make it in general for all the visual novels (i.e. Change the time the user will spend in the visual novel).

This research has still several open issues for future work, such as improving the prediction with more data as the available applications are used or if other applications are added, they would all be along the lines of visual novels. Also, we want to design a system that, depending on the abandonment predictions, makes a series of changes that modify the application to get the user's attention. Finally, to improve the prediction system we can consider including more variables obtained from the application. We can even use the variables that the user enters previously in the application to make a classification and separate them into several user profiles, depending on these profiles the application initially adapts the application to further reduce the option of abandonment of the application by the user. To do this, a user-adaptive application would be created using these user profiles.

Acknowledgements. This work has been done in the context of the projects "Reshaping Attention and Inclusion Strategies for Distinctively vulnerable people among the forcibly displaced (RAISD)" (grant 822688) supported by the European Commission in the Horizon 2020 programme and "Collaborative Design for the Promotion of the Well-Being in Inclussive Smart Cities (DColbici3)" (grant TIN2017-88327-R) supported by the Spanish Ministry for Economy, Industry, and Competitiveness.

References

1. Deng, T., Kanthawala, S., Meng, J., et al.: Measuring smartphone usage and task switching with log tracking and self-reports. Mob. Media Commun. **7**, 3–23 (2019)
2. Eirinaki, M., Vazirgiannis, M.: Web mining for web personalization. ACM Trans. Internet Technol. (TOIT) **3**, 1–27 (2003)
3. Goncalves, J., Hosio, S., Liu, Y., Kostakos, V.: Eliciting situated feedback: a comparison of paper, web forms and public displays. Displays **35**, 27–37 (2014)
4. Wilson, A., Tewdwr-Jones, M., Comber, R.: Urban planning, public participation and digital technology: App development as a method of generating citizen involvement in local planning processes. Environ. Plann. B: Urban Anal. City Sci. **46**, 286–302 (2019)
5. Rieser, L., Furneaux, B.: Share of attention: exploring the allocation of user attention to consumer applications. Comput. Hum. Behav. **126**, 107006 (2022)
6. Giua, E.M., Malavolta, I., Lago, P.: Self-adaptation in mobile apps: a systematic literature study. In: 2019 IEEE/ACM 14th International Symposium on Software Engineering for Adaptive and Self-Managing Systems (SEAMS), pp 51–62. IEEE (2019)
7. Garber-Barron, M., Si, M.: Adaptive storytelling through user understanding. In: Ninth Artificial Intelligence and Interactive Digital Entertainment Conference (2013)
8. Camingue, J., Carstensdottir, E., Melcer, E.F.: What is a visual novel? Proc. ACM Human-Comput. Interact. **5**, 1–18 (2021)
9. Goanta, C., Yohanis, A., Jaiman, V., Urovi, V.: Web monetisation. Internet Policy Rev. **11**, 1–8 (2022)
10. Appel, G., Libai, B., Muller, E., Shachar, R.: On the monetization of mobile apps. Int. J. Res. Mark. **37**, 93–107 (2020)
11. Ewieda, M., Shaaban, E.M., Roushdy, M.: Customer retention: detecting churners in telecoms industry using data mining techniques. Int. J. Adv. Comput. Sci. App. **12**, 1–10 (2021)
12. Kumar, M.R., Venkatesh, J., Rahman, A.M.J.: Data mining and machine learning in retail business: developing efficiencies for better customer retention. J. Amb. Intell. Human. Comput. **2021**, 1–13 (2021)

13. Abdulrahman, S.A., Khalifa, W., Roushdy, M., Salem, A.-B.M.: Comparative study for 8 computational intelligence algorithms for human identification. Comput. Sci. Rev. **36**, 100237 (2020)
14. NavarroAlamán, J., Lacuesta, R., GarcíaMagariño, I., Lloret, J.: EmotIoT: an IoT system to improve users' wellbeing. Appl. Sci. **12**, 5804 (2022)
15. Al-Hoqani, W.M.: Difficulties of marking decision tree diagrams. In: 2017 Computing Conference, pp. 1190–1194. IEEE (2017)
16. Athey, S., Tibshirani, J., Wager, S.: Generalized random forests. Ann. Stat. **47**(1148–1178), 31 (2019)
17. Loria, E., Marconi, A.: Exploiting limited players' behavioral data to predict churn in gamification. Electron. Commer. Res. Appl. **47**, 101057 (2021). https://doi.org/10.1016/j.elerap.2021.101057
18. GonzálezBriones, A., GarcíaMagariño, I., Gómez-Sanz, J.J., FuentesFernández, R., Pavón, J.: A collaborative platform for the detection of non-inclusive situations in smart cities. In: Alba, E., et al. (eds.) Advances in Artificial Intelligence. LNCS (LNAI), vol. 12882, pp. 206–215. Springer, Cham (2021). https://doi.org/10.1007/978-3-030-85713-4_20
19. Menze, B.H., Kelm, B.M., Masuch, R., et al.: A comparison of random forest and its Gini importance with standard chemometric methods for the feature selection and classification of spectral data. BMC Bioinform. **10**, 1–16 (2009)
20. Matthew, W.: Bias of the Random Forest out-of-bag (OOB) error for certain input parameters. Open J. Statist. **2011**, 7 (2011)

Insights from Deep Learning in Feature Extraction for Non-supervised Multi-species Identification in Soundscapes

Maria J. Guerrero[1]([⊠]) [iD], Jonathan Restrepo[1] [iD], Daniel A. Nieto-Mora[2] [iD],
Juan M. Daza[3] [iD], and Claudia Isaza[1] [iD]

[1] SISTEMIC, Facultad de Ingeniería, Universidad de Antioquia, Calle 70 No. 52-21, Medellín, Colombia
{mariaj.guerrero,jonathan.restrepor,victoria.isaza}@udea.edu.co

[2] MIRP, Máquinas Inteligentes y Reconocimiento de Patrones, Instituto Tecnológico Metropolitano, Calle 54a No 30-99, Medellín, Colombia
danielnieto152326@correo.itm.edu.co

[3] Instituto de Biología, Grupo Herpetológico de Antioquia, Universidad de Antioquia, Calle 70 No. 52-21, Medellín, Colombia
juanm.daza@udea.edu.co

Abstract. Biodiversity monitoring has taken a relevant role in conservation management plans, where several methodologies have been proposed to assess biological information of landscapes. Recently, soundscape studies have allowed biodiversity monitoring by compiling all the acoustic activity present in landscapes in audio recordings. Automatic species detection methods have shown to be a practical tool for biodiversity monitoring, providing insight into the acoustic behavior of species. Generally, the proposed methodologies for species identification have four main stages: signal pre-processing, segmentation, feature extraction, and classification. Most proposals use supervised methods for species identification and only perform for a single taxon. In species identification applications, performance depends on extracting representative species features. We present a feature extraction analysis for multi-species identification in soundscapes using unsupervised learning methods. Linear frequency cepstral coefficients (LFCC), variational autoencoders (VAE), and the KiwiNet architecture, which is a convolutional neural network (CNN) based on VGG19, were evaluated as feature extractors. LFCC is a frequency-based method, while VAE and KiwiNet belong to the deep learning area. In ecoacoustic applications, frequency-based methods are the most widely used. Finally, features were tested by a clustering algorithm that allows species recognition from different taxa. The unsupervised approaches performed multi-species identification between 78%–95%.

Keywords: Feature extraction · Deep learning · Multi-species identification · Biodiversity monitoring · Soundscape

A. C. Bicharra Garcia et al. (Eds.): IBERAMIA 2022, LNAI 13788, pp. 218–230, 2022.
https://doi.org/10.1007/978-3-031-22419-5_19

1 Introduction

Passive acoustic monitoring (PAM) is presented as an alternative for biodiversity and conservation monitoring which allows the identification of different ecosystem processes and their changes by collecting the activity present in the soundscape through acoustic recorder units (ARU) [1–3]. Soundscape studies have provided information on biodiversity conditions, population trends, and have permitted the identification of changes in ecosystem richness and composition due to natural and anthropogenic impacts [4].

Soundscape analysis can be performed through species acoustic pattern recognition. These patterns come from vocalizations that species perform to communicate with each other for reproductive, warning, and territorial purposes. PAM is a cost-efficient alternative that gives away to obtaining species data throughout the day and night, analyzing long periods, and thus collecting more information about the object of study than direct observation [3]. Moreover, compared to photo trapping, PAM offers an advantage in species monitoring because the analysis of camera trap images is more complex for cryptic and small animal species such as frogs and insects.

The species detection in soundscapes can be realized by manual human analysis using computer tools as Raven (The Cornell Lab of Ornithology) and Avisoft (Avisoft Bioacoustics). These tools allow audio analysis by spectrograms but require high user intervention. This can be an issue when large amounts of data recordings need to be analyzed. Therefore, it is necessary to design automatic recognition methods that support large-scale monitoring.

The most common methods for automatic species detection through their vocalization are based on machine learning techniques, as in the case of birds [5, 6], frogs [7, 8], and mammal species such as bats [9]. Most of these works use supervised learning techniques, requiring labeling data for the training stage, and there is no information about species in all ecosystems [10]. In a country such as Colombia which has a significant amount of fauna that is unknown to science, some species would not be correctly detected. Furthermore, most of the proposals are focused on the identification of species of a specific taxonomic group (e.g., avians, anurans, cetaceans, chiropterans) or specific target species. In general, the steps to be followed to perform multi-species recognition from audio are: (I) signal pre-processing, (II) audio segmentation, (III) feature extraction, and (IV) classification [11].

Some studies have implemented feature extraction algorithms in combination with machine learning techniques for species identification. However, most proposals are focused on determining the presence of a small number of species from a single taxonomic group. Among the most commonly used features are Mel Cepstral Coefficients (MFCC) [7], Linear Cepstral Coefficients [12], spectral centroid, mean energy, and signal time-frequency features as call length, peak frequency, and minimum and maximum frequencies [13]. Recently, deep learning methods have been used to extract discriminative features in different audio applications. These features are the inputs of a clustering algorithm that differentiates between bird species [14, 15], frog species [16], and individuals of a bird species [17]. However, the application of these techniques in bioacoustics, specifically in multi-species detection using unsupervised learning, is understudied.

This work aims to explore the feature extraction for automatic multi-species identification, which involves the analysis of all frequency bands present in the soundscape

(100 Hz–90 kHz), and the use of unsupervised learning techniques to perform species detection in different ecosystems without the need for labeled data. For feature extraction, we use the features commonly extracted in bioacoustics: cepstral coefficients and frequency features such as peak frequency and frequency range. From the deep learning area, Variational Autoencoders [14], which is an unsupervised and unapplied method in bioacoustics, was analyzed. Furthermore, we analyzed a supervised approach using a pre-trained Convolutional Neural Network architecture based on VGG19 [17] as feature extractor. These features are the input of a clustering algorithm that groups the species according to their similarities.

This paper is presented as follows. Section 2 explains the used dataset and feature extraction methodology for multi-species recognition; the clustering algorithm used will also be briefly described. Section 3 presents the proposed experiment results and discussion. Finally, in Sect. 4 conclusions and future work are presented.

2 Materials and Methods

2.1 Materials

The dataset used for this case study consists of 832 audio recordings obtained through acoustic monitoring collected in March and June 2021 from a rural area in Puerto Wilches, Santander, Colombia (7°21′52.5″N, 73°51′33.0″W). For this dataset, two recordings were used: a Song Meter Mini (Wildlife Acoustics, Inc) device programmed to collect 1-min recordings every 10 min was used with a sampling rate of 48 kHz for audible species (100 Hz–22 kHz), and a Song Meter Mini bat (Wildlife Acoustics, Inc) device programmed to collect 15 s every 15 min with a sampling rate of 384 kHz for species in high frequencies (22 kHz–90 kHz). More than 10000 segments are estimated in the audios, possibly corresponding to animal calls. Experts identified the presence/absence of species in this dataset. This information was only used to measure the performance of the clustering algorithm. It has 30 identified species in multiple frequency bands, including chiropterans (ultrasound), orthopterans (ultrasound), avians, anurans, and a primate.

As CNN architectures are a supervised method, it was required to label each species call (segments in the audio). Therefore, 315 audio recordings were randomly selected from the dataset and approximately 4000 calls were labeled in this subset. These calls were used to train the KiwiNet network.

2.2 Multi-species Identification Methodology

Generally, automatic species identification follows the methodology presented in Fig. 1. First, audio recordings pass through a pre-processing stage in which time-frequency representation is obtained using the short-time Fourier transform (STFT), thus generating the spectrogram [18]. This representation provides a detailed description of the energy present at each frequency and time and can help to differentiate between species call types. At this stage, the background noise is also removed from the audio recordings using a Gaussian filter and spectral subtraction [19]. After obtaining the time-frequency representation of the audio and removing the background noise, signal segmentation is performed. In this stage, the acoustic units or acoustic events that could be associated with the species calls are extracted. In our work, this stage is carried out by image analysis using the Acoustic Event Detection (AED) methodology proposed by Xie et al. [19] but analyzing all frequencies. The next stage consists of extracting the most relevant acoustic information for each segment found in the previous stage. Here, the ability of deep learning techniques to extract features was analyzed in contrast to the most widely used approaches in the literature, which are the cepstral coefficients and frequency features (peak frequency, minimum and maximum frequencies). This stage will be discussed in more detail in the following section. In the final stage, a clustering algorithm analyzes the previously extracted features and aggregates the segments based on the similarity of species vocalizations. Due to the type of problem where the number of species present at the site are not previously known, we selected a clustering algorithm LAMDA (Learning Algorithm for Multivariate Analysis) [7]. This algorithm does not require the number of classes as an input parameter and allows the identification of species that were not present at the training stage.

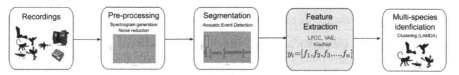

Fig. 1. Automatic multi-species identification methodology

2.3   Feature Extraction for Multi-species Identification

For automatic species identification, some methods have been applied for feature extraction. These features can be based on signal energy, zero-crossing, and time-frequency features such as peak frequency, bandwidth, call duration, and frequency range [13]. Other works have used Mel frequency cepstral coefficients (MFCC) [7, 13, 15] linear frequency cepstral coefficients (LFCC) [12] and recently, feature extraction has been performed using deep learning techniques such as Convolutional Neural Networks (CNN) [17, 18] and Variational Autoencoders (VAE) [14, 15]. Then we considered interesting to analyze the behavior of multi-species identification method using deep learning approaches.

Feature Extraction Using Linear Cepstral Coefficients
Mel-Frequency Cepstral Coefficients [20] are widely used discriminatory features in human speech and speaker recognition. They redistribute the frequency across the spectrum in a logarithmic way, Mel scale, in order to benefit specific frequency bands in which the human vocal apparatus works [20]. In some cases, the spectral range of the studied species is in the human speech spectrum (30 Hz–3 kHz) and species recognition can be made successfully; Nevertheless, generating a Mel-like scale for identify vocalizations of different taxa would not be an optimal solution. For this reason, Cepstral Coefficients

uniformly and linearly distributed across the spectrum have been used when it is not desired to benefit specific frequency bands [12, 21]. The process for calculating LFCC is described as follows:

Based on the segments previously obtained in the segmentation stage, the first step of process consists in estimating the logarithm of the energy $\mathbf{Q} \in \mathbb{R}^{N_s \times N_l}$ for each vocalization segment $\mathbf{H} \in \mathbb{R}^{N_s \times N_t}$. Where N_s and N_t are the length of the segmented call in the spectral and temporal domain respectively and N_l is the number of logarithms calculated in each window. This operation extracts the relevant acoustic information from the temporal domain and redistribute it across the spectral domain in a non-linear way.

Then, the unitary Discrete Cosine Transform (DCT) of $\mathbf{Q}$ is calculated in order to reduce the dimensionality of $\mathbf{Q}$ and set a common length for the extracted feature vector in all vocalizations. Finally, the vector $\mathbf{y}_a \in \mathbb{R}^{N_k}$ is obtained, where N_k is the number of coefficients. In our case, 24 coefficients are obtained.

These coefficients are combined with call frequency information: peak frequency, minimum and maximum frequencies which are a relevant biological information seeking to improve the detection performance.

This procedure allows the species identification across the entire spectrum without the inherent need to change its parameters according to the specific animal species.

Feature Extraction Using Variational Autoencoders (VAE)

Variational autoencoders (VAE) are an unsupervised method for feature extraction. As it is a relatively new approach [22], there are few analyses in bioacoustics. Rowe et al. [16] use them for biodiversity analysis focused on bird detection and was also used by Ntalampiras and Potamitis at [15] for detection of unknown bird species.

VAE is an improvement of the traditional autoencoders where the latent space generated by the encoding stage is used to learn a normal distribution of the features and then reconstruct the original inputs by sampling features. This method provides a principled framework for learning deep latent-variable models and corresponding inference models.

In this case, we based our architecture on the work of [23]. Our neural network is composed of three convolutional layers and their corresponding relu for encoding. Original inputs are RGB spectrogram with size 224×224, then for each layer, the input size is halved, and the number of channels increases until 64. The latent space was generated by computing the mean and standard deviation on a fully-connect layer at the end of the encoder. Finally, the reconstruction was performed by five deconvolutional layers and relu layers among them. This process can be observed in the Fig. 2.

From latent space, a vector $\mathbf{y}_b \in \mathbb{R}^{N_L}$ is obtained, where N_L is the number of features, in this case, 64. These features were evaluated initially alone and then the frequency features were added.

Feature Extraction Using a Convolutional Neural Network-KiwiNet

Recent works perform feature extraction through deep learning methods. Most of them use Convolutional Neural Network (CNN) for feature extraction and classification tasks [18]. This method presents the advantage of having pre-trained architectures such as VGG19 [24] which vary each other in the number of layers, loss function, and order

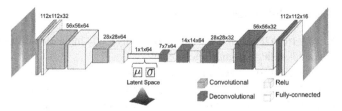

Fig. 2. Variational Autoencoder architecture used for feature extraction for multi-species identification.

of different components in their layers. However, as a supervised method, it requires a labeled dataset.

For feature extraction purposes, we used a CNN based on the architecture proposed by Bedoya and Mell [17] called KiwiNet. This network was created to identify individuals of a bird species (Great spotted kiwi/roroa) in New Zealand using as a core the VGG19 architecture [24] with two extra convolutional layers to reduce the number of filters from 512 to 32 and a global average pooling layer to generate a 1-dimensional latent space. We chose KiwiNet because its high individual identification accuracy using the species call segment and was trained for feature extraction purpose.

CNNs are characterized by requiring fine-tuning of parameters depending on the application. In our case, the KiwiNet input is 224×224 RGB spectrogram image segments containing the species vocalizations; the previous labeled calls were used for training. The first seven layers were frozen, and as presented in [17], we introduced two layers before the fully connected layer. The training was performed using stochastic gradient descent with momentum in mini-batches of eight segments, a learning rate of 1×10^{-4}, and a stopping criterion with the number of epochs. We used transfer learning by applying the KiwiNet architecture with VGG19 values. VGG18 was pre-trained with the ImageNet dataset; this allowed us to accelerate the training workflow and do the training with our small dataset. From the patterns generated by convolutional layers, a vector $\mathbf{y}_c \in \mathbb{R}^{N_f}$ is obtained, where N_f is the number of features, in this case, 32. These features were initially evaluated alone and then the frequency features were added.

2.4 KiwiNet as a Supervised Classifier

CNNs are commonly used as a classification method. We wanted to test the performance of our trained KiwiNet network to classify 30 species from 4 different taxonomic groups (aves, amphibians, mammals, and insecta).

The results section presents the performance comparison of the clustering algorithm using the extracted features and the KiwiNet network as a supervised classifier.

2.5 Clustering for Multi-species Identification

Acoustic features **y** from the previous stage were analyzed using a clustering algorithm to identify the sounds related to multiple species.

LAMDA [25] is an unsupervised fuzzy clustering methodology that has previously been used for the detection of anuran species [7] and birds individuals [17]. It does not require the number of classes as an input parameter and estimates the global adequacy degree of each datum by estimating the contribution of the input features using a non-iterative process. We use the Yager-Rivalov triple Π operator [26] as fuzzy connective to strengthen the relationship between the number of generated clusters and the number of classes associated with biophonie sounds. The detailed description of the algorithm can be found in [7, 25].

The procedure for using LAMDA for multi-species identification is divided into two steps: species learning (clustering) and species identification.

(i) Species learning: the user enters the audio recordings to be analyzed. LAMDA in unsupervised way returns the proposed clusters. The expert decides which clusters are of interest, associates them to a species, and saves the cluster (or clusters) containing the representative vocalization of the species.

(ii) Species identification: Once the species are trained, new audios can be analyzed. LAMDA recognizes the trained clusters (species clusters saved in the training stage) and if a new biophonie it is identified, it assigns them to a Non-Informative Class (NIC), showing that this methodology is able to find unexpected species that were not include in the training stage.

2.6 Evaluation Metrics

Methods performance was evaluated according to the species presence-absence detection in a recording using the recognition rate Eq. (1)

$$\text{Recognition rate } (\%) = \frac{N_d}{N_p} \tag{1}$$

where N_d is the number of audio recordings where the species was correctly identified, and N_p is the total number of audio files where the species is present according to the ground truth (expert labels).

This metric was used to evaluate the performance of the different feature extraction approaches using an unsupervised algorithm (LAMDA-clustering). Moreover, since CNN methods are widely used as a classifier, although it is a supervised method, we diced to compare the LAMDA results with the KiwiNet as a classifier.

3   Experimental Results and Discussion

In this study, we compare the multi-species identification performance using three different methods for feature extraction (LFCC, VAE, and KiwiNet) and an unsupervised clustering methodology. Here, each feature extraction method was analyzed individually and then with the conjunction of the frequency features (peak frequency, minimum and maximum frequencies). Table 1 shows the average recognition rate of 30 species in the 832 audio recordings using the feature extraction methods. Additionally, we present the results of the KiwiNet architecture used as a supervised multi-species classifier.

Our results show that multi-species recognition rate increases significantly when frequency information is used in conjunction with the three different tested methods showing that biological information is relevant to differentiating between taxa.

Feature extraction from LFCC in combination with frequency information shows that it is possible to detect species automatically in different frequency bands with high performance. This combination allows a clustering algorithm to generate an accurate cluster for species. However, as a biological problem, some species present intra-species variability due to background noise or environmental conditions, and this affects some species detection.

Table 1. Comparing the average performance for multi-species identification using different features representation. Six feature representation were evaluated with an unsupervised clustering methodology and a CNN was evaluated for feature extraction and classifier.

Features representation	Classifier	Hit rate
Frequency information	LAMDA	84%
LFCC and frequency information	LAMDA	**95%**
VAE	LAMDA	78%
VAE and frequency information	LAMDA	92%
KiwiNet-CNN	LAMDA	89%
KiwiNet-CNN and frequency information	LAMDA	90%
KiwiNet-CNN	KiwiNet-CNN	88%

VAE is able to encode and decode from the spectrogram representation of the species call, achieving high reconstruction probability as presented in Fig. 3. As an unsupervised method, it has the advantage of not requiring species labels or large datasets for accurate reconstructions. However, in this case, when the generated latent space is used as an input for clustering algorithm, some features are not relevant to group some species correctly, and the global species detection average decreases.

When these features are combined with species frequency information providing by frequency features, multi-species detection improves.

In the same way, the KiwiNet performance as a feature extraction was relatively similar to the other methods used. KiwiNet architecture was designed originally for feature extraction for individuals of the Great spotted kiwi (New Zealand bird species) [19] but with the tuning of some architecture parameters, it has proven to be useful for the multi-species approach.

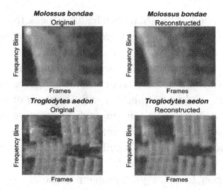

Fig. 3. Examples of original and reconstructed calls using VAE of two species in ultrasonic and audible frequencies. *Molossus bondae* (bat) and *Troglogytes aedon* (bird).

The KiwiNet, as any CNN architecture, requires an extensive and properly labeled dataset for training. Due to the diversity of species calls, the dataset is not balanced. It was necessary to manually find the most representative song segments for each species and use them for training. We used 40 vocalizations from each of the 30 species. This made it difficult to train the network.

Despite CNN being a supervised method, we decided to evaluate the performance of KiwiNet as a classifier due to CNN recently being one of the most used classifiers for species identification. This gives us a baseline to compare our unsupervised purpose. Table 1 shows the performance obtained with this technique.

Additionally, Table 2 presents time execution of complete process for multi-species identification using each feature extraction method. It is important to consider that CNN architectures training may require significant computational resources. We worked in a high-performance computer (Ryzen 5 3600, 16 gb ram, Nvidia rtx 2060 super) where CNN is notoriously the most time-consuming method.

Table 2. Comparing the execution times to perform multi-species identification in the entire dataset using different feature extraction methods.

Features representation	Execution time (seconds)
Frequency information	1053.0
LFCC and frequency information	2065.7
VAE	3986.1
VAE and frequency information	4341.6
KiwiNet-CNN	10707.9
KiwiNet-CNN and frequency information	11067.9

Finally, the identification results of the 30 analyzed species are presented in Fig. 4. These results show the difference between species detection in each tested feature extraction method. In general, high frequency species such as in the case of bats (most mammals) and orthopterans (most insecta) has a high performance, this may be since at these frequencies, the presence of background noise is low or almost null, which makes the detection improved.

Overall, species detection was above 50%. This shows that both classical methods such as LFCC and more current methods such as deep learning models in combination with species frequency information can provide relevant information about the calls to the clustering algorithm.

Table 3. Summary of species name and corresponding code for 30 analyzed species.

No.	Species-name	Code	No.	Species-name	Code	No.	Species-name	Code
1	Boana platanera	BPA	11	cf. Rhogeessa	CRA	21	Saccopteryx leptura	SLA
2	Dendropsophus microcephalus	DMS	12	Cynomops	CPS	22	Molossus bondae	MBE
3	Leptodactylus fragilis	LFS	13	Dasypterus ega	DEGA	23	Molossus molossus	MMS
4	Leptodactylus fuscus	LFUS	14	Eppia truncatipennis	ETS	24	Alouatta sp.	ASP
5	Crotophaga ani	CANI	15	Eptesicus brasiliensis	EBS	25	Neoconocephalus affinis	NAIS
6	Crypturellus soui	CSI	16	Eptesicus furinalis	EFS	26	Orthoptera sp. 1	OSP1
7	Dendroplex picus	DPS	17	Eumops glaucinus	EGS	27	Orthoptera sp. 2	OSP2
8	Nyctidromus albicollis	NAS	18	Lasiurus blossevillii	LBI	28	Orthoptera sp. 3	OSP3
9	Patagioenas cayennensis	PCS	19	Myotis keaysi	MKI	29	Orthoptera sp. 4	OSP4
10	Troglodytes aedon	TAN	20	Saccopteryx bilineata	SBA	30	Orthoptera sp. 5	OSP5

Species names in Fig. 4 were coded, the complete name of each species is presented in Table 3.

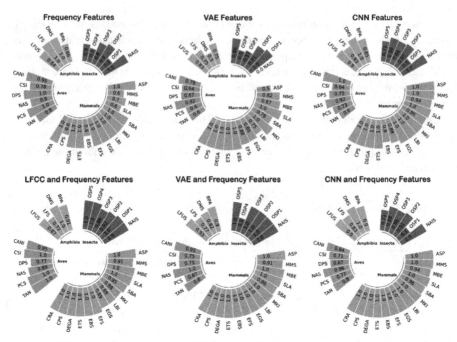

Fig. 4. Multi-species identification results using different features extraction method with LAMDA algorithm. There are six circular bar diagrams, each one representing multi-species results using each feature representation. Each bar represents the performance of every analyzed species. Species results are organized according to their taxonomic group (aves, amphibians, mammals, and insects).

4   Conclusions and Future Work

This work aims to evaluate different methods to feature extraction for multi-species identification. Deep learning models are currently the most widely used due to their high performance in different applications but have not been widely used to extract features in bioacoustics application. We decided to test them in a multi-species identification approach compared to classical feature extraction methods used in machine learning for ecoacoustics analysis.

It was found that deep learning methods for feature extraction allow the clustering algorithm to perform the species identification of different taxonomic groups. However, frequency-based methods such as LFCC showed to be faster at execution time without high performance sacrificing. Furthermore, frequency variables proved to be of significant contribution to multi-species identification.

Supervised deep learning methods such as CNN showed adequate performance in multi-species identification task. However, this method is highly dependent on a robust dataset, even with transfer learning. This can be a problem in biological applications, where a lot of audio data is collected for biodiversity monitoring and species information is not always available.

Variational Autoencoders is an unsupervised method that shows that accurate detection is possible in a multi-species approach in combination with frequency features. This method presents the advantage of working with the complete dataset due that do not require species labels. This model obtained a low reconstruction error. Nevertheless, we used the spectrogram with call segment as an image, and as an image, the latent space was appropriate for reconstruction but could be improved for clustering tasks.

Future work consists of evaluating VAE with the call segments coming directly from the time-frequency representation (spectrogram) and not taking them as an image in order to try to obtain a better feature representation in the latent space as input for the LAMDA algorithm. Besides, to analyze the generalization capacity of deep learning methods with databases from other regions since the training requires fine-tuning of parameters to find an adequate performance.

Acknowledgments. This work was supported by Universidad de Antioquia, Instituto Tecnológico Metropolitano de Medellín, Alexander von Humboldt Institute for Research on Biological Resources and Colombian National Fund for Science, Technology and Innovation, Francisco Jose de Caldas - MINCIENCIAS (Colombia) [Program No. 111585269779].

References

1. Pimm, S.L., et al.: Emerging technologies to conserve biodiversity. Trends Ecol. Evol. **30**, 685–696 (2015). https://doi.org/10.1016/j.tree.2015.08.008
2. Dumyahn, S.L., Pijanowski, B.C.: Soundscape conservation. Landsc. Ecol. **26**, 1327–1344 (2011). https://doi.org/10.1007/s10980-011-9635-x
3. Sueur, J., Farina, A.: Ecoacoustics: the ecological investigation and interpretation of environmental sound. Biosemiotics **8**(3), 493–502 (2015). https://doi.org/10.1007/s12304-015-9248-x
4. Aide, T.M., Hern, A., Campos-cerqueira, M.: Species richness (of insects) drives the use of acoustic space in the tropics. Remote Sens. Ecol. Conserv., 1–12 (2017). https://doi.org/10.3390/rs9111096
5. Ross, S.-J., Friedman, N.R., Dudley, K.L., Yoshimura, M., Yoshida, T., Economo, E.P.: Listening to ecosystems: data-rich acoustic monitoring through landscape-scale sensor networks. Ecol. Res. **33**(1), 135–147 (2017). https://doi.org/10.1007/s11284-017-1509-5
6. Ruff, Z.J., Lesmeister, D.B., Duchac, L.S., Padmaraju, B.K., Sullivan, C.M.: Automated identification of avian vocalizations with deep convolutional neural networks. Remote Sens. Ecol. Conserv. **6**, 79–92 (2020). https://doi.org/10.1002/rse2.125
7. Bedoya, C., Isaza, C., Daza, J.M., López, J.D.: Automatic recognition of anuran species based on syllable identification. Ecol. Inform. **24**, 200–209 (2014). https://doi.org/10.1016/j.ecoinf.2014.08.009
8. LeBien, J., et al.: A pipeline for identification of bird and frog species in tropical soundscape recordings using a convolutional neural network. Ecol. Inform. **59**, 101113 (2020). https://doi.org/10.1016/j.ecoinf.2020.101113
9. Ruff, Z.J., Lesmeister, D.B., Appel, C.L., Sullivan, C.M.: Workflow and convolutional neural network for automated identification of animal sounds. Ecol. Indic. **124**, 107419 (2021). https://doi.org/10.1016/j.ecolind.2021.107419
10. Stowell, D.: Computational bioacoustic scene analysis. In: Computational Analysis of Sound Scenes and Events, pp. 303–333. Springer, Cham (2018). https://doi.org/10.1007/978-3-319-63450-0

11. Xie, J., Colonna, J.G., Zhang, J.: Bioacoustic signal denoising: a review. Artif. Intell. Rev. **54**(5), 3575–3597 (2020). https://doi.org/10.1007/s10462-020-09932-4

12. Noda, J.J., David Sánchez-Rodríguez, C.M.T.-G.: We are IntechOpen, the world's leading publisher of Open Access books Built by scientists, for scientists TOP 1%. Intech **32**, 137–144 (2018)

13. Nirosha Priyadarshani, S.M., Castro, I.: Automated birdsong recognition in complex acoustic environments: a review. Avian Biol. (2018). https://doi.org/10.1111/jav.01447

14. Rowe, B., Eichinski, P., Zhang, J., Roe, P.: Acoustic auto-encoders for biodiversity assessment. Ecol. Inform. **62**, 101237 (2021). https://doi.org/10.1016/j.ecoinf.2021.101237

15. Ntalampiras, S., Potamitis, I.: Acoustic detection of unknown bird species and individuals. CAAI Trans. Intell. Technol. **6**, 291–300 (2021). https://doi.org/10.1049/cit2.12007

16. Xie, J., Hu, K., Guo, Y., Zhu, Q., Yu, J.: On loss functions and CNNs for improved bioacoustic signal classification. Ecol. Inform. **64**, 101331 (2021). https://doi.org/10.1016/j.ecoinf.2021.101331

17. Bedoya, C.L., Molles, L.E.: Acoustic censusing and individual identification of birds in the wild (2021)

18. Stowell, D.: Computational bioacoustics with deep learning: a review and roadmap. PeerJ **10**, e13152 (2022). https://doi.org/10.7717/peerj.13152

19. Xie, J., Towsey, M., Zhu, M., Zhang, J., Roe, P.: An intelligent system for estimating frog community calling activity and species richness. Ecol. Indic. **82**, 13–22 (2017). https://doi.org/10.1016/j.ecolind.2017.06.015

20. Mermelstein, P.: Distance measures for speech recognition, psychological and instrumental. Pattern Recognit. Artif. Intell. **116**, 374–388 (1976)

21. Zhou, X., Garcia-Romero, D., Duraiswami, R., Carol Espy-Wilson, S.S.: 2011 IEEE Workshop on Automatic Speech Recognition & Understanding: ASRU 2011: Proceedings, Waikoloa, Hawaii, U.S.A., 11–15 December 2011, p. 564 (2011)

22. Dong, C., Xue, T., Wang, C.: The feature representation ability of variational autoencoder. Proceedings - 2018 IEEE Third International Conference on Data Science in Cyberspace, DSC 2018, pp. 680–684 (2018). https://doi.org/10.1109/DSC.2018.00108

23. Fukumoto, T.: Anomaly detection using Variational Autoencoder (VAE) (2020). https://github.com/mathworks/Anomaly-detection-using-Variational-Autoencoder-VAE-/releases/tag/1.0.1, GitHub. Accessed 23 Apr 2022

24. Simonyan, K., Zisserman, A.: Very deep convolutional networks for large-scale image recognition. In: 3rd International Conference on Learning Representations ICLR 2015 - Conference Track Proceedings, pp. 1–14 (2015)

25. Lamrini, B., Le Lann, M.V., Benhammou, A., Lakhal, E.K.: Detection of functional states by the "LAMDA" classification technique: application to a coagulation process in drinking water treatment. Comptes Rendus Phys. **6**, 1161–1168 (2005). https://doi.org/10.1016/j.crhy.2005.11.017

26. Bedoya, C., Waissman Villanova, J., Isaza Narvaez, C.V.: Yager–Rybalov triple Π operator as a means of reducing the number of generated clusters in unsupervised anuran vocalization recognition. In: Gelbukh, A., Espinoza, F.C., Galicia-Haro, S.N. (eds.) MICAI 2014. LNCS (LNAI), vol. 8857, pp. 382–391. Springer, Cham (2014). https://doi.org/10.1007/978-3-319-13650-9_34

Evaluation of Transfer Learning to Improve Arrhythmia Classification for a Small ECG Database

Larissa Montenegro⬛, Hugo Peixoto(✉)⬛, and José M. Machado⬛

ALGORITMI/LASI, University of Minho, 4710-057 Braga, Portugal
larissa.montenegro@algoritmi.uminho.pt, {hpeixoto,jmac}@di.uminho.pt
https://algoritmi.uminho.pt/

Abstract. Deep learning algorithms automatically extract features from ECG signals, eliminating the manual feature extraction step. Deep learning approaches require extensive data to be trained, and access to an ECG database with a large variety of cardiac rhythms is limited. Transfer learning is a possible solution to improve the results of cardiac rhythms classification in a small database. This work proposes a open-access robust 1D-CNN model to be trained with a public database containing cardiac rhythms with their annotations. This study explores transfer learning in a small database to improve arrhythmia classification tasks. Overall, the 1D-CNN model trained without TL achieved an average accuracy of 91.73 % and F1-score 67.18 %; meanwhile, the 1D-CNN model with TL achieved an average accuracy of 94.40 % and F1-score of 79.72 %. The F1-score has an overall improvement of 12.54 % over the baseline model for rhythm classification. Moreover, this method significantly improved the F1-score precision and recall, making the model trained with transfer learning more relevant and reliable.

Keywords: Transfer learning · Deep learning · ECG classification · Heart rhythms

1 Introduction

Specific pathologies related to the heart can be detectable from the short-term acquisition of data in the hospital. The most commonly used non-invasive screening procedure is electrocardiography (ECG), which records the cardiac electrical activity, thereby extracting significant parameters to assess the patient's overall cardiac health [3,4]. In contrast, others require long-term by using medical devices such as ECG-Holter and wearable, leading to a large volume of acquired data. An unusual rhythm might be related toa cardiac arrhythmia. Arrhythmias are caused by different reasons, such as changes in cardiac tissue, stress, imbalance in the blood, e.g., excess or deficiency of electrolytes or hormones, a side effect of medications, chronic diseases, or problems with the heart's electrical system [3]. Automatic ECG signal analysis plays a crucial role in assisting

© The Author(s), under exclusive license to Springer Nature Switzerland AG 2022
A. C. Bicharra Garcia et al. (Eds.): IBERAMIA 2022, LNAI 13788, pp. 231–242, 2022.
https://doi.org/10.1007/978-3-031-22419-5_20

healthcare professionals, providing real-time alarms for immediate treatment in intensive care units (ICUs), and improving people's quality of life through early detection of abnormal patterns [4]. Interpreting the ECG in an automatic process is a challenge, as an ECG signal can differ among and within patients due to different physical conditions [9]. For ECG-based diagnosis, the algorithm must characterize and recognize ECG morphology and rhythm [1].

Most research has focused on automated learning algorithms involving pre-processing, feature extraction, feature selection, and the automatic classification of cardiac arrhythmias. The accessibility of computer systems with higher processing and performance has facilitated the migration to research and development of deep learning algorithms to perform cardiac arrhythmia classification tasks. Deep learning (DL) algorithms benefit from not manually extracting features from the ECG signals, as the algorithm extracts the features automatically, eliminating the manual feature extraction step. Deep learning approaches require an extensive amount of data to be trained. However, access to an ECG database with a large variety of cardiac rhythms is limited. Transfer learning is a possible solution to improve the results of cardiac rhythms classification in a small database [1,2]. Transfer learning has been widely used for image and natural language processing tasks. For this study, the focus will be on transfer learning on time-series ECG signals. The implementation of transfer learning requires a pre-trained model. The pre-trained model is an architecture previously trained with a broad database. The idea behind this technique is that the pre-trained model with an extensive database can generalize, thus effectively functioning as a generic model for a specific task. The advantage is that by reusing the pre-trained model, it is not necessary to start training the model from scratch; the features already learned by the internal layers of the model are then reused. Today, there are several pre-trained models with image data, such as VGG and AlexNet; in comparison, time-series have not been broadly explored [25].

This study explores transfer learning to improve arrhythmia classification tasks in a small database. A pre-trained model with ECG time-series signals will be used to perform this task. In the absence of a publicly available pre-trained model. A robust 1D-CNN model will be trained with a public database containing many cardiac rhythms with their annotations. Transfer Learning will be implemented to improve the classification results on a small database but with a more variety of heart rhythms. The rest of the paper is organized as follows: First, related and previous relevant works and public databases will be described in Sect. 2. Related methods based on transfer learning for ECG arrhythmia detection are briefly reviewed in Sect. 3. Then, in Sect. 4, the proposed method is explained in detail, and in Sect. 5, the results are discussed. The last Sect. 6 presents the conclusion.

2   Related Work

2.1   Public Database

Large public databases of annotated ECG signals play a fundamental role in the developed algorithms for automatic ECG classification, serving as a benchmark for comparing the validation and quantitative evaluation of algorithms from different works in the scientific community [4]. The PhysioNet Computing in Cardiology Challenge 2017 (cinc17) [5,6] and the Telehealth Network of Minas Gerais (TNMG) [8] are among the popularly used databases for heart rhythms classification. Table 1 highlights both databases, which differ in terms of the number of records, data acquisition (i.e., sampling frequency), condition, and annotations pathologies. The cinc17 database focused on Atrial Fibrillation (AF) detection by differentiating the AF from noise, normal sinus, or other rhythms. It contains 8,528 single-lead ECG recordings with a time length from 9 s to just over 60 s. The database was recorded with the AliveCor device with a frequency sample 300 Hz and filtered with a band-pass filter. The TNMG database contains six different ECG annotations: 1st degree AV block (1dAVb), Left Bundle Branch Block (LBBB), Right Bundle Branch Block (RBBB), sinus bradycardia (SB), AF, sinus tachycardia (ST), and normal (N). The Massachusetts Institute of Technology-Beth Israel Hospital (MIT-BIH) Arrhythmia database is frequently used to detect cardiac arrhythmias and heartbeats, but this study is not included.

Table 1. ECG Database Overview

Abbrev.	Sampling Frequency (Hz)	Number Records	Leads	Number of Pathologies	Annotation Type
cinc17	300	8,528	Lead I	4	rhythms
TNMG	400	827	12-Lead	6	rhythms

2.2   ECG Classification

The continuous increase in performance of processing systems has prompted researchers to evaluate more complex methods for classifying arrhythmias and heartbeats. Research has shown an increase in the number of extracted features and combinations of Machine Learning (ML) algorithms [10]. The state-of-the-art algorithms for machine learning has been broadly evaluated in recent years. Sansome et al. (2013) [26] presented a literature review describing some methods for feature extraction from ECG signals and the basic workflow for implementing ML algorithms in heartbeat and rhythm classification, with SVM and ANN algorithms being the most popular. A study published by Montenegro et al.(2022) [27] compares the performance of an SVM algorithm with that of 1D-CNN for the task of arrhythmia classification with time-series ECG signals. The results show that the SVM and 1D-CNN algorithms perform very similarly when trained and validated with the same dataset.

DL algorithms rely primarily on data representation learning techniques. Raw ECG signals can be used as input, thus obtaining QRS-complex information and P and T-wave information, thereby better representing the signal [11]. This does not imply that the raw ECG signal can be provided as the sole input. The following is a literature review of Deep Learning methods for the task of cardiac arrhythmia extraction and classification. ECG classification research can be divided into either heartbeat [11–13] or arrhythmia [1,16–18]. The databases used often are the MIT-BIH [1,11–13,18] and cinc17 [16]. The input of the model varied among research, being usually extracted features [1,16], raw time-series ECG signals [11,12,17,18] and 2-dimensional images of the signal [14,15,19,29], for example, spectrogram [13,20,21]. Transfer Learning has been used for arrhythmia classification as it works by sharing knowledge among related tasks. For instance, Van Steenkiste et al. (2020) [1] conducted a study to automatically analyze equine electrocardiography (eECG), namely equine cardiac signals. To perform this study, they proposed using Wavelet Transform (WT) for both filtering and QRS detection in eECGs, and classification was with a CNN architecture. The authors used transfer learning from a network trained with a human ECG dataset to an auto-generated equine dataset with 20,000 beats recorded in a clinical setting. The model was trained with the MIT-BIH database. The researchers included four types of beats, resulting in 237,704 heartbeats for training. The network was optimized up to an accuracy of 97.7% and 92.6% for the MIT-BIH and eECG databases, respectively. Then, after applying transfer learning from the MIT-BIH dataset to the eECG database, the average accuracy, recall, positive predictive value, and F1 score of the network increased with an accuracy of 97.1%.

A study where transfer learning was implemented for ECG classification in time series was presented by Weimann et al. (2021). They present a pre-trained model with the Icential1K dataset and then, through transfer learning, fine-tune the model for the classification of Atrial Fibrillation. Their results showed an improvement in the performance of CNNs on the target task by up to 6.57%. In our study, we will focus on ECG time series input data as opposed to other studies where the focus is on the use of ECG images. Additionally, we will not use such a broad database to pre-train the CNN model. Instead, we will pre-train the model with the cinc17 database, which, as mentioned above, is very popular in deep learning and arrhythmia classification studies. Although, the use of a broad database to pre-train the model is frequently suggested for transfer learning. We sought to evaluate whether the cinc17 database (despite being a small dataset) is sufficient to train the model to learn enough features from the ECG time series signals to be reused to fine-tune a much smaller database to classify different types of arrhythmias.

3   Methods

The main task is to explore whether transfer learning improves arrhythmia classification results in a small ECG rhythms database. The procedure requires a pre-trained deep neural network model with an extensive database for task A. Then, the pre-trained model is used to tune the deep neural network model for a smaller database (TNMG); namely, it is reused as a starting point to train the model for task B (see Fig. 1). The starting points are the weights learned by the pre-trained model A. The knowledge acquired is provided to model B, improving the learning of this model. The new data from TNMG database is expected to be better to generalize. To verify this theory, we will compare the evaluation metrics of model B with model C, which is trained with TNMG, but without transfer learning. For this research, ECG time-series signals acquired by Lead-I are the focus, and public access to pre-trained deep learning models from 1-Lead ECG signals is almost nonexistent. For this purpose, a robust CNN model was pre-trained with cinc17. Although it is not a database with a large amount of data, it is one of the most widely used databases for deep learning cardiac rhythms classification. Next, we extracted the weights from the pre-trained model and tuned the CNN model to classify six new cardiac arrhythmias from TNMG database. Model B was also trained from scratch with TNMG to compare accuracy metrics results versus the model with the transfer learning knowledge. The databases and the CNN model will be described below.

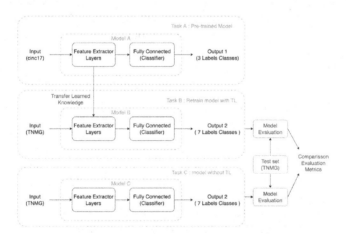

Fig. 1. Proposed Workflow Method to apply Transfer Learning (TL) to Task B and finally compare the evaluation metrics with Task C

3.1   Databases

Normalization was the only pre-processing step needed in the time-series signals from cinc17 database. TNMG database was pre-processed to meet the same conditions (e.g., removal of baseline wandering, re-sampling to a standard sampling

Fig. 2. ECG signal pre-processing steps in cinc17 database and TNMG database.

frequency) as in cinc17, see Fig. 2. Before pre-training, the cinc17 database was redesigned with three different types of rhythms, Normal (5076), Other Rhythms (3173), and Noise (279), being a total of 8528 ECG Signals. The TNMG database contains seven rhythms that include 1dAVb (20), LBBB (28), RBBB (32), SB (16), AF (13), ST (37), and Normal (30), being 176 ECG signals. Both databases have their respective heart rhythms annotations and were split randomly into 90% training dataset and 10% validation dataset.

3.2   1D-CNN Architecture Model

The 1D-CNN architecture designed by Hannun et al. [17] is open access, and its robustness is to a cardiologist-level comparison. The 1D-CNN includes 34 layers, of which 33 are 1D-convolutional, followed by a fully-connected layer with softmax. Additionally, the network contains 16 residual blocks, which are the shortcut connections that have the task of extracting deeper features by going deeper into the network. The 1D-convolutional network receives raw ECG time series as input, and the output predicts one of the heart rhythms by implementing the softmax layer. For more details on the model architecture and configuration parameters, please refer to the author's article where the model was initially developed [17].

3.3   Transfer Learning Method

The pre-trained 1D-CNN model algorithm is trained with cinc17 database. The algorithm is set to train up to 100 epochs but will stop if the validation accuracy and validation loss values stop changing. The model with the best validation accuracy and validation loss is the pre-trained model for the second task taken. Before fine-tuning the 1D-CNN algorithm for the second task, the output layer was replaced to ensure that the output would match the number of new arrhythmia categories from TNMG database and initialized with random weights. The initial layer did not need to be re-adjusted as it received the same input as the pre-trained model. The rest of the layers are with the weights extracted from the pre-trained model set. The 1D-CNN model is with a 50 epochs train. Validation metrics were obtained and compared with another model trained without transfer learning.

4   Results

The model B trained with transfer learned knowledge was compared with model C trained from scratch with TNMG database. The model C was trained with random initial weights and is considered the benchmark for task C, arrhythmias classification from TNMG database.

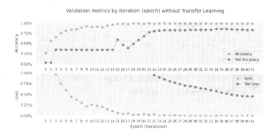

Fig. 3. Training process per epoch TNMG database without transfer learning

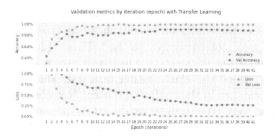

Fig. 4. Training process per epoch TNMG database with transfer learning

Figure 3 shows the training process per epoch (training iteration) for Model C with TNMG without transfer learning, and Fig. 4 shows the training process for Model B with transfer learning. The figures show four-line plots in percentage. The upper graph is for accuracy and validation accuracy and the lower for loss and validation loss. The algorithm trained with transfer learning converged to a solution of the model faster than the model trained from scratch. The validation accuracy converges to a model solution after epoch 18, where the validation accuracy stabilized around 90.00 %. In contrast, the trained model without transfer learning converged to a model solution at epoch 31, where the validation accuracy stabilized at around 80.00 %. Moreover, the model without transfer learning took much longer to stabilize at the beginning of training since it learned the signal representations from scratch. In contrast, the model with transfer learning stabilized faster since it had the knowledge acquired from the pre-trained model. The model C (without transfer learning) has a final accuracy of 99.95 %, a validation accuracy of 84.80 %, a loss of 0.85 %, and a validation

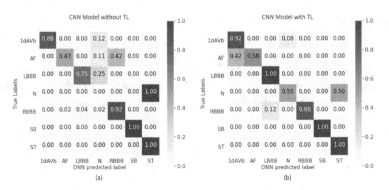

Fig. 5. (a) Confusion Matrix of CNN Model without Transfer Learning. (b) Confusion Matrix of CNN Model with Transfer Learning

loss of 46.25 %. The model B (with transfer learning) has a final accuracy of 99.95 %, a validation accuracy of 89.22 %, a loss of 0,78 %, and a validation loss of 27.71 %. Overall, the model B with transfer learning improves the validation accuracy up to 4.42 % and reduces validation loss to −18.54 %. Figure 5 shows the confusion matrix for the CNN model trained with and without transfer learning. The diagonal values of the confusion matrix show the true-positives values; the rest are those that were not correctly classified. These values serve to compute precision, recall, and F1-score metrics, which will give a better description of how the model performed the classification task individually for each rhythm category. The performance in the confusion matrix, it is observed that only four cardiac arrhythmias were true-positive predicted in the model C (without TL). However, The model B (with TL) shows that five cardiac arrhythmias were true-positive predicted in its confusion matrix. Evaluation metrics are calculated for each arrhythmia to get a better perspective of the performance of both models.

Table 2 shows the evaluation metrics for both models. At first glance, the model C without transfer learning shows overall good accuracy results. However, the F1-score per cardiac rhythm was not as high, except for arrhythmia 1dAVb and SB, which means that this model's prediction performance per individual cardiac rhythm is not optimal. In contrast, the model B trained with transfer learning shows improvement and increase in the F1-score for five of the seven cardiac rhythms. The F1-scores improved 9.82 % for AF, 11.72 % for LBBB, 44.64 % Normal Sinus, 31.67 % for RBBB and 13.33 % for ST. As the precision and recall as well increased, the prediction results are more relevant and reliable in the model B with transfer learning. Overall, the 1D-CNN model trained without TL achieved an average accuracy of 91.73 % and F1-score 67.18 %; meanwhile, the 1D-CNN model with TL achieved an average accuracy of 94.40 % and F1-Score 79.72 %. The F1-score has an overall improvement of 12.54 % over the baseline model for rhythm classification.

Table 2. Classification summary report per heart rhythm.

Model	ECG rhythm	Accuracy (%)	Precision (%)	Recall (%)	F1-Score (%)	Support
CNN Model	1dAVb	97.88	88.00	100.00	93.61	3
Without TL	AF	84.18	47.00	98.32	63.60	5
	LBBB	95.11	75.00	96.57	84.43	3
	N	84.12	0.00	0.00	0.00	1
	RBBB	86.71	92.00	46.70	61.95	2
	SB	100.00	100.00	100.00	100.00	2
	ST	94.12	100.00	50.00	66.67	1
CNN Model	1dAVb	86.24	**92.00**	56.79	70.23	3
With TL	AF	**87.65**	**58.00**	**100.00**	**73.42**	5
	LBBB	**98.59**	**100.00**	92.60	**96.15**	3
	N	**92.71**	**33.33**	**67.57**	**44.64**	1
	RBBB	**98.58**	88.00	**100.00**	**93.62**	2
	SB	100.00	100.00	100.00	100.00	2
	ST	**97.06**	100.00	**66.67**	**80.00**	1

5   Discussion

This study showed that a robust CNN model does not necessarily require being trained with an extensive ECG time-series database to be used for transfer learning on a smaller database. The fine-tuned model with the pre-trained model weights improved its performance considerably on model B compared to model C trained with TNMG database from scratch, reducing the training time and enhancing its precision and F1-score. In multiple classification tasks, the overall accuracy validation does not indicate the accuracy for each rhythm category. Therefore, the evaluation metrics and confusion matrix are valuable when evaluating a model. Interestingly, the normal sinus had the worst performance in both models because there was only a single normal signal in the evaluation dataset. The model mix it with sinus tachycardia due is morphology. Depending on the type of ECG rhythm to be predicted, it can be confused with normal sinus since its characteristics in the time domain are similar and vary in frequency.

The transfer learning model improved the prediction accuracy; however, ideally, there should have been more than a single rhythm sample at the evaluation time to observe a better performance of the model. Most cardiac rhythms improved their values, showing better measurement quality and quantity in their predictions. The pre-trained model helped perform transfer learning despite not having such an extensive database because of the robustness and depth of the model layers. The study presented by Weinmann et al. [29] used a wider database to train the pre-training model. After TL, the model showed a 4.92 % F1-score improvement in heart rhythm classification over their baseline model. In our case, we used fewer data in the pre-training model but used a model with a much deeper convolutional layered configuration. We still obtained promising results compared to the baseline model, with an F1-Score improvement of 12.54 %. The model learned more specific characteristics of the signals as the model trained on deeper layers. This knowledge of the signals, especially the more distinctive features, is probable the model trained with transfer learning trained faster and obtained improved performance results in contrast to the model without transfer learning prior knowledge. Since all the convolution layers of the model were fine-tuned, it would be interesting to test the above theory by reducing the number of convolutional layers that are fine-tuned with the weights of the pre-trained model. Instead of using the 33 convolutional layers with the weights, set fewer layers and let the remaining ones be initialized with random weights and compare the results.

6   Conclusion

This study used transfer learning to enhance convolutional neural networks (CNNs) trained to classify cardiac rhythm from a database with few data samples. Pre-trained CNN models for time-series ECG signals are inaccessible compared to pre-trained CNN models for images. This is due to limited access to large-high-quality ECG databases since most are private for security reasons or the cost of accessing them is very high, especially for independent researchers. Using public databases for this purpose would be a solution to get pre-trained models, however databases differ in the number of records, data acquisition (i.e., sampling frequency), number of leads and pathology annotations. The existence of large public databases of annotated ECG signals plays a fundamental role in design open access pre-trained models of time-serie signals. The access to large databases is a recurrent problem faced by the research community. Although, the continuous development of computerized ECG systems has increased the collection of more data both at the clinic and in remote settings [??]

Systems based on ECG classification with transfer learning could be explored in parallel with cardiovascular disease prediction systems based on Data Mining. Studies based on data mining have shown positive results in the prediction of cardiovascular disease by using clinical data collected during medical examinations [23,24]. Implementing such applications in reliable mobile monitoring systems would benefit patients with limited access to continuous care. Overall, This study demonstrated first, that the cinc17 database could be used to pretrain a robust and successful CNN model, and second, transfer learning improves classification performance on a small database and speeds up the training time for the CNN model.

Acknowledgements. Funding: This work has been supported by FCT-Fundação para a Ciência e Tecnologia within the R&D Units Project Scope: UIDB/00319/2020 and the project "Integrated and Innovative Solutions for the well-being of people in complex urban centers" within the Project Scope NORTE-01-0145-FEDER- 000086.

References

1. Van Steenkiste, G., van Loon, G., Crevecoeur, G.: Transfer learning in ECG classification from human to horse using a novel parallel neural network architecture. Sci. Rep. **10**, 186 (2020). https://doi.org/10.1038/s41598-019-57025-2
2. Isin, A., Ozdalili, S.: Cardiac arrhythmia detection using deep learning. Proc. Comput. Sci. **120**, 268–275 (2017). https://doi.org/10.1016/j.procs.2017.11.238, ISSN 1877–0509
3. National Heart Lung and Blood Institute Web page. Arrhythmia. https://www.nhlbi.nih.gov/health-topics/arrhythmia. (Accessed May 2022)
4. Sansone, M., Fusco, R., Pepino, A., Sansone, C.: Electrocardiogram pattern recognition and analysis based on artificial neural networks and support vector machines: a review. J Healthc Eng. **4**(4), 465–504 (2013). PMID: 24287428. https://doi.org/10.1260/2040-2295.4.4.465
5. Clifford, G.D., et al.: AF Classification from a Short Single Lead ECG Recording: the PhysioNet/Computing in Cardiology Challenge 2017. Comput. Cardiol **44** (2017) https://doi.org/10.22489/CinC.2017.065-469
6. Clifford, G.D., et al.: AF Classification from a Short Single Lead ECG Recording: the PhysioNet/Computing in Cardiology Challenge (2017). https://physionet.org/content/challenge-2017/1.0.0/. (Accessed May 2022)
7. Goldberger, A.L., et al.: PhysioBank, PhysioToolkit, and PhysioNet: components of a new research resource for complex physiologic signals. Circulation. 13, 101(23), E215–20 (2000). PMID: 10851218. https://doi.org/10.1161/01.cir.101.23.e215
8. Ribeiro, A.H., Ribeiro, M.H., Paixão, G.M.M., et al.: Automatic diagnosis of the 12-lead ECG using a deep neural network. Nat. Commun. **11**, 1760 (2020). https://doi.org/10.1038/s41467-020-15432-4
9. Banerjee, S., Mitra, M.: Application of cross wavelet transform for ECG pattern analysis and classification. IEEE Trans. Instrum. Meas. **63**, 326–333 (2014)
10. Mondéjar-Guerra, V., Novo, J., Rouco, J., Penedo, M.G., Ortega, M.: Heartbeat classification fusing temporal and morphological information of ECGs via ensemble of classifiers. Biomed. Signal Process. Contr. **47**, 41–48 (2019). https://doi.org/10.1016/j.bspc.2018.08.007, ISSN 1746–8094
11. Xu, S.S., Mak, M.-W., Cheung, C.-C.: Towards End-to-End ECG classification with raw signal extraction and deep neural networks. IEEE J. Biomed. Health Inform. **23**(4), 1574–1584 (2019). https://doi.org/10.1109/JBHI.2018.2871510
12. Kiranyaz, S., Ince, T., Gabbouj, M.: Real-time patient-specific ECG classification by 1-D convolutional neural networks. IEEE Trans. Biomed. Eng. **63**(3), 664–675 (2016). https://doi.org/10.1109/TBME.2015.2468589
13. Hao, C., Wibowo, S., Majmudar, M., Rajput, K.S.: Spectro-temporal feature based multi-channel convolutional neural network for ECG beat classification. In: 2019 41st Annual International Conference of the IEEE Engineering in Medicine and Biology Society (EMBC), pp. 5642–5645 (2019). https://doi.org/10.1109/EMBC.2019.8857554
14. Venton, J., Aston, P.J., Smith, N.A.S., Harris, P.M.: Signal to image to classification: transfer learning for ECG. In: 2020 11th Conference of the European Study Group on Cardiovascular Oscillations (ESGCO), pp. 1–2 (2020). https://doi.org/10.1109/ESGCO49734.2020.9158037
15. Pal, A., Srivastva, R., Narain Singh, Y.: CardioNet: an efficient ecg arrhythmia classification system using transfer learning. Big Data Res. **26**, 100271 (2021). https://doi.org/10.1016/j.bdr.2021.100271, ISSN 2214–5796

16. Ghiasi, S., Abdollahpur, M., Madani, N., Kiani, K., Ghaffari, A.: Atrial fibrillation detection using feature based algorithm and deep convolutional neural network. Comput. Cardiol. (CinC) **2017**, 1–4 (2017). https://doi.org/10.22489/CinC.2017. 159-327

17. Hannun, A.Y., Rajpurkar, P., Haghpanahi, M., et al.: Cardiologist-level arrhythmia detection and classification in ambulatory electrocardiograms using a deep neural network. Nat. Med. **25**, 65–69 (2019). https://doi.org/10.1038/s41591-018-0268-3

18. Yıldırım, Ö., Pławiak, P., Tan, R.S., Acharya, U.R.: Arrhythmia detection using deep convolutional neural network with long duration ECG signals. Comput. Biol. Med. **102**, 411–420 (2018). https://doi.org/10.1016/j.comp.biomed.2018.09.009

19. Jun, T.J., Nguyen, H.M., Kang, D., Kim, D., Kim, D., Kim, Y.: ECG arrhythmia classification using a 2-D convolutional neural network. ArXiv, abs/ arXiv: 1804.06812 (2018)

20. Mousavi, S., Afghah, F., Razi, A., Acharya, U.R.: ECGNET: learning where to attend for detection of atrial fibrillation with deep visual attention. In: IEEE EMBS International Conference on Biomedical & Health Informatics (BHI), vol. 2019, pp. 1–4 (2019). https://doi.org/10.1109/BHI.2019.8834637

21. Zihlmann, M., Perekrestenko, D., Tschannen, M.: Convolutional Recurrent Neural Networks for Electrocardiogram Classification, arXiv:1710.06122 (2018)

22. Ferreira, D., Silva, S., Abelha, A., Machado, J.: Recommendation system using autoencoders. Appli. Sci. **10**(16), 5510. 659 MDPI (2020). 660

23. Martins, B., Ferreira, D., Neto, C., Abelha, A., Machado, J.: Data Mining for Cardiovascular Disease Prediction. J. 661 Med, Syst. **45**(1), 662 (2021)

24. Aqra, I., Abdul Ghani, N., Maple, C., Machado, J., Sohrabi Safa, N.: Incremental Algorithm for Association Rule Mining under 663 Dynamic Threshold. Appli. Sci. **9**(24). MDPI (2019)

25. Weber, M., Auch, M., Doblander, C., Mandl, P., Jacobsen, H.: Transfer learning with time series data: a systematic mapping study. IEEE Access **9**, 165409–165432 (2021). https://doi.org/10.1109/ACCESS.2021.3134628

26. Sansone, M., Fusco, R., Pepino, A., Sansone, C.: Electrocardiogram pattern recognition and analysis based on artificial neural networks and support vector machines: a review. J Healthc. Eng. **4**(4), 465–504 (2013) PMID: 24287428. https://doi.org/ 10.1260/2040-2295.4.4.465

27. Montenegro, L., Abreu, M., Fred, A., Machado, J.M.: Human-Assisted vs. deep learning feature extraction: an evaluation of ECG Features extraction methods for arrhythmia classification using machine learning. Appli. Sci. **12**(15), 7404 (2022). https://doi.org/10.3390/app12157404

28. Gajendran, M.K., Khan, M.Z., Khattak, M.A.K.: ECG classification using deep transfer learning. In: 2021 4th International Conference on Information and Computer Technologies (ICICT), pp. 1–5 (2021). https://doi.org/10.1109/ICICT52872. 2021.00008

29. Weimann, K., Conrad, T.O.F.: Transfer learning for ECG classification. Sci. Rep. **11**, 5251 (2021). https://doi.org/10.1038/s41598-021-84374-8

A General Recipe for Automated Machine Learning in Practice

Hernan Ceferino Vazquez[✉]

MercadoLibre Inc., Applied Machine Learning Research, Buenos Aires, Argentina
`hernan.vazquez@mercadolibre.com, hernan.c.vazquez@gmail.com`

Abstract. Automated Machine Learning (AutoML) is an area of research that focuses on developing methods to generate machine learning models automatically. The idea of being able to build machine learning models with very little human intervention represents a great opportunity for the practice of applied machine learning. However, there is very little information on how to design an AutoML system in practice. Most of the research focuses on the problems facing optimization algorithms and leaves out the details of how that would be done in practice. In this paper, we propose a frame of reference for building general AutoML systems. Through a narrative review of the main approaches in the area, our main idea is to distill the fundamental concepts in order to support them in a single design. Finally, we discuss some open problems related to the application of AutoML for future research.

Keywords: Machine learning · Automated machine learning · AutoML · Applied machine learning · Applied AutoML

1 Introduction

In last decade, machine learning has been applied in various fields and used to solve many challenging business problems. This has led to a growing demand for data scientists with solid knowledge and experience to harness massive amounts of data and create business-impacting machine learning solutions [4]. However apply machine learning to business problems is labor-intensive and human experts are scarce and heavily demanded in organizations.

Automated machine learning (AutoML) has become an area of growing interest for machine learning researchers and practitioners. AutoML groups together many techniques and methods that can be used to automate the tasks that constitute the process of applying machine learning. This has led the researchers to propose many literature reviews that try to summarize the area from different perspectives and propose many reusable components to solve the different AutoML challenges.

In this work we take advantage of these perspectives to propose a general recipe that brings them closer to the practice of applied machine learning. Based

A. C. Bicharra Garcia et al. (Eds.): IBERAMIA 2022, LNAI 13788, pp. 243–254, 2022.
https://doi.org/10.1007/978-3-031-22419-5_21

on findings in several AutoML reviews, we describe a design based on learning loops that attempts to provide the flexibility to incorporate main AutoML methods. Furthermore, we propose a new way to approach the goal of AutoML systems from a multi-objective perspective that takes into account time and computational resources.

2   Methodology

The paper aims to gather information on AutoML, especially when it comes from literature reviews. This is because the main objective is to identify the most important parts for the comprehensive application of AutoML systems in practice. To carry out the literature review, the narrative and scoping literature review approaches have been adopted [6,28], and a research search strategy has been developed [5,30]. Figure 1 shows the process flow for the systematic literature review.

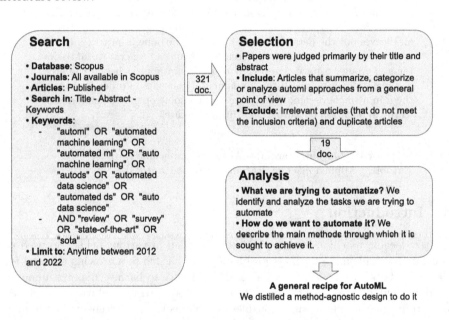

Fig. 1. Process flow for AutoML literature review.

The articles on AutoML were identified from the Scopus database to find the most relevant published articles or in press articles. We search within the title, abstract and key words for various terms such as "automated machine learning", "automl", "automated data science" and "autods". The search is then narrowed to documents that also contain either in the title or the abstract or in the keywords, the terms "review", "survey", "state-of-the-art" or "sota". With this we seek to keep all the articles that summarize various methods in the area. In order to focus on recent literature, the search is limited to articles published in the last decade. The search was carried out on May 11, 2022 and retrieved 321 documents. After manual screening, e.g. removing duplicate or irrelevant articles, 19 articles remained, which form the core of this review.

3   Results Analysis

In this section the results are presented. We focus on answering two main questions. The first question is: What we are trying to automatize? With this question we are trying to identify and analyze the tasks we are trying to automate. The second question is: How do we want to automate it? With this question we are looking for describing the main methods through which it is sought to achieve the automation.

3.1   What We Are Trying to Automatize?

Most of the articles describe AutoML as a process in which tasks that would normally be performed by a data scientist are automated. Table 1 describes an overall process and maps each part of the process to the articles. It is evident that the majority of articles are concentrated around *Model Selection* and *Hyperparameter Optimization*. Moreover, the least explored areas are *Task Formulation* and *Prediction Engineering*.

Table 1. Machine learning process phases identified in the reviewed articles.

Phases	Reviews
Task Formulation	$[7, 31]$
Prediction Engineering	$[7, 31]$
Data Preparation	$[4, 7, 10, 15, 21, 22, 26, 27, 40]$
Feature Engineering	$[2, 4, 7, 10, 15, 21, 22, 24\text{--}27, 31, 32, 36, 40]$
Model Selection	$[2, 4, 7, 9, 10, 15, 21, 22, 24\text{--}27, 31\text{--}34, 36, 37, 40]$
Hyperparameter Opt.	$[2, 4, 7, 9, 10, 15, 21, 22, 24\text{--}27, 31\text{--}34, 36, 37, 40]$
Model Estimation	$[2, 9, 10, 15, 21, 32, 36, 40]$
Results Summarizing	$[4, 7, 31]$

Task Formulation. This is the process through which a machine learning task is formulated that could help solve a business problem. Only two works of those reviewed incorporate this phase within the scope of AutoML. Santu et al. [31], highlights in this phase the interaction between domain experts and data scientists, while De Bie et al. [7] relates it more to an EDA (Exploratory Data Analysis) process. Generally, the output of this task are available data sources, verified hypotheses and the main business metric to impact.

Prediction Engineering. This is the phase in which the business problem is framed as a machine learning problem. This includes deciding between different frameworks, for example, a ranking problem can be solved as a scoring problem (point-wise), a binary classification problem (pair-wise) or a position assignment problem (list-wise). According to Santu et al. [31] This phase also involves constructing and assigning labels to data points according to the goal prediction task. The output of this phase is generally the framing of the problem represented by the data points and targets, and a refinement of the business metric into a proxy metric to be optimized.

Data Preparation. Many jobs incorporate data preparation as part of the tasks that can be automated. The data preparation process consists of performing operations on the defined dataset to make it ready for the next process. Within this process we define two types of data preparation, those that increase the number of data points (e.g. data collection, data augmentation) [4,7,10,15,22,27] and those that do not (e.g. data cleaning, data inputation, data standardization) [4,7,15,21,22,26,27,40]. The output of this phase is typically a curated dataset ready to be used for feature engineering.

Feature Engineering. Feature Engineering task aims to maximize the extraction of features from raw data for use by algorithms and models [15]. In this context, raw data could be structured data, such as tabular and relational datasets [4], or unstructured data, such as text and images. Feature Engineering consists mainly of two sub-task: feature selection and feature transformation (e.g. feature extraction, feature construction). Feature Engineering is one of the most explored task due to its impact on the performance of the model [40] since data and features determine the upper bound of ML, and that models and algorithms can only approximate this limit.

Model Exploration and Hyper-Parameter Tuning. This task is one of the most explored in the literature as it was one of the places where researchers started looking for automation [13]. With the advancement in computing power, a growing wave of machine learning methods and techniques became available to data scientists. Being able to explore different models with different hyper-parameters automatically is something that usually saves machine learning practitioners a lot of time. This problem is generally approached from two sub-problems: The definition of a search space of possible models to be explored and the search method to be used to traverse that space.

Model Estimation. Evaluating models is an expensive process, since it usually requires a series of training and test stages, usually in a cross-validation scenario using all the data. Because of this, researchers have focused over the years on creating methods to estimate the performance of models in less expensive ways [15]. This is usually done in roughly two ways. Either reducing the amount of data needed to evaluate the model (e.g. multi-fidelity approaches [11]) or modeling the performance of the models in a way that allows us to predict it without the need to evaluate it (e.g. surrogate models [8], relative landmarks [14]).

Results Summarizing/Recommendation. The last part of the procedures is to summarize all the findings and recommend the most useful/promising solution to the stakeholders. There is very little information about this task in the literature. Santu et al. [31] consider that the recommendations are made at the model, function or computational overhead level. This part is still mostly done manually without any systematic structure. However, some AutoML tools automatically select the best solution from the target metric, while others allow the data scientist to select an option from a ranking of available options.

3.2   How Do We Want to Automate It?

The general way researchers have found to automate the process is to think of it as a search problem. Every possible decision within the machine learning application process becomes a configuration variable. Thus, the problem is reduced to finding the best configuration among all possible configurations. The main methods to carry out this search according to the review of the literature are described below.

Random Search and Grid Search. Random Search and Grid Search are the most widely used strategies for automatically explore the search space for hyperparameter optimization [3]. Random search consists of exploring the search space randomly. Usually this search is restricted to a fixed number of attempts. Grid Search, consists of exploring the search space as if it were a grid. For this it is necessary to discretize the values of the continuous numerical variables in order to fit them into the grid. Both methods are widely used but do not have any kind of optimization when it comes to exploring the space efficiently.

SMBO/SMAC. Sequential Model-Based Optimization (SMBO) involves tuning a model of the predictive performance at the same time as configurations are explored [19]. It then uses it to make decisions about which configurations are most promising to evaluate. The classical implementation of this model is using Bayesian optimization and a surrogate model based on Gaussian processes. In Sequential Model-based Algorithm Configuration (SMAC) [16] Hutter et al. generalize this model in an attempt to overcome some of its limitations. They do this mainly by using random forests as surrogate models.

Reinforcement Learning. Reinforcement Learning has been widely used as a search method [15]. It mainly consists of a controller model, usually a recurrent neural network (RNN) [2,39]. The controller executes an action at each step to sample a new configuration from the search space and receives an observation of the state together with a reward from the environment to update the controller's sampling strategy. Here environment refers to the application of the configuration to the training procedure to train and evaluate the solution generated by the controller, after which the corresponding predictive performance (such as accuracy) are returned.

Evolution Based Methods. Evolution-based optimization methods follow a process inspired by biological concepts related to evolution [26]. Generally the most used is the one based on genetic programming. In this method, it first creates a random population of possible configurations from the search space. Then each individual (configuration) of the population is evaluated to know its fitness function (predictive performance). Based on this aptitude, the best builds have a higher chance of passing through to the next generation and interbreeding with others. Generally this process is repeated until the performance is not improved or until a certain number of generations is reached.

Bandit-Based Methods. Bandit-based methods consist of dividing the search space and evaluating many options in parallel to then decide how to proceed [9]. The two most popular strategies are Successive Halving [17] and HyperBand [23]. On the one hand, Successive Halving consists of first evaluating all configurations with a small data set. Configurations are then ranked based on their performance and the worst half is eliminated. Finally, the data is doubled and the process is repeated until only one configuration remains. On the other hand, HyperBand uses the same technique, but instead of eliminating less promising configurations, it assigns them a lower chance of being selected in the next iteration.

Adaptive Methods. Adaptive methods are those that aim to adapt the configuration during training. This type of method is commonly used in neural architecture search (NAS) to learn the best network architecture while learning its parameters [15]. For example, self-tuning networks (STN) and population based training (PBT) fall into this category. Furthermore, in deep learning, another widely used method is to adapt the learning rate during the training of a network [38].

Meta-Learning. Meta-learning is learned from prior experience in a systematic, data-driven way [35]. It is a process which can be found in many reviews about AutoML and that aims to improve the process itself from learning obtained after the application in many tasks. It generally consists of two problems. The first is how to represent and collect the prior knowledge, usually through meta-features. The second problem is how to learn from this data to extract and transfer knowledge that guides the process of finding an optimal solution for new tasks. Meta-learning techniques can generally be roughly categorized into three broad groups [9]: learning based on the properties of the task, learning from evaluations of previous models, and learning from already trained models.

4   A General Recipe for AutoML

There are several AutoML methods in literature that could be used to search the best machine learning solution to an specific problem. Most of this methods rely on a feedback loop to explore efficiently the search space. In particular, we identify the necessity of three main loops in which the search of the best machine learning solution can be decomposed (Fig. 2). Each of these learning loops are described below.

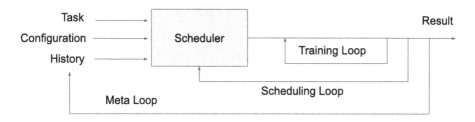

Fig. 2. Context diagram of the main feedback loops in AutoML.

Another important component of AutoML systems is the objective function. The objective function is the function that the system aims to maximize or minimize. We will describe these components in more detail in the following sections.

4.1   Scheduling Loop

AutoML systems rely on the possibility of evaluating a possible good configuration, analyzing the results and being able to decide which is the next best configuration to evaluate. This is the basis for most AutoML methods. We will call the component that is responsible for making the decision the Scheduler. Algorithm 1 shows the pseudo-code for the general operation of the Scheduler. The Scheduler is not only responsible for deciding which are the best configurations to explore but also for defining an evaluation plan to efficiently explore the search space. This evaluation plan is composed of two abstractions, the steps and the stages. A step can be defined as the evaluation of one configuration. A stage can be defined as a set of independent steps that are likely to be parallelized. For example, Random Search and Grid Search only contain steps. SMBO and SMAC only contain 1-step Stages. Reinforcement learning techniques are also commonly 1-Step Stages. Bandit-based and Evolution-based methods, require the evaluation of a set of stages made up of independent steps. In real-world scenarios, the Scheduler may also have to decide on which computational resources to perform the evaluation and for how long (for those methods that may not converge) [12].

Algorithm 1. AutoML Learning Algorithm

Input: Tasks, Configurations, History, Resources
for task in Tasks **do**
    Create an Evaluation Plan consisting of a set of stages,
      with each stage being made up of independent steps.
      with each independent step assigned to a resource
    for stage in Evaluation Plan **do**
      for step in stage **do**
        Evaluate the step
        Add step results to the stage results
      end for
      Update the Evaluation Plan with stage results
    end for
    Summarize results
    Add results to History
end for

4.2   Meta Loop

Another important capability of an AutoML system is to be able to learn from its own experience and thus become more and more efficient in exploration. We describe this learning ability as the Meta-Loop. The Meta-Loop allows us to learn transversely about the problems we are solving. This loop is potentially exploited by meta-learning techniques that learn from other tasks. Something important to enable this learning is to define how the information will be stored, what will be the way in which the tasks will be described (meta features) and how this information will be consumed by the Scheduler. Many authors have studied how to represent tasks for use in a machine learning process. On the one hand, some works have made an effort to identify the best characteristics that describe a data set [29]. On the other hand, others have chosen to create distributed representations [1,18].

4.3   Training Loop

Finally, it is possible see a third loop, the training loop. This loop occurs in training time, and it is the basis for adaptive methods that change configurations in a single trial, like adaptive learning rate [38]. In adaptive learning rate, the value is selected in a dynamic way using information in training time. This attempts to alleviate the task of choosing the best learning rate before training. In practice, the learning information may not return to the Scheduler until the training is complete. This is due to the overhead that can be caused if the

Scheduler and the training are running on different processes or even machines. Because of this, it is very important to consider that the Scheduler will only see, for example, that adaptive learning rate was activated (as a binary configuration) and then see the results of this after training.

4.4   Objective Function

The AutoML problem is generally defined as a combined algorithm selection and hyperparameter search (CASH) problem [20]. In CASH the objective is to find a pipeline ($\mathcal{M}$) and a set of hyperparameters (λ) that minimizes the generalization error (GE) of a particular task ($\mathcal{D}$). Feurer et al. extend this definition to generalize it to many tasks and thus include the idea of meta-learning in the optimization problem [12]. It also proposes incorporating time and computational resources as constraints on how much we are willing to invest (T) as shown in Eq. 1.

$$\mathcal{M}_{\lambda^*} \in \underset{\lambda \in \Lambda}{\operatorname{argmin}} \ \widehat{GE}(\mathcal{M}_\lambda, \mathcal{D}) \quad \text{s.t.} \quad \left(\sum t_{\lambda i}\right) < T \tag{1}$$

where $\mathcal{M}_{\lambda^*}$ denotes the best pipeline configuration, and $t_{\lambda i}$ denotes the time and computational resources used to evaluate the i-th configuration λ_i of a particular pipeline.

This definition is very useful at the experimental level. However, in practice, two solutions can achieve the same or similar predictive performance and consume far fewer resources. That solution would probably be the best. In that case, seeing the budget as a constraint is not useful. Based on the works reviewed, we believe that it is most convenient to model the problem as a multi-objective optimization problem, in which the aim is to minimize the generalization error together with the time and computational resources used.

$$\mathcal{M}_{\lambda^*} \in \underset{\lambda \in \Lambda}{\operatorname{argmin}} \ \left(\widehat{GE}(\mathcal{M}_\lambda, \mathcal{D}) \wedge \left(\sum t_{\lambda i}\right)\right) \quad \text{s.t.} \quad \left(\sum t_{\lambda i}\right) < T \tag{2}$$

Equation 2 tries to synthesize the purpose of the three learning cycles presented in the previous sections. In essence, what we pursue in the general learning process is to be more and more efficient in the search for the best machine learning solutions.

5   Discussion

AutoML is an area that has gained importance in recent years and has led to the appearance of numerous literature reviews. In particular, we took into account only those whose sources are indexed by Scopus, leaving out gray literature that could be important. We believe that this helped us to better define the scope of this work and to define a clear methodology. In addition, we indirectly consider the references of the analyzed works where some references to gray literature were found.

The loops described in this work are important to visualize where learning occurs. On one hand, we believe that the boundaries between these loops are permeable in terms of hyper-parameters. For example, one hyper-parameter could be initialized as a range from knowledge in the meta loop and then refined in the other loops. On the other hand, this boundaries are clearly defined in terms of the execution of the loops. For example, each iteration of a higher level loop might depend on the set of iterations of lower level loops to complete.

Another interesting point to discuss is that when we consider the entire data science process for the application of machine learning, *Task Formulation* and *Prediction Engineering* are two of the most difficult data science task to automate. The main difficulty lies in the fact that these tasks involve a lot of back and forth, where data scientists, domain experts and other stakeholders have to consider multiple possibilities and check whether the required data and business conditions are suitable each time before to make a decision. However, we believe that this general view of AutoML could support some parts of these tasks as long as the decisions made can be coded as configurations.

In addition, this article proposes to address CASH problems from a multi-objective perspective. This brings with it the new challenge of having to define the trade-off between maximizing predictive performance or minimizing time and resource consumption. This may be difficult to determine in practice and further research is required to define suitable criteria.

6   Conclusions

In this paper, we propose a general recipe for AutoML systems in practice generated from the findings of a systematic literature review. In particular, we describe the main tasks in the process of apply machine learning and the main methods used to automate it. After the review, we describe a general design for AutoML systems from the perspective of feedback loops necessary for learning. Additionally, we propose a multi-objective function as the general purpose for AutoML systems in practice that takes into account time and computational resources. Despite the recentness of the AutoML area, we hope this work would be helpful for research scholars and practitioners of machine learning, to understand and integrate the latest research efforts related to AutoML into your own systems.

References

1. Achille, A., et al.: Task2vec: Task embedding for meta-learning. In: Proceedings of the IEEE/CVF International Conference on Computer Vision, pp. 6430–6439 (2019)
2. Baymurzina, D., Golikov, E., Burtsev, M.: A review of neural architecture search. Neurocomputing **474**, 82–93 (2022)
3. Bergstra, J., Bengio, Y.: Random search for hyper-parameter optimization. Journal Mach. Learn. Res. **13**(2) (2012)
4. Bouneffouf, D., et al.: Survey on automated end-to-end data science? In: Proceedings of the International Joint Conference on Neural Networks (2020). www.scopus.com, cited By :2

5. Creswell, J.W., Creswell, J.D.: Research design: Qualitative, quantitative, and mixed methods approaches. Sage publications (2017)
6. Cronin, P., Ryan, F., Coughlan, M.: Undertaking a literature review: a step-by-step approach. Br. J. Nursing **17**(1), 38–43 (2008)
7. De Bie, T., De Raedt, L., Hernández-Orallo, J., Hoos, H.H., Smyth, P., Williams, C.K.: Automating data science: Prospects and challenges. arXiv preprint arXiv:2105.05699 (2021)
8. Eggensperger, K., Hutter, F., Hoos, H.H., Leyton-Brown, K.: Surrogate benchmarks for hyperparameter optimization. In: MetaSel@ ECAI, pp. 24–31 (2014)
9. Elshawi, R., Sakr, S.: Automated machine learning: techniques and frameworks. In: Kutsche, R.-D., Zimányi, E. (eds.) eBISS 2019. LNBIP, vol. 390, pp. 40–69. Springer, Cham (2020). https://doi.org/10.1007/978-3-030-61627-4_3
10. Escalante, H.J.: Automated Machine Learning-A Brief Review at the End of the Early Years. Natural Computing Series (2021). www.scopus.com
11. Fernández-Godino, M.G., Park, C., Kim, N.H., Haftka, R.T.: Review of multi-fidelity models. arXiv preprint arXiv:1609.07196 (2016)
12. Feurer, M., Eggensperger, K., Falkner, S., Lindauer, M., Hutter, F.: Auto-sklearn 2.0: The next generation. arXiv preprint arXiv:2007.04074 24 (2020)
13. Feurer, M., Hutter, F.: Hyperparameter optimization. In: Automated machine learning, pp. 3–33. Springer, Cham (2019), https://doi.org/10.1007/978-1-4899-7687-1_100200
14. Fusi, N., Sheth, R., Elibol, M.: Probabilistic matrix factorization for automated machine learning. In: Advances in Neural Information Processing Systems 31 (2018)
15. He, X., Zhao, K., Chu, X.: Automl: A survey of the state-of-the-art. Knowledge-Based Systems 212 (2021). www.scopus.com, cited By :155
16. Hutter, F., Hoos, H.H., Leyton-Brown, K.: Sequential model-based optimization for general algorithm configuration. In: Coello, C.A.C. (ed.) LION 2011. LNCS, vol. 6683, pp. 507–523. Springer, Heidelberg (2011). https://doi.org/10.1007/978-3-642-25566-3_40
17. Jamieson, K., Talwalkar, A.: Non-stochastic best arm identification and hyperparameter optimization. In: Artificial Intelligence and Statistics, pp. 240–248. PMLR (2016)
18. Jomaa, H.S., Schmidt-Thieme, L., Grabocka, J.: Dataset2vec: Learning dataset meta-features. Data Min. Knowl. Disc. **35**(3), 964–985 (2021)
19. Jones, D.R., Schonlau, M., Welch, W.J.: Efficient global optimization of expensive black-box functions. J. Global Optim. **13**(4), 455–492 (1998)
20. Kotthoff, L., Thornton, C., Hoos, H.H., Hutter, F., Leyton-Brown, K.: Auto-WEKA: automatic model selection and hyperparameter optimization in WEKA. In: Hutter, F., Kotthoff, L., Vanschoren, J. (eds.) Automated Machine Learning. TSSCML, pp. 81–95. Springer, Cham (2019). https://doi.org/10.1007/978-3-030-05318-5_4
21. Kulbach, C., Philipp, P., Thoma, S.: Personalized automated machine learning. Frontiers in Artificial Intelligence and Applications, vol. 325 (2020). www.scopus.com
22. Lakshmi Patibandla, R.S.M., Srinivas, V.S., Mohanty, S.N., Ranjan Pattanaik, C.: Automatic machine learning: An exploratory review. In: 2021 9th International Conference on Reliability, Infocom Technologies and Optimization (Trends and Future Directions), ICRITO 2021 (2021). www.scopus.com

23. Li, L., Jamieson, K., DeSalvo, G., Rostamizadeh, A., Talwalkar, A.: Hyperband: A novel bandit-based approach to hyperparameter optimization. J. Mach. Learn. Res. **18**(1), 6765–6816 (2017)
24. Li, Y., Wang, Z., Ding, B., Zhang, C.: Automl: A perspective where industry meets academy. In: Proceedings of the ACM SIGKDD International Conference on Knowledge Discovery and Data Mining. pp. 4048–4049 (2021). www.scopus.com
25. Li, Y., Wang, Z., Xie, Y., Ding, B., Zeng, K., Zhang, C.: Automl: From methodology to application. In: International Conference on Information and Knowledge Management, Proceedings. pp. 4853–4856 (2021). www.scopus.com, cited By :1
26. Nagarajah, T., Poravi, G.: An extensive checklist for building automl systems. In: CEUR Workshop Proceedings, vol. 2360 (2019). www.scopus.com
27. Nagarajah, T., Poravi, G.: A review on automated machine learning (automl) systems. In: 2019 IEEE 5th International Conference for Convergence in Technology, I2CT 2019 (2019). www.scopus.com, cited By :10
28. Paré, G., Trudel, M.C., Jaana, M., Kitsiou, S.: Synthesizing information systems knowledge: A typology of literature reviews. Inf. Manag. **52**(2), 183–199 (2015)
29. Rivolli, A., Garcia, L.P., Soares, C., Vanschoren, J., de Carvalho, A.C.: Characterizing classification datasets: a study of meta-features for meta-learning. arXiv preprint arXiv:1808.10406 (2018)
30. Robson, C.: Real world research: A resource for social scientists and practitioner-researchers. Wiley-Blackwell (2002)
31. Santu, S.K.K., Hassan, M.M., Smith, M.J., Xu, L., Zhai, C., Veeramachaneni, K.: Automl to date and beyond: Challenges and opportunities. ACM Comput. Surv. **54**(8) (2022). www.scopus.com, cited By :2
32. Tuggener, L., et al.: Automated machine learning in practice: State of the art and recent results. In: Proceedings - 6th Swiss Conference on Data Science, SDS 2019, pp. 31–36 (2019). www.scopus.com, cited By :18
33. Vaccaro, L., Sansonetti, G., Micarelli, A.: Automated machine learning: prospects and challenges. In: Gervasi, O., et al. (eds.) ICCSA 2020. LNCS, vol. 12252, pp. 119–134. Springer, Cham (2020). https://doi.org/10.1007/978-3-030-58811-3_9
34. Vaccaro, L., Sansonetti, G., Micarelli, A.: An empirical review of automated machine learning. Computers **10**(1), 1–27 (2021). www.scopus.com, cited By :7
35. Vanschoren, J.: Meta-learning. In: Automated Machine Learning, pp. 35–61. Springer, Cham (2019). https://doi.org/10.1007/978-1-4419-9863-7_613
36. Waring, J., Lindvall, C., Umeton, R.: Automated machine learning: Review of the state-of-the-art and opportunities for healthcare. Artifi. Intell. Med. **104** (2020). www.scopus.com, cited By :125
37. Weng, Z.: From conventional machine learning to automl. J. Phy. Conf. Ser. **1207** (2019). www.scopus.com, cited By :9
38. Zeiler, M.D.: Adadelta: an adaptive learning rate method. arXiv preprint arXiv:1212.5701 (2012)
39. Zoph, B., Vasudevan, V., Shlens, J., Le, Q.V.: Learning transferable architectures for scalable image recognition. In: Proceedings of the IEEE Conference On Computer Vision And Pattern Recognition. pp. 8697–8710 (2018)
40. Zöller, M., Huber, M.F.: Benchmark and survey of automated machine learning frameworks. J. Artifi. Intell. Res. **70**, 409–472 (2021). www.scopus.com, cited By :30

Semi-supervised Hierarchical Classification Based on Local Information

Jonathan Serrano-Pérez$^{(\boxtimes)}$ and L. Enrique Sucar

Instituto Nacional de Astrofísica, Óptica y Electrónica, Puebla, Mexico
{js.perez,esucar}@inaoep.mx

Abstract. In this work, a semi-supervised hierarchical classifier based on local information (SSHC-BLI) is proposed. SSHC-BLI is a semi-supervised learning algorithm that can be applied to hierarchical classification, that is, it can handle labeled and unlabeled data in scenarios where the labels are arranged in a hierarchical structure. SSHC-BLI tries to pseudo-label each unlabeled instance using information of its nearest labeled instances. It uses a similarity function to determine whether the unlabeled instance is similar to its nearest labeled instances to assign it a label; if it is not, then it continues unlabeled. A heuristic similarity function of an instance with a set of instances was proposed to determine similitude. The method was tested in several datasets from functional genomics and compared against a hierarchical supervised classifier and two state of the art methods, showing in most cases superior performance, with statistical significance in *accuracy* and *hierarchical F-measure*.

Keywords: Hierarchical classification · Semi-supervised learning · Local distance

1 Introduction

A common problem in supervised classification is the scarcity of labeled data. This may be because hand-labeling data is time consuming and costly or just hard to label [4]. Hence, training a classifier with few labeled data could produce an unreliable classifier. However, large amounts of unlabeled information can be obtained from different sources, such as the Internet. Moreover, unlabeled information could be required in scenarios where instances can be associated to multiple labels [5], like hierarchical classification [11], which makes more challenging making use of unlabeled data. Strategies that take advantage of that information are required.

A suitable Semi Supervised Hierarchical Classification algorithm, that considers the hierarchy, trained with labeled and unlabeled data, can produce a hierarchical classifier with better performance than using only the few labeled data. The proposed method, *semi-supervised hierarchical classification based on local information* (SSHC-BLI), can handle tree hierarchies and predicts a single path of labels. Its main idea is to pseudo-label the unlabeled data, which are later used to train a hierarchical classifier. The nearest labeled instances to each

A. C. Bicharra Garcia et al. (Eds.): IBERAMIA 2022, LNAI 13788, pp. 255–266, 2022.
https://doi.org/10.1007/978-3-031-22419-5_22

unlabeled instances are used to build a pseudo-label, then, the similitude of the unlabeled instance with its nearest labeled instances is estimated, if they are similar, the unlabeled instance is pseudo-labeled, else it is kept unlabeled. A heuristic similitude function, *similarity of an instance with a set of instances* (SISI), is proposed to measure this similitude. Experiments on a subset of the Functional Catalogue (FunCat) datasets show promising results; SSHC-BLI outperforms the baseline, a supervised hierarchical classifier (Top-Down) trained only on the labeled instances; as well as two other state-of-the art semi-supervised hierarchical classifiers.

The main contributions of this manuscript are: (i) a semi-supervised hierarchical classifier that can handle hierarchies of tree type and predicts a single path of labels that always reaches a leaf node, which makes use of local information to pseudo-label the unlabeled instances and use them to train a classifier; and (ii) a heuristic function, *similarity of an instance with a set of instances (SISI)*, which, as it name says, indicates if an instance is similar or not to a set of instances.

The document is organized as follow. Section 2 summarizes fundamentals of hierarchical classification and semi-supervised learning. Section 3 reviews related work. Section 4 presents the proposed method. Section 5 presents the experiments and results. Finally, in Sect. 6, conclusions and some ideas for future work are given.

2   Fundamentals

Hierarchical Classification. The hierarchy or *hierarchical structure* (HS) is denoted with the graph notation: $HS = (L, E)$, where L is the set of labels (nodes), E is the set of edges that link the labels (nodes), and HS is a DAG.

Furthermore, the subset of labels for an instance has to fulfill the *hierarchical constraint*. The hierarchical constraint states that if an instance x is associated to the label $l \in L$ then x has to be associated to the ancestors of l, $Anc(l)$, given by the HS:

$$\forall x \in l \rightarrow x \in z, \forall z \in Anc(l) \tag{1}$$

Therefore, a *valid* or *consistent path* is a subset of the labels that complies the hierarchical constraint.

Silla and Freitas [11] described the hierachical problems as a 3-tuple (Υ, Ψ, Φ), where Υ specifies the hierarchical structure type, T if it is a tree or DAG if it is a Direct Acyclic Graph; Ψ specifies whether an instance can be associated to a single or multiple paths of labels (SPL, MPL); and Φ indicates the depth of the paths of the instances, Full Depth (FD) or Partial Depth (PD).

Semi-Supervised Learning. (SSL) can be seen as the branch of machine learning that combines supervised and unsupervised learning [2,15]. Semi-supervised classification methods are appropriate to scenarios where labeled data is scarce, and a reliable classifier could be hard to obtain.

There are some assumptions that are the foundation of most semi-supervised learning algorithms, which depend on one or more of them being satisfied [2]. They are briefly described next: *Smoothness assumption*: It states that, for two input points $x_i, x_j \in X$ that are close by in the input space, the corresponding labels y_i, y_j should be the same. This assumption can be applied transitively to unlabeled data. *Low-density assumption*: It states that the decision boundary of a classifier should preferably pass through low-density regions in the input space. *Manifold assumption*: It states that the input space is composed of multiple lower dimension manifolds on which all data points lie, therefore, data points on the same manifold must have the same label.

Van Engelen and Hoos [4] proposed a taxonomy for the SSL methods which are divided into two main groups, inductive and transductive, the former produce a classification model, while the second only label the unlabeled data points.

3   Related Work

Metz and Freitas [5] proposed the first method for semi-supervised hierarchical classification. It is essentially a Top-Down classifier that can handle hierarchies of tree type and predicts a single path of labels that always reaches a leaf node. It trains *decision trees* as binary classifier for each node, then each classifier is self-trained following one of three strategies. Their results are not superior in a statistically significant way against a supervised hierarchical classifier.

Santos and Canuto [7] proposed Hierarchical Multi-label classification using Semi-Supervised Label Powerset (HMC-SSLP). First, a HMC-Label Powerset (HMC-LP) [1] is trained with labeled data, then it is used to pseudo-label a predefined proportion of the unlabeled data that will be added to the training set. HMC-LP [1] combines all the classes of an example to generate a new hierarchy, however, examples of how to combine paths of different lengths are not shown. On the other hand, the positives instances for each new class could be too few, which can result on unreliable classifiers.

Hierarchical Multi-label using Semi-Supervised Random k-Labelsets (HMC-SSRAkEL) is the semi-supervised version of HMC-RAkEL [7]. This method trains a RAkEL classifier [12] as local classifier per parent node. Then, a Top-Down procedure is used to pseudo-label the predefined proportion of unlabeled data that will be added to the training set. Santos and Canuto [8] also proposed Hierarchical Multi-label Classification using Semi-Supervised Binary Relevance (HMC-SSBR), that trains support vector machines as a local classifier per node which are self-trained [9]. The Top-Down procedure for labeling instances in HMC-SSRAkEL and HMC-SSBR can predict multiple labels in the *same* level. On the other hand, the methods HMC-SSRAkEL, HMC-SSBR and HCM-SSLP [7] pseudo-label a predefined proportion of unlabeled data in each iteration, which are added to the training set, that is, these methods do not select the instances with the most confident predictions but add all of them.

Later, Xiao et al. [14] proposed the Path Cost-Sentive Algorithm with Expectation Maximization (PCEM) which consists on the following steps. First, the

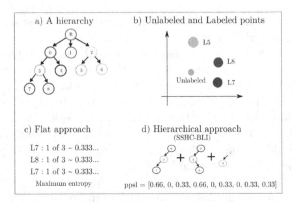

Fig. 1. Example of pseudo-labeling an instance based on the smoothness assumption. The labeled points in b) are associated to the path that ends in the corresponding label of the hierarchy in a). In a flat approach, as the 3 instances are different, the instance will stay unlabeled. On the other hand, in the hierarchical approach, the internal nodes {0,3} got the highest score, so the instance could be pseudo-labeled. (Best seen in color).

base classifier PCNB [14] is trained with the labeled data; then it pseudo-labels all the unlabeled instances and trains the PCNB with labeled and pseudo-labeled instances; this is iterated until convergence. PCEM was proposed for hierarchical text classification, so it is no straightforward to apply it in non-text domains.

The proposed method pseudo-labels instances making use of a labeling strategy which considers the local information of the unlabeled data, and selects those with the most confident predictions for training in the next iterations; that is, it does not pseudo-label a predefined amount of data in each iteration. Furthermore, pseudo-labels may be updated or removed, in order to correct *wrong* pseudo-labels. Additionally, the method can be applied to multiple domains.

4   SSHC Based on Local Information

Making use of the flat approach (only considering the leaf nodes and ignoring the rest of the hierarchy) to pseudo-label the unlabeled instances is not the best way, because unlabeled instances that may be partially pseudo-labeled will stay unlabeled. An example is shown in Fig. 1c, as the 3 instances are different, the unlabeled point will stay unlabeled. On the other hand, making use of an hierarchical approach a partial path, that fulfills the hierarchical constraint could be obtained. For instance, in Fig 1d, the number of times a label is seen is averaged for each label, in this way, the internal nodes 0 and 3 got the highest score, hence, the unlabeled instance could be pseudo-labeled with those labels.

The proposed method, SSHC-BLI, is mainly based on the smoothness assumption, that is, neighboring instances must have the same or similar paths of labels. SSHC-BLI begins building pseudo-labels for each unlabeled instance

---

Algorithm 1. SSHC-BLI algorithm

Require: (X, Y): labeled data, U: unlabeled data, k: number of nearest neighbors, THR: similitude threshold, $t2label$: threshold to pseudo-label an instance, HS hierarchy, $maxIterations$: maximum number of iterations.

Ensure: f_{sshc}: semi supervised hierarchical classifier

1:	$T \leftarrow 1$	▷ Iteration
2:	$LD \leftarrow X$	▷ LD: Labeled data
3:	$CL \leftarrow L$	▷ Labels of labeled data
4:	**while** $True$ **do**	
5:	**for each** $u_j \in U$ **do**	
6:	$IND_j \leftarrow getKNN(k, u_j, LD)$	▷ Get k-nearest neighbors
7:	$PSL_j \leftarrow getPseudoLabel(IND_j, LD, t2label)$	▷ Pseudo label for u_j
8:	**end for**	
9:	**for each** $u_j \in U$ with valid PSL_j **do**	
10:	**if** $similitude(u_j, IND_j) < THR$ **then**	
11:	$PSL_j = \emptyset$	▷ Invalid pseudo-label
12:	**end if**	
13:	**end for**	
14:	**if** $(T > maxIterations)$ or $(PSL^T == PSL^{T-1})$ **then**	
15:	**break** loop (while)	
16:	**else**	▷ join labeled data with valid pseudo-labeled data
17:	$CL \leftarrow Y \cup valid(PSL)$	
18:	$LD \leftarrow X \cup U[valid(PSL)]$	
19:	**end if**	
20:	$T \leftarrow T + 1$	
21:	**end while**	
22:	$f_{SSHC} \leftarrow trainHC(LD, CL, HS)$	▷ Train a hierarchical classifier

---

using its neighboring labeled instances, but if the unlabeled instance is not similar to its labeled neighbors, it stays unlabeled; this process iterates until all the pseudo-labels do not change.

Algorithm 1 shows the general steps. It is an iterative method where the unlabeled data is pseudo-labeled using its nearest labeled neighbors (lines 6–7), details of how pseudo-labels are built (line 7) are shown in Subsect. 4.1. Also, the similitude of the unlabeled point with its neighbors (line 10) is considered, if they are not *similar* the unlabeled point looses its pseudo-label, details of how the similitude is estimated are shown in Subsect. 4.2. The method finishes when the pseudo-labels for the unlabeled data do not change from an iteration to the next, or when the maximum number of iterations is reached.

There are three variants of the SSHC-BLI algorithm:

- **Variant 1:** It corresponds to the version described in Algorithm 1.
- **Variant 2:** In each iteration the pseudo-labels for the unlabeled set are re-estimated, so after the first iteration an instance that was added to the training set, it will have itself as one of the k-nearest neighbors. This results in a *biased* pseudo-label, because the instance is contributing to itself. In order

to avoid this, the function *getKNN* (line 6) is modified so that it guarantees that none of nearest neighbors is the instance itself.

- **Variant 3:** Considering that the number of instances in the training set could increase in each iteration, it may be interesting to increase the number of nearest neighbors. In this way, variant 3 increases the value of k after a predefined number of iterations.

4.1 Pseudo-Label an Instance

To pseudo-label an instance, labeled instances *close* to it are required. Let $Y = [y_1, ..., y_k]$ be the labels of k instances close to the unlabeled instance x, where $y_i \in \{0, 1\}^{|L|}$, that is, $y_{i,j}$ is 1 if the i-th instance is associated to the j-th label, 0 otherwise. So the probabilities for each individual label can be estimated in the following way:

$$ppsl_j = \left\{ \frac{\sum_{i=1}^{k} y_{i,j}}{k} \right., \forall j \in \{1, ..., |L|\} \tag{2}$$

A threshold, *t2label*, is used to determine if an instance is associated to the label:

$$psl_j = \begin{cases} 1 : ppsl_j >= t2label \\ 0 : ppsl_j < t2label \end{cases}, \forall j \in \{1, ..., |L|\} \tag{3}$$

$$0 <= t2label <= 1 \tag{4}$$

Finally, *psl* is the pseudo label for x, but if *psl* is *zero* for all labels, then it is an invalid path, therefore, x remains unlabeled.

4.2 SISI: Similarity of an Instance with a Set of Instances

A *similarity* function to estimate how similar is an instance to a small set of instances is required, and the result should be in the interval $[0, 1]$; that is, 1 if the instance is similar to the set of instances, and 0 in the opposite case. Nevertheless, it is not known[1] a *similarity* function that complies the previous requisites to the best of our knowledge.

Therefore, the heuristic function *Similarity of an Instance with a Set of Instances* (**SISI**) is proposed. This is a *local* measure, because it does not consider the complete data distribution, but just the instance of interest, p, and a set of instances, A. SISI takes into account the distances among the set of instances A, *lavg*; and the distances of an instance p with the instances of set A, *uavg*:

$$lavg = \frac{\sum_{i=1}^{k} \sum_{j=i+1}^{k} d(x_i, x_j)}{\frac{k(k-1)}{2}} \tag{5}$$

$$uavg = \frac{\sum_{i=1}^{k} d(p, x_i)}{k} \tag{6}$$

---

[1] The Mahalanobis distance estimates the distance between a point and a distribution. Furthermore, this measure does not comply with the interval result, $[0, 1]$.

where $x_i \in A$, k is the length of A and $d(A, B)$ is any metric. Additionally, two assumptions are made:

Assumption 1: It is assumed that if $uavg$ is equal or lower than $lavg$, the instance p is similar to the set of instances A with score 1.

Assumption 2: It is assumed that if $uavg$ is greater than n times $lavg$, with $n > 1$, the instance p is not similar to the set of instances A, that is, score 0.

Hence, from Assumptions 1 and 2, the equation of the line that passes through points $(lavg, 1)$ and $(n * lavg, 0)$ is defined as:

$$score = \frac{lavg - uavg}{(n-1)lavg} + 1 \tag{7}$$

That is, Eq. 7 scores the similitude of the point p with the set of instances A in interval $(lavg, n * lavg)$. Finally, from assumptions 1,2 and Eq. 7, function $SISI$ is defined in Eq. 8.

$$SISI(p, A) = \begin{cases} 1 & uavg <= lavg \\ 0 & uavg >= n * lavg \\ \frac{lavg - uavg}{(n-1)lavg} + 1 & otherwise \end{cases} \tag{8}$$

SISI requires a parameter, n, that was set to 3, and the *Euclidean distance* was used as the metric; the experiments were made with this configuration. However, the method is not limited to the Euclidean distance; it can be replaced for other metric according to the problem of interest.

5   Experiments and Results

The experiments are focused on showing that using unlabeled data may help to improve the performance of a hierarchical classifier trained only on labeled data.

5.1   Evaluation Measures

In order to evaluate the performance of the compared methods, two common evaluation measures in hierarchical classification are used. Let N be the number of instances in the test set, let Y be the real subset of labels to which an instance is associated and let $\widehat{Y}$ be the subset of predicted labels. The evaluation measures [6,11] are accuracy, Eq. 9; and hierarchical F-measure (hF), Eq. 10, that is composed by hierarchical precision (hP) and hierarchical recall (hR).

$$Accuracy = \frac{1}{N} \sum_{i=1}^{N} \frac{\left|Y_i \cap \widehat{Y}_i\right|}{\left|Y_i \cup \widehat{Y}_i\right|} \tag{9}$$

$$hF = \frac{2 * hP * hR}{hP + hR} \; ; \; hP = \frac{\sum_{i=1}^{N} \left|Y_i \cap \widehat{Y}_i\right|}{\sum_{i=1}^{N} \left|\widehat{Y}_i\right|} \; ; \; hR = \frac{\sum_{i=1}^{N} \left|Y_i \cap \widehat{Y}_i\right|}{\sum_{i=1}^{N} |Y_i|} \tag{10}$$

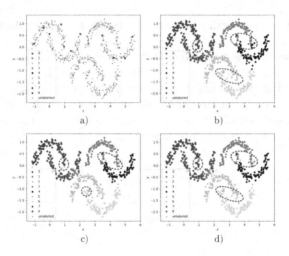

Fig. 2. Pseudo-labels for the artificial dataset, labeled points are shown in the color of the deepest node of their path and unlabeled instances are shown in gray. a) Initial labeled and unlabeled data. Pseudo labels of variants 1-b), 2-c) and 3-d). (Best seen in color.) (Color figure online)

5.2 Artificial Dataset

To illustrate the differences of the three variants of the SSHC-BLI, they were applied to a simple artificial dataset. The parameters for the three variants are: nearest neighbors, $k = 3$; similitude threshold, $THR = 0.5$; threshold to positively label an instance, $t2label = 0.5$; for variant 3 k increase each 5 iterations. Finally, a Top-Down (TD) classifier is trained with labeled and pseudo-labeled data, the TD trains support vector machines (regularization parameter: 1000, kernel: rbf, gamma=$1/n\_features$) as local classifier per node, and the *less inclusive* policy is used to select positive and negative instances in each node.

Figure 2 shows an example of how the variants pseudo-label the unlabeled data. Variant 1 pseudo-labeled the whole unlabeled data, most of wrongly pseudo-labeled instances are found at the ends of the half moons (circled in blue) as can be seen in Fig. 2 b). Variant 2 got better results at the ends of the half moons (circled in blue), see Fig. 2 c). However, there are unlabeled points that are surrounded by points with the same labels, so its natural to think that they should have the same set of labels, but they were no pseudo-labeled because the estimation of similitude with its nearest neighbors. Finally, variant 3 seems to smooth the results obtained by the 2nd variant, see Fig. 2 d).

Results of the SSHC-BLI variants trained with the labeled and pseudo-labeled data are shown in Table 1. Column TD corresponds to a hierarchical classifier trained only on labeled data. Making use of the unlabeled data helped to obtain a better performance than training only on the labeled data.

Table 1. Results of SSHC-BLI variants (1, 2 and 3) and the supervised classifier, Top-Down (TD), for the artificial dataset. In bold the best score.

	TD	V1	V2	V3
Accuracy	0.4662	0.8928	**0.8951**	0.8930
h Recall	0.7692	**0.9492**	0.9477	**0.9492**
h Precision	0.7082	0.9551	0.9595	**0.9596**
hF	0.7375	0.9522	0.9536	**0.9544**

5.3   Real World Datasets

In order to evaluate the performance of the methods, five challenging hierarchical data sets from the field of functional genomics (FunCat) [13] were used. The preprocessing applied to the datasets is the same than Serrano-Pérez et al. [10] (Tree, SPL, FD); that is, the hierarchies of the datasets are trees and the instances are associated to a single path of labels which always reach a leaf node. Description of datasets is shown in Table 2. A stratified 5-fold cross validation was carried out, 80% for training and 20% for test. Additionally, the training set was stratified split in labeled and unlabeled sets in the following way:

- Labeled: $\{10, 30, 50, 70, 90\}\%$.
- Unlabeled: $\{90, 70, 50, 30, 10\}\%$, complement with respect to labeled.

The training set was divided 3 times, so the results are the average of 15 (3×5) executions. The parameters for the variants (V1, V2, V3) are: nearest neighbors, $k = 3$; similitude threshold, $THR = \{0.7, 0.5, 0.3\}$; threshold to positively label an instance, $t2label = 0.5$; maximum number of iterations, 45; variant 3 increase k each 10 iterations. A TD classifier is trained with labeled and pseudo-labeled data; the TD trains random forest classifiers (number of trees: 100, criterion: *gini*, bootstrap: *True*) as local classifiers per node, and the *balanced bottom-up* policy is used to select positive and negative instances in each node. For each variant, the configuration that results in the highest average rank is selected: variant 1, $THR = 0.5$; variant 2, $THR = 0.3$; and variant 3 $THR = 0.7$.

The proposed method is compared against two related methods. First, HMC-SSBR [8] that uses support vector machines (regularization parameter: 1, kernel:

Table 2. Description of the FunCat datasets. MD: Maximum Depth

Dataset	Instances	Attr.	Classes	MD
cellcycle_FUN	2339	77	36	4
derisi_FUN	2381	63	37	4
eisen_FUN	1681	79	25	3
gasch1_FUN	2346	173	36	4
gasch2_FUN	2356	52	36	4

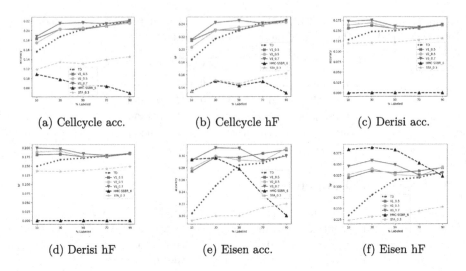

(a) Cellcycle acc.   (b) Cellcycle hF   (c) Derisi acc.

(d) Derisi hF   (e) Eisen acc.   (f) Eisen hF

Fig. 3. Results of FunCat datasets. Cellcycle: 3a and 3b. Derisi 3c and 3c. Eisen: 3e and 3f. (Best seen in color.)

Table 3. Average rank of each classifier in the FunCat datasets. In bold the best (lower is better).

	TD	V1	V2	V3	HMC-SSBR	STA
Accuracy	3.72	2.52	2.52	**1.76**	5.12	5.36
hF	3.84	2.8	2.72	**2.0**	4.0	5.64

rbf, gamma=$1/n\_features$) as local classifier per node; the method was iterated $\{2,3,6\}$ times, the configuration with the highest average rank was selected: 6 iterations. Second, the method self-training A (STA) [5], that uses decision tree classifiers (criterion: $gini$, minimum number of intances per leaf: 1, pruning: $None$) as local classifier per node; three thresholds were used $\alpha = \{0.3, 0.5, 0.7\}$, the configuration with the highest average rank was selected: $\alpha = 0.3$. Additionally, a supervised hierarchical classifier, TD, was trained only on the labeled data, with the same parameters than the used by the SSHC-BLI.

The results in terms of accuracy and hF are shown in Figs. 3 and 4. Each graph compares the three variants of SSHC-BLI with the related methods and the baseline, TD, on different percentages of labeled-unlabeled data. Table 3 summarizes the results in terms of the average rank of each method.

5.4   Statistical Analysis and Discussion

To estimate if there is statistical difference among the variants of the proposed method and the related methods (HMC-SSBR, STA, and TD), the Wilcoxon test was used [3]. Given n independent samples (x_i, y_i) from paired samples,

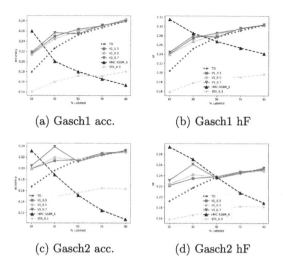

(a) Gasch1 acc.   (b) Gasch1 hF

(c) Gasch2 acc.   (d) Gasch2 hF

Fig. 4. Results of FunCat datasets. Gasch1 4a and 4b. Gasch2 4c and 4d. (Best seen in color.)

the differences $d_i = x_i - y_i$ are computed, where x_i and y_i are the proposed and related methods scores, respectively. The null hypothesis (one-side) states that the median of the differences is negative against the alternative that it is positive. 25 experiments, 5 datasets with 5 divisions each, were considered, with $\alpha = 0.05$. For each SSHC-BLI variant against each related method: the null hypothesis can be rejected in favor of the alternative, that is, the median is greater than zero (positive) for both evaluation measures, accuracy and hF. In other words, the different SSHC-BLI variants got better performance than all the related methods (HMC-SSBR, STA, and TD) with statistical significance.

6   Conclusions and Future Work

In this manuscript was proposed a semi-supervised hierarchical classifier based on local information. The method is based in the *smoothness assumption*, so it finds the nearest labeled instances to each unlabeled instance in order to generate a pseudo-label. Also, a heuristic similarity function, *similarity of an instance with a set of instances (SISI)*, was proposed, which is used to evaluate whether an unlabeled instance is similar to its labeled neighbors.

The three variants of method were evaluated on a subset of the Functional Catalogue datasets and the results were compared against different related methods. The proposed method outperforms the rest of the classifiers, mainly when there is only few labeled data. Additionally, the Wilcoxon test showed that the results of the proposed approach are significantly better than those of the related methods. As future work, the proposed method will be extended to handle any hierarchy of directed acyclic graph type, and to predict multiple paths of labels.

Acknowledgements. This work was sponsored in part by CONACYT, project A1-S-43346. J. Serrano-Pérez acknowledges the support from CONACYT scholarship number (CVU) 84075.

References

1. Cerri, R., de Carvalho, A.F.A.: Comparing local and global hierarchical multilabel classification methods using decision trees, January 2009
2. Chapelle, O., Schlkopf, B., Zien, A.: Semi-Supervised Learning1st edn. The MIT Press, Cambridge (2010)
3. Demšar, J.: Statistical comparisons of classifiers over multiple data sets. J. Mach. Learn. Res. **7**, 1–30 (2006)
4. van Engelen, J.E., Hoos, H.H.: A survey on semi-supervised learning. Mach. Learn. **109**(2), 373–440 (2019). https://doi.org/10.1007/s10994-019-05855-6
5. Metz, J., Freitas, A.A.: Extending hierarchical classification with semi-supervised learning. In: Proceedings of the UK Workshop on Computational Intelligence, pp. 1–6 (2009)
6. Nakano, F.K., Pinto, W.J., Pappa, G.L., Cerri, R.: Top-down strategies for hierarchical classification of transposable elements with neural networks. In: 2017 International Joint Conference on Neural Networks (IJCNN), pp. 2539–2546, May 2017
7. Santos, A., Canuto, A.: Applying semi-supervised learning in hierarchical multi-label classification. Expert Syst. Appl. **41**(14), 6075–6085 (2014)
8. Santos, A., Canuto, A.: Applying the self-training semi-supervised learning in hierarchical multi-label methods. In: 2014 International Joint Conference on Neural Networks (IJCNN), pp. 872–879 (2014)
9. Santos, A.M., Canuto, A.M.P.: Using semi-supervised learning in multi-label classification problems. In: The 2012 International Joint Conference on Neural Networks (IJCNN), pp. 1–8 (2012)
10. Serrano-Pérez, J., Sucar, L.E.: Hierarchical classification with Bayesian networks and chained classifiers. In: Proceedings of the Thirty-Second International Florida Artificial Intelligence Research Society Conference, Sarasota, Florida, USA, 19–22 May 2019, pp. 488–493 (2019)
11. Silla, C.N., Freitas, A.A.: A survey of hierarchical classification across different application domains. Data Min. Knowl. Disc. **22**(1), 31–72 (2011)
12. Tsoumakas, G., Katakis, I., Vlahavas, I.: Random k-labelsets for multilabel classification. IEEE Trans. Knowl. Data Eng. **23**(7), 1079–1089 (2011)
13. Vens, C., Struyf, J., Schietgat, L., Džeroski, S., Blockeel, H.: Decision trees for hierarchical multi-label classification. Mach. Learn. **73**(2), 185 (2008)
14. Xiao, H., Liu, X., Song, Y.: Efficient path prediction for semi-supervised and weakly supervised hierarchical text classification. In: The World Wide Web Conference, pp. 3370–3376. WWW 2019, Association for Computing Machinery, New York (2019)
15. Zhu, X.: Semi-supervised learning literature survey. Technical report, University of Wisconsin-Madison, July 2008

The Impact of Allostatic Load on Machine Learning Models

William da Rosa Fröhlich[1]($\boxtimes$), Sandro José Rigo[1], Marta Rosecler Bez[2], Daiane Rocha de Oliveira[3], and Murilo Ricardo Zibetti[3]

[1] Applied Computing Graduate Program, UNISINOS University,
São Leopoldo, Brazil
william_r_f@hotmail.com
[2] Creative Industry Graduate Program, Feevale University, Novo Hamburgo, Brazil
[3] Psychology Graduate Program, UNISINOS University, São Leopoldo, Brazil

Abstract. Stress is a social problem affecting society in different ways. Obtaining an accurate diagnosis of stress is complex because the symptoms of stress are very similar to the symptoms of many other illnesses. Some studies have used wearable sensors for psychophysiological signal acquisition and artificial intelligence approaches for automatic stress pattern detection. Nevertheless, the literature does not present specific group analyses among different participants' characterizations. This work evaluates the impact of the allostatic load on Machine Learning models considering clinical patients compared to non-clinical patients. We created and used a new dataset that uses a stress-inducing protocol based on repetitive negative thoughts. This dataset contains 27 participants with clinical and non-clinical profiles. The acquired data are electrocardiogram, respiration patterns, electrodermal activity, weight, height, and Body Mass Index. The classifier that showed the best results was Random Forest. The model using only non-clinical participants obtained 94.54% accuracy against 92.08% of clinical ones. Results obtained with all participants had slight improvement, with 92.72%. Comparing clinical and non-clinical patients showed an accuracy mean difference of 2.5%, indicating aspects of the impact of the allostatic load on the model generalization for people outside the dataset.

Keywords: Machine learning · Wearable sensors · Allostatic load

1 Introduction

Even though stress is part of daily life, prolonged periods of stress harm our health and can lead to the development of illnesses [1]. According to O'Connor et al. [5], stress is a common risk factor in about 75% of diseases, including those that cause the most morbidity and mortality. Giannakakis et al. [9] explain that several studies show promising results for identifying stress using wearables to acquire biosignals and applying Artificial Intelligence (AI) to detect stress

© The Author(s), under exclusive license to Springer Nature Switzerland AG 2022
A. C. Bicharra Garcia et al. (Eds.): IBERAMIA 2022, LNAI 13788, pp. 267–278, 2022.
https://doi.org/10.1007/978-3-031-22419-5_23

patterns. Wearables have many applications, and with the improvement of the sensors' quality, the use of wearables is expanding [18].

The impact of allostatic load on stress identification has been studied in recent years but still is considered a research gap. Allostatic load is the accumulation of chronic stress load [14], which tends to impact people's physiological signals directly. Individuals who go through long periods of chronic stress tend to have a disorder in their allostatic load, leading to an allostatic overload [14,21]. Several studies have evaluated the impact of allostatic load on stress monitoring and the differences in physiological signs in stress cases. They point out that further studies are needed to explore comparisons with individuals with a normal allostatic load [7,8,10].

Based on this context, this work aims to study and compare the impact of allostatic load on AI models of stress identification in clinical and non-clinical patients based on physiological signals acquired through wearables. This work used a new Stress-Inducing Features (SIF) dataset. This SIF dataset follows inducing stress through repetitive negative thinking protocol [11,15,16]. The SIF dataset has two kinds of participants, the clinical and the non-clinical. Clinical participants are those who have received regular psychological support. Moreover, the non-clinical participants do not have any psychological follow-up or diagnosis.

The SIF dataset is composed of multiple physiological signals. Several studies pointed out the multiple physiological signals as a reliable alternative for detecting stress [17,19,23]. Because stress affects different aspects of the human body, affecting different signs. To support overall stress effects identification, the dataset has signals such as the electrocardiogram (ECG), respiration pattern (RSP), and electrodermal activity (EDA). The levels of common physiological signs of each individual are affected by several factors, such as age, height, weight, and Body Mass Index [6,12].

The interest in investigating clinical and non-clinical participants is motivated by the need to assess the behavior of physiological signals in these two sets of participants. The main motivation of this study is to evaluate the impact of allostatic load on pattern prediction using artificial intelligence, since allostatic load directly affects the behavior of the individual's vital signs. According to allostatic load theory, clinical participants tend to have a different response to stressful situations [10,14]. Thus, it may be that a model intended for use in clinical participants may not be helpful in non-clinical participants due to the unpredictability of the data. The central differential of this work is the in-depth analysis of the impact of clinical individuals on the possible generalization of a Machine Learning model with a focus on stress identification. The remainder of this text is organized as follows. Section 2 presents a literature review. In Sect. 3, the new dataset is described in detail. Section 4 presents the methodology, and the results obtained are evaluated in Sect. 5. The conclusions and future work indications are presented in Sect. 6.

2   Literature Review

We conducted a review to identify recent work results on stress detection with wearables and the presence of allostatic load joint studies. We have surveyed the CAPES database, Elsevier, IEEE Explore, and ACM. For each repository, we used the following strings: "wearable AND stress"; "wearable AND artificial intelligence"; "artificial intelligence AND stress"; "artificial intelligence AND allostatic load"; "allostatic load AND stress detection". We have filtered the papers published from 2017 to 2022. Bellow, we discuss the most relevant aspects present in the area.

Han et al. [17] carried out a work to detect stress; for it, they applied a stress induction protocol, the Montreal Imaging Stress Task (MIST). The authors collected the signs using an unspecified wearable. The experiment had 39 participants who measured RSP and ECG data. The Random Forest (RF) and Support Vector Machine (SVM) combined presented the best result among all the algorithms. They obtained an accuracy of 84% for the stress classification with three classes and 94% for binary classification.

Ahuja and Banga [19] carried out an experiment in which they monitored young people during exam preparation and while using the internet. This work aimed to detect mental stress. The experiments had 206 students' data, and for the AI experiments, the authors used Linear Regression (LR), Naïve Bayes (NB), RF, and SVM algorithms.

Corrigan et al. [20] highlight the use of heart rate variability as a way to monitor stress and allostatic load. The allostatic load consists of metabolic energy composed of biological measures. It captures the dysregulation of multiple physiological systems as a result of chronic exposure to stress [14,23]. HRV decreases in response to a stressful situation, whether physically or cognitively [20]. They observed that HRV recovery is usually slow in response to higher magnitude stressors. They also conclude that people with allostatic load disorder usually have a lower standardization HRV response. The authors emphasize that further studies are needed to further the usefulness of HRV to assess allostatic load [10].

The use of AI for stress detection has several limitations and uncertainties that we need to deepen. Research indicates that experiments in controlled laboratory environments have high precision for detecting stress [2,13]. The stress level is often significantly different from the induced stress level outside of controlled environments. They also emphasize that wearables are excellent means to carry out these experiments. The participants tend to accept easily non-intrusive and discreet devices. Based on the state-of-the-art, we can conclude that there is still a lot to be explored in stress detection.

Several works have been studying how to make a reliable stress detection using wearables as a means of data acquisition due to their reliability and practicality. Our study pointed out the uses of classifiers such as DL and RF as the best alternatives for stress detection. In addition, we found that the use of multiple physiological signals tends to be more reliable in detection. As mentioned, stress detection still has limitations, including real-time stress detection in environments outside of laboratories and the use of a single dataset for people who have not participated in the experiments. Another relevant issue is the relationship between allostatic load and the generalization of a machine learning model. Based on the literature, it is evident that the allostatic load impacts physiological signs. Nevertheless, it is necessary to investigate further how people with disturbances in the allostatic load impact a model generalization.

3   Dataset Composition

This section describes the dataset used in the experiments, which is a new dataset obtained in experiments based on the Repetitive Negative Thoughts protocol [11,15,16]. We denominate it Stress-Inducing Features (SIF) Dataset, and it comprises a set of data from 27 participants. The dataset contains data for ECG, EDA, RSP, age, height, weight, clinical status, and Body Mass Index (BMI), which is an international measure used to calculate whether a person has an ideal weight [4].

We used a proprietary architecture to acquire this dataset. The architecture's name is ATHENA I, an acronym for "Architecture for Healthcare reinforced by Artificial Intelligence". The developed architecture can acquire sensor data, process it, and store it to compose a dataset. The architecture incorporates wearable sensors and a single board computer (SBC). The wearable used for the architecture was PsychoBIT from the BITalino company. PsychoBIT focuses on acquiring signs related to the mental and psychological state, which is ideal for collecting stress data. The sensors used collected ECG, EDA, and RSP data. The SBC used for the architecture was the Raspberry Pi 4 Model B. We choose this device for its processing power, as it needs to perform several functions in parallel, collecting, processing, and storing data in real-time during long periods.

This section will present how the dataset acquisition process was in practice. To perform this experiment, we followed a stress induction protocol. We have performed this practical experiment in a multi-discipline partnership with the Post Graduate Program in Psychology. The ethics committee (CAAE number 40555420.0.0000.5344) had approved the experiment to further studies regarding the effects of Repetitive Negative Thoughts on the body.

With the stress induction protocol followed, it is possible to obtain three classes of data: pre-stress, stress, and post-stress (relaxing). We aim to evaluate the heart rate variability in clinical and non-clinical patients (without a prior diagnosis). Moreover, verify the behavior of this signal throughout a process that leaves the normal state, with orientation to concentrate, going to a moment of stress, and then returning to the normal state, trying to avoid the stress. The protocol seeks to assess stress induction in clinical and non-clinical patients, comparing the body's responses to the effect of stress. We have considered clinical participants who have some psychological follow-up previously diagnosed. Whereas non-clinical participants who have nothing diagnosed. We have not made any previous tests to assess whether non-clinical patients have something to be diagnosed.

The first stage of the protocol is the coupling of wearable sensors. We decided to place the electrodes at the beginning of the experiment. This way, the participants would get used to using the sensors. We intended to reduce the impact of the bother of the sensors during the experiment. In the next step, participants filled in sociodemographic data and relevant data for inclusion in the model as the social context and physiological information tend to impact the physiological stress response. After completing the sociodemographic data, the pre-stress meditation process begins, in which the participant remains looking at a white screen, oriented to try not to think about anything specific. The participant remains that way for five minutes. After completing the five pre-stress minutes, the white screen changes to a screen with the phrase "Think about a recent situation that has made you anguished or upset ...". In this stage, the induction of stress occurs through negative thinking. The participant spends 1 min thinking about the current situation that has made him go through some unpleasant moments.

After completing this minute of stress induction, the screen returns to a white screen, in which the participant again tries to avoid thinking, constituting a post-stress stage. In this stage, as in pre-stress, it occurs for 5 min to return to a neutral state. When completing the period time of the post-stress stage, the participant performs a self-report. We have performed the experiments with 27 participants. Among the participants, 12 are non-clinical, and 15 are clinical cases. Of these 27 participants, 66.67% participants are women aged 18 and 38. Male participants represent 33.33%, aged between 19 and 38 years.

4 Pattern Detection Methodology

This section describes the data preparation, processing, and filtering steps, the training process, and the Machine Learning approach. The data filtering process took place using the Python library for physiological signals. We performed specific techniques to filter each signal. We applied a Hamilton method for ECG extraction. Firstly, the library applies a Butterworth low-pass filter. The filter has order four and a cutoff frequency 25 Hz. The next step is to apply a high-pass filter, Butterworth of order 4. We defined a cutoff frequency 3 Hz. Lastly, a derivative function obtains a moving average of 80 ms. Finally, peak detection and QRS complex detection are applied.

Related to the EDA filter, the first step is to apply a Butterworth low-pass filter of order four, with a cutoff frequency of five Hz. A moving average function of N terms is applied to smooth out the distortions. Finally, the library applies differential functions with an analysis in frequency, and it extracts the intrinsic values through zeros of the transfer functions.

Regarding the processing of the RSP, a second-order Butterworth band-pass filter is applied, with cutoff frequencies set to 0.1 Hz 35 Hz. Then, the library uses a zero-crossings function, which checks the points at which the signal intercepts the axis.

We have 11 min of data, of which the initial 5 min refer to pre-stress, 1 min of stress, and the last 5 min of post-stress. To balance the data, we defined that the data that we would use for pre-and post-stress would be the third minute because in an evaluation carried out, these points correspond better to the objective of the classifier step. In other words, of the 11 min collected in the experiment, we selected the third, sixth, and ninth minutes. Thus, the data we used during the AI experiment totaled 3 min of collection per participant, with 27 participants.

In sequence, we applied the windowing process, which aims to reinforce physiological signals' characteristics. For this, the data, after going through the filtering process, are arranged in matrices. Each column is a signal, and the lines its instant in time. We defined 600 ms windows for each of the signs with an overlapping of 50% for each window, according to Fig. 1.

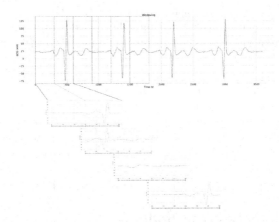

Fig. 1. Windowing

As seen in Fig. 1, we split the data into 600 ms windows, and windowing occurs with the overlapping of these windows by sliding one over the other. For example, the first window occurs from 0 ms to 600 ms, so we relocate these selected data to the new dataset. The second window has an overlapping of 50%. This overlapping of 50% means that the window advances 300 ms, thus forming a time window of 300 ms to 600 ms. We relocate this second window again into the new dataset, repeating 300 ms of data. This technique tends to reinforce the characteristics of the trained data. This process repeats for the entire dataset.

After the windowing process, we performed the ML experiments with variations regarding the participants and the classification objectives. We performed experiments with all and part of the dataset dedicated to identifying the clinical and non-clinical behavior. We also evaluated classification with the three classes (pre-stress, stress, and post-stress) and with some binary combinations. We described the complete set of experiments in the following.

We performed a series of ML experiments using five different classification algorithms. Among the algorithms used, we can mention DT, kNN, RF, Ada Boost, and MLP. We separated these experiments into four sets. The first experiment had all participants and classification using the three classes. We also performed experiments separating the dataset into clinical and non-clinical participants. For each of these sets, we carried out experiments to detect stress.

Furthermore, it is essential to evaluate broadly the results obtained. In the evaluation step, we used the accuracy, precision, F1, and Recall metrics. In addition to the before-mentioned metrics, we also used the confusion matrix to assess how the model classified the data. Moreover, we use a cross-validation method to evaluate the possible overfitting. We selected the kFold method for the cross-validation.

5 Results and Analysis

We performed a series of Machine Learning training experiments evaluating classification approaches and exploring the results with the collected dataset. The new dataset has two different groups of participants, clinical and non-clinical participants, so we also sought to evaluate them in isolation. According to the allostatic load, theory [10,20], people with some mental disorders may present different behavior in stressful situations. Furthermore, clinical people may not feel anything in a stressful situation. Alternatively, after starting a stressful situation, they can take more time to return to a normal state or even do not return.

The first set of tests performed involved all participants, classifying three classes. The predicted classes are pre-stress, stress, and post-stress. The results verified (Table 1) in this set of experiments showed significantly lower accuracy when compared with the first experiment. However, we consider this result as expected, considering that the classification went from two to three classes. In this set of tests, the algorithms that stand out are still Decision Tree and Random Forest, with 90% and 92% accuracy, respectively.

Table 1. Results from all participants - classes: pre-stress, stress, and post-stress

Algorithm	Accuracy	Precision	Recall	kFold
DT	90.522	90.521	90.522	86.341
kNN	72.046	72.136	72.046	69.992
RF	**92.720**	**92.725**	**92.725**	**89.697**
Ada Boost	57.605	57.273	57.605	58.042
MLP	33.551	33.357	33.551	33.806

In the second and third Machine Learning experiments performed, we divided the dataset into non-clinical and clinical participants, using the three classes. This division aimed to identify whether any of these two sets had any impact on the trained values obtained. According to the allostatic load theory, we expected lower results with the clinical participant set. The second experiment, with non-clinical participants, showed better values when compared to training values with all participants. The accuracy for DT and RF was above 93%.

In the third experiment we carried out with clinical participants, the values obtained were lower than non-clinical ones. We expected this range of values, as the signs of clinical participants tend not to respond in a standard way in stressful situations. Therefore the model presents difficulties in classifying these individuals correctly. Even though the values obtained are lower than those of the previous experiments, the accuracy and precision values for Decision Tree and Random Forest continued to be above 90%. We can compare both results in the Table 2.

Table 2. Results of non-clinical and clinical participants separately - classes: pre-stress, stress, and post-stress

Algorithm	Clinical	Accuracy	Precision	Recall	kFold
DT	No	93.035	93.037	93.035	90.532
kNN	No	80.861	80.962	80.861	79.536
RF	**No**	**94.564**	**94.564**	**94.564**	**92.845**
Ada Boost	No	64.917	64.817	64.917	65.157
MLP	No	41.428	43.532	41.428	39.561
DT	Yes	90.081	90.083	90.081	86.547
kNN	Yes	70.276	70.322	70.276	68.407
RF	**Yes**	**92.083**	**92.083**	**92.081**	**88.940**
Ada Boost	Yes	60.140	60.071	60.140	59.046
MLP	Yes	36.344	35.039	36.344	34.242

The values obtained in Cross-Validation are a means of varying the model's generalization. Thus, in these experiments, all values obtained are close to the accuracy and precision values for the data sets and the algorithms, with an average of 3% difference.

Another way to validate the values obtained in training is using confusion matrices. We selected the algorithm Random Forest Classifier with three classes for this verification. We first analyzed the results with all participants. Next, we compared the set of non-clinical and clinical participants. In Fig. 2a, we look at the predicted values in each class in percentage terms using DT. As we can see, the classification occurred adequately, with the Post-Stress classifications with the highest number of correct predictions. In the Fig. 2b, we check the confusion matrix with absolute values. By analyzing the matrix, we verified that the correctly predicted values in each class were very close.

On the other hand, the model made fewer predictions between post-stress and stress. We can justify it because the final period of participant stress is at its maximum stress level. The post-stress used has already passed 2 min of the participant trying to get back to his basal state.

We can analyze the impact of allostatic load on non-clinical and clinical patients by comparing the confusion matrix. In Fig. 3a, we have the confusion

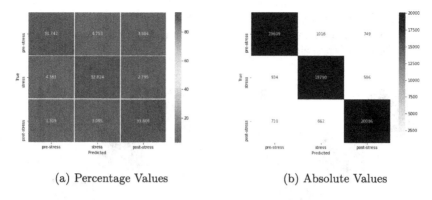

(a) Percentage Values   (b) Absolute Values

Fig. 2. Confusion matrix of all participants

matrix in percentage values for non-clinical patients with a three-class classification. In Fig. 3b, we see the confusion matrix for clinical patients, also in percentage values for three classes. Both matrices are obtained from the training using RF. Comparing the values of the two matrices, we verified an average decrease of 2.5% in the model accuracy with clinical patients compared to non-clinical patients. The most significant difference found is in post-stress, in which the classification for non-clinical patients reached 95.49% and dropped to 92.71% with clinical patients.

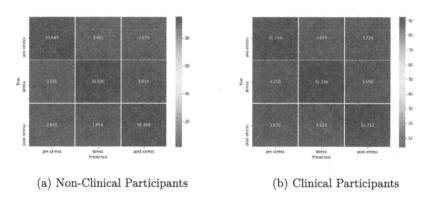

(a) Non-Clinical Participants   (b) Clinical Participants

Fig. 3. Confusion matrix percentage values of all categories

If we analyze the matrices of non-clinical (Fig. 4a) and clinical (Fig. 4b) patients in absolute values, we verify the magnitude of the difference in predictions. Before performing this direct comparison of absolute values, it is necessary to highlight the number of participants in each group in the dataset. The dataset has 12 non-clinical and 15 clinical participants, totaling three more participants for the set of clinical participants. One of the possible comparisons between the two groups is the significant increase in post-stress classification errors. This aspect aligns with our expectations because clinical participants have greater difficulty returning to their normal state.

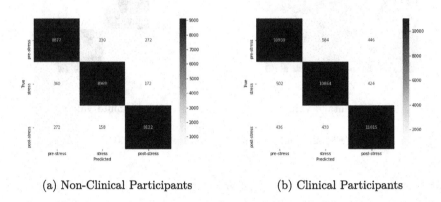

(a) Non-Clinical Participants   (b) Clinical Participants

Fig. 4. Confusion matrix absolute values of all categories

We performed a brief analysis of the means of some clinical and non-clinical participants in a comparative way. Using ECG data as a basis, we can verify the behavior of the signs. It justifies the drop in the accuracy value. By analyzing the non-clinical participants, we verified the expected behavior in the stress-inducing protocol. Starting with lower heart signals, during the stress period, an increase in the signals, and during the stress period, again a relaxation, thus decreasing the heart rate levels.

However, if we analyze the clinical participants' heart rates, we see different behaviors in the signs. For example, participant 12 has an average of 82.81 bpm in pre-stress, the heart hates rises to 83.53 bpm, and in post-stress, instead of lower, the heart rate rises to 84.95 bpm. Participant 17, on the other hand, has a heartbeat behavior that is the opposite of what we expected, starting with 85.90 bpm, decreasing to 84.15 bpm, and rising again to 86.26 bpm. Analyzing participant 18, we verified that the behavior is as expected but at much higher levels when compared to the other participants.

6   Conclusion

This work aimed to obtain a Machine Learning model that performs stress detection based on stress induction through repetitive negative acquired through wearables sensors. Our approach seeks to differentiate clinical and non-clinical participants. This research considers the load theory, which defines that patients with a psychological diagnosis exhibit behavior different from physiological signs. Clinical people can show different behaviors at stressful moments, such as not showing any difference in the levels of physiological signs.

The study showed good values in detecting stress in clinical and non-clinical individuals using ML algorithms for classification. Among the algorithms used, we can highlight the Random Forest with 92.72% accuracy in the prediction using three classes and the cross-validation test with 89.69%. The three classes we used for detection were pre-stress, stress, and post-stress. For binary classification, the model using RF Classifier obtained accuracy and cross-validation of 98.20%.

Another conclusion highlighted is the impact that clinical patients have on the model. When comparing the training model with the set of clinical and non-clinical participants, the values obtained with non-clinical participants were significantly better. Nonetheless, the values are in line with expectations. According to the allostatic load theory, clinical patients tend to have a different behavior of physiological sign than non-clinical people. Clinical participants may have a naturally higher basal (non-stress) level. Alternatively, their physiological signal levels do not return to normal without intervention when going through a stressful situation.

As the main contribution of this work, we can mention the acquisition of a dataset aimed at inducing stress through repetitive negative thinking, with the possibility of evaluating differences in the behavior of physiological signs in clinical and non-clinical patients. Thus, the dataset helps assess allostatic load's impact on stress and uses artificial intelligence to detect stress. Another contribution is the Machine Learning experiments performed to obtain a model for stress detection, evaluating in parallel the impact of allostatic load on the model accuracy.

In future work, we intend to expand the dataset for broader coverage of the data regarding the variety of physiological signs of the participants. Each person has physiological and biological characteristics, which makes physiological signals different. It is not easy to generalize the model to individuals outside the study, thus requiring a tuning for each new participant. Furthermore, we foresaw the practical application of the model obtained for validation in the actual application. Since stress-inducing protocols tend to generate well-defined signals, but outside these controlled environments, the values tend to be different.

Acknowledgement. We acknowledge Fapergs for supporting this work by means of Grant PQG, 5/2019.

References

1. Ghaderi, A., Frounchi, J., Farnam, A.: Machine learning-based signal processing using physiological signals for stress detection. In: 22nd Iranian Conference on Biomedical Engineering(ICBME 2015), pp. 25–27 (2015)
2. Pinge, A., et al.: A comparative study between ECG-based and PPG-based heart rate monitors for stress detection. In: 14th International Conference on COMmunication Systems and NETworkS (COMSNETS) (2022)
3. Carreiras, C., et al.: BioSPPy: biosignal processing in Python (2015)
4. Centers for Disease Control and Prevention, U.S., "Body Mass Index (BMI)" (2020)

5. O'Connor, D.B., Thayer, J.F., Vedhara, K.: Stress and health: a review of psychobiological processes. Annu. Rev. Psychol **72**, 663–688 (2021)

6. Dias, D., Cunha, J.P.S.: Wearable health devices-vital sign monitoring. Syst. Technol. Sens. **18**, 2414 (2018)

7. Mauss, D., Jarczok, M.N.: The streamlined allostatic load index is associated with perceived stress in life - findings from the MIDUS study. Int. J. Biol. Stress **24**, 404–412 (2021)

8. Whelan, E., et al.: Evaluating measures of allostatic load in adolescents: a systematic review. Psychoneuroendocrinology **131**, 105324 (2021)

9. Giannakakis, G., et al.: Review on psychological stress detection using biosignals. IEEE Trans. J. **13**(1), 440–460 (2019)

10. Parker, H.W., et al.: Allostatic load and mortality: a systematic review and meta-analysis. Am. J. Prev. Med. (2022)

11. Chalmers, J.A., et al.: Worry is associated with robust reductions in heart rate variability: a transdiagnostic study of anxiety psychopathology. BMC Psychol. **4**, 1–9 (2016)

12. Früh, J., et al.: Variation of vital signs with potential to influence the performance of qSOFA scoring in the Ethiopian general population at different altitudes of residency: a multisite cross-sectional study. Plos One **19**, 1–13 (2021)

13. Can, J.Y.S., Arnrich, B., Ersoy, C.: Stress detection in daily life scenarios using smart phones and wearable sensors: a survey. J. Biomed. Inform. **92**, 103139 (2019)

14. Guidi, J., et al.: Allostatic load and its impact on health: a systematic review. Psychother. Psychosom. **90**, 11–27 (2021)

15. Kelly, K.M.: Examining the effect of perseverative thinking on physiological activation in response to stress. Honors Theses (2018)

16. Carnevali, L., et al.: Heart rate variability mediates the link between rumination and depressive symptoms: a longitudinal study. Int. J. Psychophysiol. **131**, 131–138 (2018)

17. Han, L., et al.: Detecting work-related stress with a wearable device. Comput. Ind. **90**, 42–49 (2017)

18. Jacobsen, M., et al.: Wearable technology: a promising opportunity to improve inpatient psychiatry safety and outcomes. J. Diabetes Sci. Technol. **15**, 34–43 (2021)

19. Ahuja, R., Banga, A.: Mental stress detection in university students using machine learning algorithms. Procedia Comput. Sci. **152**, 349–353 (2021)

20. Corrigan, S.L., et al.: Monitoring stress and allostatic load in first responders and tactical operators using heart rate variability: a systematic review. BMC Public Health **21**, 1701 (2021). https://doi.org/10.1186/s12889-021-11595-x

21. Doan, S.N.: Allostatic load: developmental and conceptual considerations in a multi-system physiological indicator of chronic stress exposure. Dev. Psychobiol. **63**(5), 825–836 (2021)

22. Russell, S., Norvig, P.: Artificial Intelligence: A Modern Approach, 3rd edn. Pearson - Prentice Hall, Hoboken (2009)

23. Suganthi, V., Punithavalli, M.: A review of deep learning and machine learning methods for analyzing Covid-19 Stress. Int. J. Innovative Sci. Res. Technol. (2022)

Natural Language Processing

Antonymy-Synonymy Discrimination in Spanish with a Parasiamese Network

Juan Camacho[✉], Juan Cámera, and Mathias Etcheverry

Universidad de la República, Montevideo, Uruguay
{juan.camacho,juan.camera,mathiase}@fing.edu.uy

Abstract. Antonymy-Synonymy Discrimination (ASD) is a challenging NLP task that has been tackled mainly for English. We present a dataset for ASD in Spanish, built using online dictionaries and Wordnet in Spanish. To evaluate the quality of the dataset, we performed two manual annotations on a random sample of the dataset, obtaining 0.89 of kappa score between them. Additionally, each annotator obtained 0.9 of agreement according to the dataset. The dataset is split into train, val and test using three algorithms: (1) randomly, (2) Shwartz's split [20], and (3) graph split. The last two are without lexical intersection between its parts, and (3) is based on the topology of the antonymy relation depicted from data. Finally, we report results using a parasiamese neural network [6] in our dataset, one of the best performing approaches in the literature.

Keywords: Antonymy-Synonymy Discrimination · Parasiamese neural network · Word embeddings

1 Introduction

The discrimination of antonyms and synonyms (ASD) is a task of lexical semantics that plays an important role in NLP. Synonyms refer to words with the same (or almost the same) meaning, for example *cute* and *adorable*, and on the other hand, antonymy can be roughly defined as words with opposite meanings, such as *hot* and *cold*. ASD consists of distinguishing synonyms and antonyms from a set of pairs of words. Although ASD is fairly easy for humans, it represents a challenging problem to be resolved automatically. Antonymic and synonymic words tend to occur in similar contexts and thus leads to have close word embeddings [15], which can lead to errors in end systems.

In ASD, as in hypernymy detection [19], there are two main approaches: distributional and pattern based. Pattern based approaches are rooted in Hearst's patterns for hyponymy-hypernymy acquisition [10]. They consist in syntactic patterns on which the related elements co-occur (e.g. "from $<ant>$ to $<ant>$" as in "from *beginning* to *end*"). Pattern based approaches can achieve high precision, but they can suffer from low coverage since the candidate pairs need to co-occur in the same context to be detected. Pattern based approaches have

A. C. Bicharra Garcia et al. (Eds.): IBERAMIA 2022, LNAI 13788, pp. 281–292, 2022.
https://doi.org/10.1007/978-3-031-22419-5_24

been mostly used in hypernymy detection, directly [19] or combined with distributional information [20]. Regarding ASD, Nguyen et al. [17] introduced a pattern based approach in which the paths between the candidate pair are given to a neural network to detect antonymy.

Distributional approaches address ASD relies in the distributional information of each word independently, for example, using word embeddings. Due to the similar distributional information on synonyms and antonyms, word embeddings may seem problematic. However, good performance has been obtained with purely distributional approaches. The ATTRACT-REPEL method [21] attracts synonyms and repels antonyms, showing that word embeddings can be fine-tuned to distinguish antonyms and synonyms. Ali et al. (2019) project word embeddings into two sub-spaces: one for synonymy and the other one for antonymy, using two supervised encoders [1]. The parasiamese network [6] is another distributional approach, inspired in siamese networks, adapted to consider the anti-transitive property of antonyms. Recently, MoE-ASD [22], a model based on mixture-of-experts (MoE) [11], has been introduced to take into account the different dimensions on which antonyms oppose; since antonyms are said to be similar in most dimensions of meaning, excepting a few dimensions in which they oppose, and those dimensions may vary between different cases of antonymy.

ASD research has focused primarily in English and as far as we know there is no substantial research in Spanish[1]. A well-known publicly available source of information for lexical relations that has a version in Spanish is WordNet [7,16] from the Multilingual Central Repository [9], that is a translation from the original WordNet in English. In this work, in addition to WordNet we consider online dictionaries to create a more exhaustive dataset in Spanish for ASD.

Following the direction of ASD, the dataset contains tagged pairs of words indicating if they are synonyms or antonyms. We impart 3 versions of partitions of the dataset: (1) random, (2) lexical split using the algorithm of Shawrtz [20] and (3) lexical split using the information of the relations graph (see Fig. 1).

2   Dataset

The dataset is a collection of pairs of words, each labeled as either *synonyms* or *antonym*. Despite Cardellino vocabulary is bigger than WordNet (3225303 vs 36681), the datasets obtained from each one were very similar, Cardelino's resulted in 1105441 WordNet's resulted in 977408 (less than 10% difference between them). After merging the different sources, the symmetric pairs were removed to reduce redundancy. Once the partitions were complete, a new set of partition was created by gene.

---

[1] [8] addressed lexical relations detection and performed a manually curated automatic translation of the dataset to spanish and german and only consists of 1850 pairs of antonyms.

It was created using WordNet in Spanish [7] and three sites of synonyms and antonyms. An extraction process was performed on each resource and the results were combined. We present 3 versions of the dataset regarding different partition criteria for train, val and test.

This section explains the process for creating the dataset and its partitions. In Sect. 2.1 we explain the process used to extract the tuples from the different sources. Once the information from each source was obtained it was merged as described in Sect. 2.1. We study the antonymy and synonymy relations using a graph representation of them in Sect. 2.2. Then we proceeded to define the dataset partition criteria, as explained in Sect. 2.3. Three variations are presented: random-split, Shwartz-split and graph-split

2.1   Data Sources

Four sources were used for synonyms and antonyms extraction. Firstly, we used the Spanish version of WordNet and we extracted all its words with their respective antonyms and synonyms.

Then, we considered the following online dictionaries:

- wordreference.com
- antonimos.net
- sinónimo.es

For each source we have built a scraper, that has extracted each word with its respective antonyms and synonyms. Due to the nature of the online dictionaries an initial vocabulary was required to scrape them. Each scraper uses this initial vocabulary to query the online dictionary. The query result is processed by the scraper to generate tagged tuples with the queried word and the ones in the response. If a word in the response is not present in the vocabulary that word is added to it, expanding the current vocabulary. The initial vocabulary was from WordNet in Spanish. After that we also used the vocabulary from Cardellino corpus [4] in a second run of the scrapers.

Data Unification. As a result of the data harvesting we had 2 datasets for each online dictionary, one generated with WordNet vocabulary and the other with Cardellino's; and additionally we had the pairs from WordNet.

Despite Cardellino vocabulary is bigger than WordNet (3225303 vs 36681), the datasets obtained from each one were very similar, resulting in 1105441 and 977408 tuples, respectively, i.e. around 10% difference between them. After merging the different sources, the symmetric pairs were removed to reduce redundancy. Once the partitions were complete, a new set of partition was created by generating the symmetrical tuples. So, two versions of each dataset were made regarding symmetry.

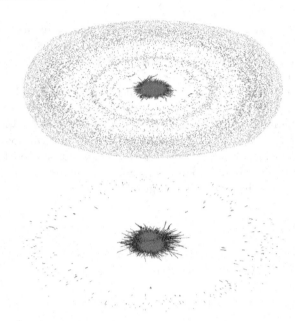

Fig. 1. Synonymy (top) and Antonymy (bottom) graphs, representing words as nodes and the relations as edges.

2.2   Antonymy and Synonymy Graphs

An analysis of the topology in the antonymy and synonymy relations was performed using Networkx [2]. We draw each relation as an undirected graph, considering words as nodes, and the relations between words as the edges.

In Fig. 1 we present the graph representation of the antonymy relation. It can be observed that there is a massive connected component which contains 16651 nodes. This means that through the antonymy relation there are 16651 words connected between them in the graph. The remaining of the graph contains 298 components ranging between 2 and 11 nodes each. In the disposition of the graph drawn it seems like a kernel and a nebula that surrounds it. Regarding synonymy, similar to the antonymy graph, there is a component that contains 44576 nodes and 9770 components containing between 2 and 20 nodes.

With the information provided by the graph representation, we can conclude that the words in Spanish are highly related to each other in the antonymy and synonymy relations, due to the existence of paths in those relations that connect the majority of them. This can be the reason why the Cardellino and WordNet vocabularies generate similar datasets.

2.3 Dataset Splits

Once the dataset was completed we proceeded to partition it into three different sets for training, validation and test. This partitions were made with different criteria: one randomly preserving the proportion between antonyms and synonyms, and two lexical splits. A partition with lexical split means that the intersections between the vocabularies of training, validation and test sets are empty (i.e. each word can only occur in one of them).

We decided to make partitions with lexical split for two reasons, the first one is to evaluate the impact of the "lexical memorization" presented in [14] on the ASD task, the second reason is to evaluate the performance of a classifier with previously unknown words. Because the train, val and test sets have no intersection in their vocabulary, this greatly increases the difficulty of the ASD task. The classifier now needs to generate an abstraction of the antonym relation that is general enough to perform well and to also apply to completely unknown words. The "lexical memorization" occurs when a model learns an independent property of one of the elements in the tuple. For example if we had the tuples (cold, hot), (cold, warm) and (cold, friendly) as positive examples in the train set, a model could learn that if the word cold is present in the left side of the tuple then is a positive example. This phenomenon is much more relevant in hypernymy detection, because the relation is prone to having one word repeat many times in one side of the tuple.

We decided to make partitions with lexical split to evaluate the performance of a model on words that have not been seen during training nor hyperparameter tuning. Another motivation is the "lexical memorization" phenomenon [14], this occurs when a model learns an independent property of one of the elements in the tuple. For example if we had the tuples (*cold, hot*), (*cold, warm*) and (*cold, friendly*) as positive examples in the train set, and the model learns that if the word *cold* is present in the left side of the tuple, then it is a positive example. This phenomenon seems much more relevant in hypernymy detection because the relation is prone to having one word repeated many times in one side of the tuple (e.g. a massive hyperonym such as *animal*).

Random Split (Non-lexical). To create the random partition the library scikit-learn [18] was used. This library has methods to create random stratified partitions of datasets, thus we decided to use this methods to generate the random partition of the dataset. As a result of the partition we obtained the train, val and test sets the number of tuples in each set is shown in Table 1.

Lexical Split Shwartz. This partition of the dataset is inspired on Shwartz et al. work [20]. We iterate over the dataset and each tuple is assigned to train, val or test trying to preserve the 70%, 10% and 20% of the original dataset, respectively. If a tuple could not be assigned to one of the three sets because it would break the lexical split, then that tuple is discarded.

Lexical Split Graph. For the creation of this partition we focused primarily in the antonymy relation graph. We created an algorithm based on its structure to attempt to reduce the amount of discarded tuples. First we take the small components of the graph and add them into the test set, because these components contain less than the 10% of the antonyms. Then, to divide the big component of the graph, we perform a breadth-first search (BFS) starting on the node with most edges and stops once it reaches the target size of the test set. The reached nodes are removed from the graph and the edges from the removed nodes that were not added to the test set are discarded. This process is repeated for the validation set and the remaining of the graph forms the training set. The test and validation sets have 20% and 10%, respectively, of the antonymy tuples, and the training set contains less than 70% due to the discarded tuples. Finally, synonyms are added, preserving lexical split, to each set to achieve a 25% ratio of antonyms/synonyms.

2.4   Dataset Quality

In order to study the quality of the dataset a sample of 200 tuples (100 antonyms and 100 synonyms shuffled) was taken and then were manually classified by two native speakers, looking for the meaning of unknown words[2].

Once we have the tuples classified, we used Cohen Kappa Score [5] to calculate the concordance between the classifications obtaining a score of 0.8999, which means a high concordance between the annotators.

When analysing the annotation mismatches, we found that many of them were cases of tuples that could be classified as antonyms or a synonyms depending on the meaning of the words that the annotator considered. An example of this is (*inhospito, selvático*) in which one annotator considered *selvático* as a place hostile to life and the other as a place full of life. Other examples are (*enervar, espabilar*), (*lateral separado*) and (*entregarse, ocuparse*)

On the other hand, in tuples like (*idiotizar, embelecerse*), or (*perversión, conversión*), the missmatch was caused due to the connotation considered by the annotators. For example, if *idiotizar* or *conversión* were considered with positive or negative connotation.

Finally, the tuple (*corrección, propiedad*), at first glance, were considered as unrelated. However, when we searched their meanings we agreed that they can be considered as synonyms due to phrases like "*hablar con corrección*" and "*hablar con propiedad*", being this meaning for *corrección* is weird in our region.

---

[2] The meanings were obtained through the *Real Academia Española* dictionary (https://dle.rae.es/).

Table 1. Quantity of tuples in the datasets without symmetrics

		Ant	Syn	Total
Rand	Train	47720	244378	292098
	Val	3592	18394	21986
	Test	12828	65693	78521
Lex Shwartz	Train	28571	132772	161343
	Val	3684	18803	22487
	Test	7498	37503	45001
Lex Graph	Train	27124	77200	104324
	Val	6647	18919	25566
	Test	13357	38017	51374

3   Experiments

In this section we explain the experiments conducted. All the experiments are based on neural networks and word embeddings. A random search was performed to get a suitable hyper-parameter configuration for each model. Finally we show and discuss the obtained results.

3.1   Models

For our experiments we considered three models: (1) a feed-forward neural network receiving as input the concatenation of the word embeddings, (2) a siamese network [3], and (3) a parasiamese network [6].

We used FastText [12] word embeddings with a vector dimension of 300. The publicly available pretrained model of FastText was used, which was trained on a Spanish Wikipedia dump.

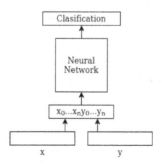

Fig. 2. Concat model.

Concat: The input of this model is the concatenation of the vectors of the words in the tuple, that is, if the tuple is the pair (x, y), the input of the network is the vector $(x_1, .., x_{300}, y_1, .., y_{300})$. Figure 2 shows the structure used for this network.

Siamese: Due to the design of siamese networks, its input consists of two vectors, one per each word in the tuple. One instance of the base neural network is used in the branches of the siamese network. The result provided by the network is the square of the euclidean distance between the output vectors. To obtain the classification an acceptance threshold is used, if the distance provided by the result is less than that value, is classified as a positive pair (antonyms), otherwise it is classified as a member of the negative class (synonyms). Figure 3 shows the structure of the network used in this model.

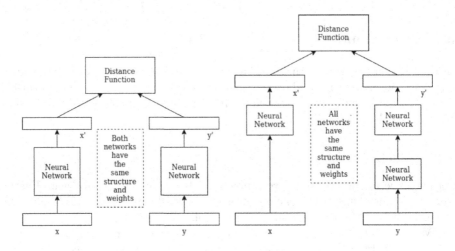

Fig. 3. Siamese (left) and parasiamese (right) neural networks.

Parasiamese: This model was proposed in [6] to deal with antitransitive relations such as antonymy. Figure 3 shows the architecture of the network. The loss and distance functions were the same used as in the siamese model, as well as the same base network. This model has two variants, parasiamese left and right, that applies the base network twice in the left and right input vector, respectively.

Pretrained Parasiamese: We have also consider a pretrained version of the aforementioned model. The pretrain consists in training a siamese network with synonyms as the positive class. Once the siamese network is trained, its weights are used as the initial weights in the parasiamese network. Then the parasiamese network is trained like the non pretrained version.

3.2   Random Search

For each dataset variant and model, a random search was performed to obtain a suitable hyperparameter configuration. This was done using the train and validation datasets with and without symmetric tuples, but in all cases the test set with symmetric tuples was used.

The hyperparameters search is stated as follows:

- Number of layer of the network in the range $[1, 4] \in N$.
- Layer size $[100, 700] \in N$. A value for every layer is selected independently.
- Learning rate: one of 1e-2, 1e-3, 1e-4, 1e-5
- Activation function: [ReLU, Tanh, Sigmoid]
- Batch size one of: 64, 128, 256, 512, 1024, 2048, 4096
- Margins. These are used by the contrasting loss. This loss requires a positive margin and a negative margin. Additionally an intermediate margin is selected as the classification threshold. This values are taken from a range $[0.01, 0.02, ..., 7]$.
- In every execution we used Adam [13] for training.

In total, 33 random searches with 200 trials each were executed. Table 3 contains the results for the executions performed using train and validation datasets without symmetrics, and Table 2 contains the results for the executions with symmetrized train and validation datasets. Both tables report precision (P), recall (R) and F1-score (F1) scores.

Table 2. Results on the three versions of the dataset with symmetric tuples in train and validation partitions. The name of the models are abbreviated where Siam means siamese, PSiam means parasiamese, R means right and L means left. PSiam Pre means pretrained parasiamese.

Models	Random			Lex Shwartz			Lex Graph		
	P	R	F1	P	R	F1	P	R	F1
Concat	0.80	0.76	0.78	**0.58**	0.50	0.53	**0.64**	0.33	0.44
Siam	0.41	0.58	0.48	0.30	0.45	0.36	0.26	**1.00**	0.41
PSiam R	0.88	0.85	0.87	0.55	0.57	0.56	0.61	0.45	0.52
PSiam L	0.89	0.87	0.88	0.47	0.65	0.55	0.58	0.40	0.47
PSiam Pre R	0.92	0.88	**0.90**	0.46	**0.66**	0.55	0.61	0.50	**0.55**
PSiam Pre L	0.88	**0.89**	0.89	0.56	0.58	**0.57**	0.60	0.49	0.54

Table 3. Results for the three versions of the dataset on the test partition with symmetric tuples and without symmetric tuples in train and validation. The name of the models are abbreviated where Siam means siamese, PSiam means parasiamese, R means right and L means left. PSiam Pre means pretrained parasiamese.

Models	Random			Lex Shwartz			Lex graph		
	P	R	F1	P	R	F1	P	R	F1
Concat	0.69	0.66	0.68	**0.59**	0.41	0.48	0.59	0.34	0.43
Siam	0.41	0.58	0.48	0.30	0.45	0.36	0.26	**1.00**	0.41
PSiam R	0.85	0.83	0.84	0.57	0.51	0.53	0.57	0.40	0.47
PSiam L	0.79	0.84	0.81	0.48	**0.62**	0.54	**0.62**	0.37	0.46
PSiam Pre R	**0.93**	0.87	**0.90**	0.52	**0.62**	0.56	0.59	0.47	0.52
PSiam Pre L	0.91	**0.88**	0.89	0.58	0.58	**0.58**	0.58	0.48	0.53

3.3   Results Analysis

As we can see in Table 3 the best model in terms of F1 score is the Parasiamese model across all the dataset partitions. The left or right versions of the parasiamese model are very close to each other in terms of F1-scores, this indicates that the side of the double application of the base network does not significantly affect the results. It can be noticed that the pretrained parasiamese variants perform better than its non pretrained counterparts, with an improvement in F1 score, ranging from 0.03 and 0.08.

In the Table 2 we can see that almost all the models benefited from the dataset containing the symmetric tuples, the only exception being the pretrained version of parasiamese model. Due to the inherently symmetric nature of the Siamese model the random search was not done in the symmetrized datasets, the results are presented in Table 2. The parasiamese model continues to be the best model in terms of F1-score. The versions of the parasiamese models with or without pretraining, are more close in terms of F1-score than in the Table 3. The performance gained by the pretrained variants ranges from 0.01 to 0.03 in most cases, excepting the graph based partition that shows an improvement of 0.07.

Comparing the results of the same model in both tables, we can see that both concat and non pretrained parasiamese perform better when are trained with symmetric tuples. The reason for this is that those models are not symmetric. Parasiamese will apply the base network twice only in one side of the tuple, so if we have the symmetric tuple the model will generate a new pair of transformed vector for the same pair of vectors. In the concat model the explanation is similar, as it uses the concatenation of the words vectors to create a new vector which is the networks input, so the symmetrical is a different input vector.

On the other hand, pretrained parasiamese seems to not improve when symmetric tuples are included. We think that this can be explained in how we are pretraining the model. As it is being pretrained with a synonyms Siamese network as explained in [6], and those models are symmetric by nature, we think that it does not need the symmetric tuple to learn it.

4 Conclusion

We studied ASD problem in the Spanish language and introduce a new dataset of tagged tuples, created from different online resources and the WordNet in Spanish. Using the dataset content, we perform an analysis of these relationships using a graph representation of them. The new dataset was splitted into train, validation and test through three different methods: random, Shwartz's based and graph split. The latter two reach a dataset split without lexical intersection, being the graph split designed using the graph topology analysis. To ensure the quality of the dataset, a manual classification of a random selection of tuples was performed by two annotators independently. This manual validation obtains a Cohen's kappa score of 0.89 between the annotators. Finally, we used the newly created dataset with a parasiamese network, one of the best performing models used in English, obtaining results of 0.9 of F1-score at best.

References

1. Ali, M.A., Sun, Y., Zhou, X., Wang, W., Zhao, X.: Antonym-synonym classification based on new sub-space embeddings. In: Proceedings of the AAAI Conference on Artificial Intelligence, vol. 33, pp. 6204–6211 (2019)
2. Hagberg, A., Dan Schult, P.S.: Networkx, July 2005. https://github.com/networkx/networkx#readme
3. Bromley, J., et al.: Signature verification using a "Siamese" time delay neural network. Int. J. Pattern Recogn. Artif. Intell. **07**(04), 669–688 (1993). https://doi.org/10.1142/S0218001493000339
4. Cardellino, C.: Spanish Billion Words Corpus and Embeddings, August 2019. https://crscardellino.github.io/SBWCE/
5. Cohen, J.: A coefficient of agreement for nominal scales. Educ. Psychol. Measur. **20**(1), 37–46 (1960). https://doi.org/10.1177/001316446002000104
6. Etcheverry, M., Wonsever, D.: Unraveling antonym's word vectors through a Siamese-like network. In: Proceedings of the 57th Annual Meeting of the Association for Computational Linguistics, pp. 3297–3307. Association for Computational Linguistics, Florence, Italy, July 2019. https://doi.org/10.18653/v1/P19-1319, https://aclanthology.org/P19-1319
7. Fernández-Montraveta, A., Vázquez, G., Fellbaum, C.: The Spanish version of wordnet 3.0, September 2008. https://doi.org/10.1515/9783110211818.3.175
8. Glavaš, G., Vulić, I.: Discriminating between lexico-semantic relations with the specialization tensor model (2018)
9. Gonzalez-Agirre, A., Laparra, E., Rigau, G.: Multilingual central repository version 3.0. In: Proceedings of the Eighth International Conference on Language Resources and Evaluation (LREC 2012), pp. 2525–2529. European Language Resources Association (ELRA), Istanbul, Turkey, May 2012. https://www.lrec-conf.org/proceedings/lrec2012/pdf/293_Paper.pdf

10. Hearst, M.A.: Automatic acquisition of hyponyms from large text corpora. In: COLING 1992 Volume 2: The 15th International Conference on Computational Linguistics (1992)

11. Jacobs, R.A., Jordan, M.I., Nowlan, S.J., Hinton, G.E.: Adaptive mixtures of local experts. Neural Comput. **3**(1), 79–87 (1991). https://doi.org/10.1162/neco.1991.3.1.79

12. Joulin, A., Grave, E., Bojanowski, P., Mikolov, T.: Bag of tricks for efficient text classification. arXiv preprint arXiv:1607.01759 (2016)

13. Kingma, D.P., Ba, J.: Adam: a method for stochastic optimization (2014). https://doi.org/10.48550/ARXIV.1412.6980, https://arxiv.org/abs/1412.6980

14. Levy, O., Remus, S., Biemann, C., Dagan, I.: Do supervised distributional methods really learn lexical inference relations? In: Proceedings of the 2015 Conference of the North American Chapter of the Association for Computational Linguistics: Human Language Technologies, pp. 970–976 (2015)

15. Mikolov, T., Chen, K., Corrado, G., Dean, J.: Efficient estimation of word representations in vector space (2013). https://doi.org/10.48550/ARXIV.1301.3781, https://arxiv.org/abs/1301.3781

16. Miller, G.A.: Wordnet: a lexical database for English. Commun. ACM. **38**(11), 39–41 (1995). https://doi.org/10.1145/219717.219748

17. Nguyen, K.A., Walde, S.S., Vu, N.T.: Distinguishing antonyms and synonyms in a pattern-based neural network. arXiv preprint arXiv:1701.02962 (2017)

18. Pedregosa, F., et al.: Scikit-learn: machine learning in Python. J. Mach. Learn. Res. **12**, 2825–2830 (2011)

19. Roller, S., Kiela, D., Nickel, M.: Hearst patterns revisited: Automatic hypernym detection from large text corpora. arXiv preprint arXiv:1806.03191 (2018)

20. Shwartz, V., Goldberg, Y., Dagan, I.: Improving hypernymy detection with an integrated path-based and distributional method. In: Proceedings of the 54th Annual Meeting of the Association for Computational Linguistics (Volume 1: Long Papers), pp. 2389–2398. Association for Computational Linguistics, Berlin, Germany, August 2016. https://doi.org/10.18653/v1/P16-1226, https://aclanthology.org/P16-1226

21. Vulić, I.: Injecting lexical contrast into word vectors by guiding vector space specialisation. In: Proceedings of The Third Workshop on Representation Learning for NLP, pp. 137–143. Association for Computational Linguistics, Melbourne, Australia, July 2018. https://doi.org/10.18653/v1/W18-3018, https://aclanthology.org/W18-3018

22. Xie, Z., Zeng, N.: A mixture-of-experts model for antonym-synonym discrimination. In: Proceedings of the 59th Annual Meeting of the Association for Computational Linguistics and the 11th International Joint Conference on Natural Language Processing (Volume 2: Short Papers), pp. 558–564. Association for Computational Linguistics, Online, August 2021. https://doi.org/10.18653/v1/2021.acl-short.71, https://aclanthology.org/2021.acl-short.71

LSA-T: The First Continuous Argentinian Sign Language Dataset for Sign Language Translation

Pedro Dal Bianco[1,3]([✉]) [iD], Gastón Ríos[1,3] [iD], Franco Ronchetti[1,2] [iD],
Facundo Quiroga[1,4] [iD], Oscar Stanchi[1] [iD], Waldo Hasperué[1,2] [iD],
and Alejandro Rosete[5] [iD]

[1] Instituto de Investigación en Informática LIDI - Universidad Nacional de La Plata,
La Plata, Argentina
pdalbianco@lidi.info.unlp.edu.ar
[2] Comisión de Investigaciones Científicas de la Pcia. de Bs. As. (CIC-PBA),
La Plata, Argentina
[3] Becario Doctoral UNLP, La Plata, Argentina
[4] Becario Postdoctoral UNLP, La Plata, Argentina
[5] Universidad Tecnológica de La Habana José Antonio Echeverría,
Havana, Cuba

Abstract. Sign language translation (SLT) is an active field of study
that encompasses human-computer interaction, computer vision, natural
language processing and machine learning. Progress on this field could
lead to higher levels of integration of deaf people. This paper presents,
to the best of our knowledge, the first continuous Argentinian Sign Language (LSA) dataset. It contains 14,880 sentence level videos of LSA
extracted from the CN Sordos YouTube channel with labels and keypoints annotations for each signer. We also present a method for inferring
the active signer, a detailed analysis of the characteristics of the dataset,
a visualization tool to explore the dataset and a neural SLT model to
serve as baseline for future experiments.

Keywords: Sign Language Translation · Computer vision · Big data ·
Sign language dataset · Deep learning

1 Introduction

Sign language is one of the main tools that the deaf community has both to
communicate and to access information. Sign Language Recognition (SLR) has
been an active research field for the last two decades [4]. However, each sign
language has its own specific linguistic rules [17] and there is no direct correspondence between a sequence of words in a traditional spoken language and a
sequence of signs. Therefore, an approach based on SLR, as a special case of a
gesture recognition problem, is not enough to close the gap between deaf and
hearing people. The problem of translating directly from sign language videos

A. C. Bicharra Garcia et al. (Eds.): IBERAMIA 2022, LNAI 13788, pp. 293–304, 2022.
https://doi.org/10.1007/978-3-031-22419-5_25

(a) "Resumen semanal"   (b) "Ley federal LSA"

(c) "Último momento"   (d) "CN Sordos ecología"

Fig. 1. Screenshots of videos from the different channel playlists.

to text in the corresponding language is known as Sign Language Translation (SLT). Compared to other translation tasks, SLT presents unique challenges [4]:

- The signs used and their meanings vary in different countries and even in different regions, meaning that for a model to be able to translate a specific sign language, for example, Argentinian Sign Language (LSA), it needs to be trained on data from that specific language.
- A model for SLT needs to learn not only the meaning of each sign but also the temporal dynamics and dependences between signs. Therefore, it has to be trained on videos that contain sequences of (continuous sign language or CSL). Datasets that contain isolated signs (videos containing one sign each) are not suitable for SLT.
- There are much fewer labeled sign language videos than in other domains (such as audio transcription) and there are also relatively few sign language users in the general population, which makes it difficult to find interpreters to label videos.
- Each sign is made of many components such as hand shapes, positions, movements, body poses, facial expressions, and lip movements. A model able to perform SLT should be able to use all these features.
- Typically, available videos of sign language share the same backgrounds, interpreters, or topics. This is specially true for laboratory-made datasets or, for example, videos from news channels that have a live interpreter. Models trained on this data might be biased toward these features and unable to generalize to other people or backgrounds.

Because of the aforementioned, a dataset that allows training models for video translation from LSA to text (and also generating LSA videos from text, altought not the focus of this work) is thought to be especially useful in order to improve the communication and inclusion of deaf people.

1.1 Contributions

The main contribution of this work is the presentation and analysis of LSA-T: the first continuous LSA dataset. The dataset contains:

- More than 20 h of video extracted from CN Sordos [16], a YouTube channel that presents the latest news in LSA.
- The corresponding Spanish translation for the sign sentences in each video.
- The keypoints for each signer, computed using AlphaPose [7].
- An estimation of the active signer in the videos where multiple people appear.

Figure 1 shows images of some of its videos. We also present a visualization tool for LSA-T that allows to explore the videos present in the dataset. Finally, a model for SLT trained on this dataset is presented as a baseline. The model uses the keypoints to infer the corresponding Spanish sentence. Details of the model and its resulting metrics are described in Sect. 4.

2 Related Work

Sign language datasets can be categorized into isolated or word-level, where each video matches a specific sign, and continuous or sentence-level, where videos contain many signs that correspond to an entire sentence. As mentioned before, while isolated datasets can be used for recognition they cannot be used for translation [3]. Since the main goal of this work is to present a dataset for SLT, only CSL datasets will be considered.

One of the most relevant SLT datasets is RWTH-Phoenix-Weather 2014 T T [4]. It contains videos of German Sign Language (GSL) extracted from German public TV weather forecasts. This dataset is used today as the main benchmark for SLT and, having a vocabulary of over 1000 signs, is considered the only resource for large-scale continuous sign language worldwide [12]. As the present work, it has the peculiarity of having been recorded in real life conditions, which may result in a more challenging dataset than if it was laboratory-made. Most of the other available SLT datasets have been recorded in laboratory conditions and contain a set of frequent or relevant sentences for its respective language. This is the case of the SIGNUM dataset [19] of GSL, the Chinese Sign Language dataset (CSL) [9], the Greek Sign Language (GSL) dataset [1] and the KETI dataset [11] of Korean Sign Language.

Table 1 shows the main features of the mentioned datasets alongside LSA-T. Many datasets also provide information about glosses translation (an intermediate translation between sign language and spoken language where each gloss

Table 1. Main features of the studied datasets.

	PHOENIX*	SIGNUM	CSL	GSL	KETI	LSA-T
Language	German	German	Chinese	Greek	Korean	Spanish
Sign language	GSL	GSL	CSL	GSL	KLS	LSA
Real life	**Yes**	No	No	No	No	**Yes**
Signers	9	25	50	7	14	**103**
Duration [h]	10.71	55.3	**100+**	9.51	28	21.78
# samples	7096	**33,210**	25,000	10,295	14,672	14,880
# unique sentences	5672	780	100	331	105	**14,254**
% unique sentences	79.93%	2.35%	0.4%	3.21%	0.71%	**95.79%**
Vocab. size (w)	2887	N/A	178	N/A	419	**14,239**
# singletons (w)	1077	0	0	0	0	**7150**
% singletons (w)	37.3%	0%	0%	0%	0%	**50.21%**
Vocab. size (gl)	**1066**	450	-	310	524	–
# singletons (gl)	**337**	0	-	0	0	–
# singletons (gl)	**31.61%**	0%	–	0%	0%	–
Resolution	210×260	776×578	1920×1080	848×480	1920×1080	1920×1080
Fps	25	**30**	**30**	**30**	**30**	**30**

*Data was not available for the whole PHOENIX dataset, so the table show its train set statistics.

corresponds to one sign) so we differentiated between gloss (gl) and word (w) level. There is no information about glosses in LSA-T because it only contains sentence-level Spanish translation. There is no gloss information about CLS dataset either, as there is no gloss notation in CSL, signs map directly to words.

It is interesting to notice that there is a significant difference between datasets generated in real life conditions and those generated in laboratory conditions regarding the amount of unique sentences, the vocabulary size and the amount of singletons (tokens that appear only once in the training set). As laboratory-made dataset first choose a number of sentences and then record them many times, the set of unique sentences is a small fraction of the total amount of samples (between 0 and 5 percent). On the other hand, as it is not as common to find an exact sentence repeated many times in realistic environments, datasets generated in real life conditions present a much higher rate of unique sentences: 79.93% for the PHOENIX dataset and 95.79% for LSA-T. This rate was expected to be lower in PHOENIX than in LSA-T as the videos in PHOENIX were taken only from the weather forecast, while videos in LSA-T feature a wide range of topics. This directly impacts the size of the vocabulary and the amount of unique words in each dataset. While laboratory-made datasets have a small vocabulary with no singletons (all sentences appear multiple times), datasets collected in real-life environments have large vocabulary sizes with a significant amount of unique words (37.3% in PHOENIX and 50.21% in LSA-T). As a consequence, datasets recorded in real-life conditions, and particularly the one presented in this work, are highly challenging for translation models.

2.1 LSA

LSA is the sign language used by the deaf community in Argentina. In 2018, 0.02% of the Argentinian population older than 6 years old presented a hearing difficulty, half of them being deaf [5]. Counting teachers, relatives and friends of deaf people, there are 2 million LSA users [6].

Regarding LSA datasets, LSA64 [14,15] is, to the best of our knowledge, the only other dataset available. It contains 3200 videos of 64 isolated signs recorded by a team of researchers from the Universidad Nacional de La Plata. As it has been generated under laboratory conditions, the background scenarios are similar. Although it has been the only source of data for training LSA sign recognition models, it cannot be used for translation.

2.2 SLT

Regarding SLT, works that use keypoints data as part of the input for a translation model can be found in [2] and [20]. In those cases keypoints data were given to the model as extra information together with the corresponding video. Furthermore, the models presented in [11] and [10] have a similar approach to the one presented in this work. Also, they only use keypoints information as input for a neural network model. In both cases the authors train a recurrent network on the KETI dataset. Particularly, in [10] no gloss information is employed, neither as input or output.

3 LSA-T Dataset

LSA-T was built from videos from the YouTube channel CN Sordos, a news channel created in 2020 by deaf people and deaf people's relatives. The hosts use LSA to communicate the news. Subtitles in Spanish are provided by the authors. There are two main hosts in most of the videos. There are other regular hosts in charge of special sections like ecology or movies, but these appear much less often. Therefore, the great majority of the 103 signers present on the dataset are guests who appear once or twice for an interview or a specific piece of news. There is gender parity among the signers and videos contain different locations, backgrounds and lighting conditions.

LSA-T was generated in two stages. In the first stage, we collected the videos and cropped them according to its subtitles. For each valid sentence in the subtitles of a video (annotations like "[background music]" were discarded), we cropped the corresponding video clip. This process resulted in 14,880 video segments. In the second stage, in those segments where there is more than one person, we inferred who is the signer with a certain score of confidence. Section 3.3 describes how this inference was done.

Finally, we also made available a visualization tool so that researchers can easily explore the dataset. The tool was developed using Fifty One [13], which provides useful features such as allowing to filter samples by label, video, playlist, or by the confidence score of the signer inference.

Full code for the development of this dataset, the analysis described below, and the start-up of the visualization tool can be found in the project's repository[1].

3.1   Statistics and Analysis

LSA-T was built from 63 different videos of the CN Sordos Channel. We only used videos that had been annotated with subtitles by the authors (LSA experts). The videos resulted in 14,880 segments that span 21.78 h of video. Each sample of the database contains a sentence taken from the subtitles provided and its respective video segment. The generated segments have a mean duration of 5.27 s and their labels a mean of 4.78 words per sample (Fig. 2).

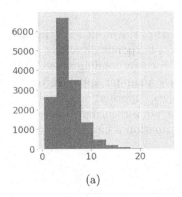
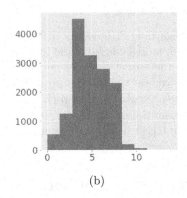

(a)   (b)

Fig. 2. Histograms of segments duration (a) and amount of words per sentence (b).

As shown in Table 1, a peculiarity of the dataset is the significant proportion of unique sentences ($\sim$ 95%) and the proportion of unique words with respect to the vocabulary size ($\sim$ 50%) (3a). Therefore, we consider the dataset to be highly challenging for a SLT model.

A common preprocessing step in the natural language processing field consist on removing sentences with words that appear very few times in the dataset so they do not add noise to the training process. Histograms of word frequencies after removing samples with singletons and words with frequencies lower than 5 are shown in Figs. 3b and 3c, respectively. After removing the corresponding subset of labels, the histograms do not show such a heavy left tail, despite the fact that other words now appear with a lower frequency than before.

Finally, Figs. 4a and 4b show the 30 most commons words and trigrams respectively. As many of the most common words correspond to articles or connectors we believe the trigrams can provide a more representative visualization of the dataset content. Additionally, we expect traditional models without special training schemes to be able to learn these common trigrams.

---

[1] https://github.com/midusi/LSA-T.

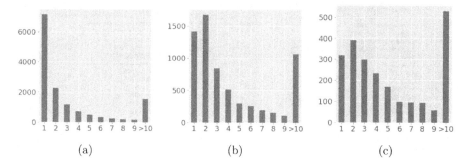

Fig. 3. Histograms of word frequency over the whole dataset (a), after removing samples with singletons (b) and and after removing samples with words with frequency lower than 5 (c).

3.2   Default Train-Test Sets

Alongside the dataset, we provide two versions of the train and test sets: a full version and a reduced version. We made the reduced version by excluding labels that contain words with frequency lower than 5. This version is expected to be easier for a SLT model to learn. In both versions we used 80% of the samples for training and 20% for testing. Samples with a signer-confidence score lower than 0.5 were discarded (Sect. 3.3).

Also, as videos labels are subtitles instead of classes, traditional stratified Train-Test splits cannot be performed. Since samples from the same video are expected to share topics and therefore vocabulary, both train and test sets were generated so that segments from the same video are included in both.

Table 2. Comparison between the two versions of the train-test sets.

	LSA-T	Full version		Reduced version	
		Train	Test	Train	Test
# sentences	14,880	11,065	2735	3767	910
% unique sentences	95.79%	96.64%	92.78%	96.88%	98.35%
Vocab. size	14,239	12,385	5546	2694	1579
% singletons	50.21%	52.01%	61.9%	23.2%	48.83%
% sentences with singletons	34.97%	40.98%	67.97%	14.36%	54.29%
% sentences with words not in train vocabulary	–	–	59.2%	–	84.5%

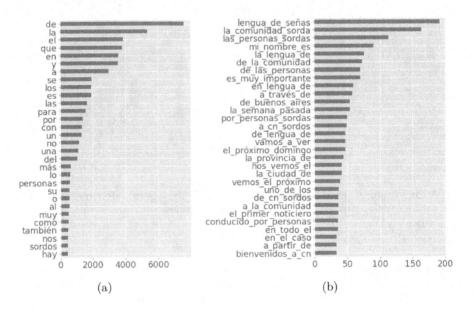

Fig. 4. Most common words (a) and trigrams (b) in the dataset

Table 2 shows a comparison between the different versions. As it can be seen, there is no significant difference between them in terms of percentage of unique sentences, but there is in terms of percentage of singletons. While in the full version around 50% of the training vocabulary are singletons, only 23% of the train vocabulary of the reduced version are. Furthermore, while 40% of the labels on the full version train train contain singletons, only 14% of the labels of the reduced version train do. Apart from this, it is also worth noticing that, in both versions, there is an important amount of sentences (∼60%) and (∼85%) respectively) of the test set that contains tokens that were not present in the train set. This must be considered when evaluating a SLT model over the dataset as it wont be able to learn about those tokens and its usage from the train set.

3.3 Signer Inference

Many videos in LSA-T feature more than one person. In most cases, they appear at the same distance from the camera and in similar positions (Fig. 1), so identifying who is the active signer is not trivial. Figure 5a presents an histogram of samples by number of people detected. Most of the videos have two people in them and only a third of the samples feature one person. Identifying the signer is particularly useful as it allows to define a region of interest (ROI) that models could focus on while performing SLT.

To infer the signer, we first obtained the keypoints and bounding boxes of the people in each video. This was done using AlphaPose with the Halpe full-body keypoints format [7]. This format has 136 keypoints of which 42 are hand

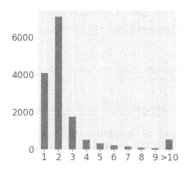

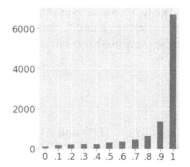

(a) Histogram of amount of people detected by AlphaPose.

(b) Histogram of signer-confidence scores (rounded to one decimal).

Fig. 5. Amount of signers and signer scores histograms.

keypoints, which makes them suitable for hand gesture recognition. Then, we transformed each keypoints' coordinates to a relative position with respect to the person's center. For each person in the video, we computed the total amount of movement of their hands. The person chosen as the signer is the one who has the largest amount of hand movement, with a high confidence (Fig. 5b).

The resulting keypoints computed by AlphaPose, the result of the signer inference and the amount of movement of each person used to compute the confidence score of the inference are provided for each sample of the dataset.

4 Baseline Experiments

4.1 Proposed Approach

A SLT model trained on LSA-T is proposed to serve as baseline for future comparisons. The model used is similar to the one presented in [10], using only the keypoint data extracted from the videos as input. However, instead of using recurrent layers we use a transformers as the main building blocks of the network.

The proposed model then takes as input a list of F frames and predicts the words of its corresponding sentence one at a time. Each frame is a matrix of size $2K$ that has the x and y position for the K keypoints and each sentence is encoded as a list of the embeddings of its words.

4.2 Metrics

Given the difficulty of LSA-T dataset caused by having many words that appear few times, we proposed a new metric called WER-N, based on the traditional Word Error Rate (WER). WER-N mimics the top-N accuracy for classification models by not penalizing the model when it misses words from the target that were seen less than N times in the training set. The metric can be used in any problem that present tokens with low frequency.

WER-N can be computed using a variant of the Edit Distance algorithm. For two strings s and t, the distance d between the sub-strings $s[0:i]$ and $t[0:j]$ can be computed recursively using the formula:

$$d[i,j] = min(d[i-1,j] + C(I)$$
$$d[i,j-1] + C(D^*) \qquad (1)$$
$$d[i-1,j-1] + C(S^*))$$

where $C(I)$, $C(D^*)$ and $C(S^*)$ represent the cost of an insertion, deletion and substitution respectively. The WER-N metric modifies the traditional definitions of C(D) and C(S) by not penalizing some errors as follows.:

$$C(D^*) = \begin{cases} 0 & \text{if frequency of token } t[j] \text{ is lower than } N \\ 1 & \text{otherwise} \end{cases}$$

$$C(S^*) = \begin{cases} 0 & \text{if } s[i] = t[j] \\ 0 & \text{if frequency of token } t[j] \text{ is lower than } N \\ 1 & \text{otherwise} \end{cases} \qquad (2)$$

4.3   Experimental Results

We performed two sets of experiments, using the two versions of the train and test sets presented in Sect. 3.1. We implemented the proposed model using PyTorch and trained the model for 30 epochs in each experiment. We used the Adam optimizer and a learning rate of 0.0001. We used 75 frames per sample, obtaining an average of ∼15 FPS over all samples in the dataset.

The size of both the keypoint embeddings and the word embeddings was set to 64. The parameters for defining the transformer network were the same as the proposed in the original paper [18]. We used the Spacy Spanish [8] tokenizer for the sample labels.

Only the 42 hands keypoints were used as the model input as those were the ones considered relevant for SLT. Finally, we used a threshold of $T = 0.3$ for the keypoint confidence. Full code for the model development, training and experimentation can be found in its GitHub repository[2].

Table 3 shows the result of evaluating the model over the two versions of the train-test sets. We used WER and the proposed WER-N with $N = 5$ and $N = 10$ as evaluation metrics. As it can be seen, the model does not perform well, although similar approaches had obtained decent results in other datasets like [11] or [10] on KETI dataset. It is interesting to notice that [10] also tried the same model over RWTH Phoenix dataset, that as Table 1 showed, has a bigger similarity to ours than the laboratory made datasets as KETI. Their model achieved a BLEU score of ∼85 over KETI, but when evaluated over Phoenix it obtained a BLEU score of ∼13 across all their experiments. Then it is not surprising that given the complexity of the presented dataset the models did not obtained good results.

---

[2] https://github.com/midusi/keypoint-models.

Table 3. Metrics for the proposed model over the two train-test sets

	Full version		Reduced version	
	Train	Test	Train	Test
WER	0.9387	0.9392	0.9207	0.957
WER-5	0.8116	0.7892	0.7982	0.68
WER-10	0.7547	0.7154	0.7193	0.5904

5 Conclusions and Future Work

We have presented a new dataset from the Argentinian Sign Language. It is, to the best of our knowlegde, the first continuous dataset that will make it possible to train SLT models for LSA.

The dataset was built using videos from the channel CN Sordos and shows a wide range of topics, signers, backgrounds and lightning conditions. It contains clips extracted from the videos, their corresponding label, the keypoints for each person in the video and, in case there is more than one person, an estimation of who is the one signing. Also a visualization tool was made available.

We have also proposed two possible versions of train and test sets and trained on them a translation model that uses keypoint information to serve as baseline.

As future works, we intend to train different state-of-the-art models over the dataset and compare their performance with the one achieved in other datasets such as Phoenix-RWTH. This models could take both video and keypoints information. Also, we plan on using data augmentation and one-shot learning techniques that will allow us to extract more information from the tokens or sentences that seldom appear on the dataset. Finally, we are working on a web tool that will allow people to use the translation model over a video recorded by themselves using a web cam.

References

1. Adaloglou, N., et al.: A comprehensive study on sign language recognition methods. arXiv preprint arXiv:2007.12530 2(2) (2020)
2. Borg, M., Camilleri, K.P.: Phonologically-meaningful subunits for deep learning-based sign language recognition. In: Bartoli, A., Fusiello, A. (eds.) ECCV 2020. LNCS, vol. 12536, pp. 199–217. Springer, Cham (2020). https://doi.org/10.1007/978-3-030-66096-3_15
3. Bragg, D., et al.: Sign language recognition, generation, and translation: an interdisciplinary perspective. In: The 21st International ACM SIGACCESS Conference on Computers and Accessibility, pp. 16–31 (2019)
4. Camgoz, N.C., Hadfield, S., Koller, O., Ney, H., Bowden, R.: Neural sign language translation. In: Proceedings of the IEEE Conference on Computer Vision and Pattern Recognition, pp. 7784–7793 (2018)
5. de Estadística y Censos, I.N.: Estudio nacional sobre el perfil de las personas con discapacidad: Resultados definitivos 2018 (2018)

6. y Empleos, E.B.: La lengua de señas argentina (lsa). https://idiomas. becasyempleos.com.ar/lengua-de-senas/, Accessed 28 May 2022

7. Fang, H.S., Xie, S., Tai, Y.W., Lu, C.: RMPE: regional multi-person pose estimation. In: Proceedings of the IEEE International Conference on Computer Vision, pp. 2334–2343 (2017)

8. Honnibal, M., Montani, I.: spaCy 2: natural language understanding with Bloom embeddings, convolutional neural networks and incremental parsing (2017). to appear

9. Huang, J., Zhou, W., Zhang, Q., Li, H., Li, W.: Video-based sign language recognition without temporal segmentation. In: Proceedings of the AAAI Conference on Artificial Intelligence, vol. 32 (2018)

10. Kim, Y., Kwak, M., Lee, D., Kim, Y., Baek, H.: Keypoint based sign language translation without glosses. arXiv preprint arXiv:2204.10511 (2022)

11. Ko, S.K., Kim, C.J., Jung, H., Cho, C.: Neural sign language translation based on human keypoint estimation. Appl. Sci. **9**(13), 2683 (2019)

12. Koller, O.: Quantitative survey of the state of the art in sign language recognition. arXiv preprint arXiv:2008.09918 (2020)

13. Moore, B.E., Corso, J.J.: Fiftyone. GitHub. Note. https://github.com/voxel51/ fiftyone (2020)

14. Ronchetti, F., Quiroga, F., Estrebou, C., Lanzarini, L., Rosete, A.: Sign languague recognition without frame-sequencing constraints: a proof of concept on the argentinian sign language. In: Montes-y-Gómez, M., Escalante, H.J., Segura, A., Murillo, J.D. (eds.) IBERAMIA 2016. LNCS (LNAI), vol. 10022, pp. 338–349. Springer, Cham (2016). https://doi.org/10.1007/978-3-319-47955-2_28

15. Ronchetti, F., Quiroga, F., Estrebou, C.A., Lanzarini, L.C., Rosete, A.: Lsa64: an argentinian sign language dataset. In: XXII Congreso Argentino de Ciencias de la Computación (CACIC 2016). (2016)

16. Sordos, C.: Cn sordos youtube channel (2020). https://www.youtube.com/ channel/UCTi9woRHA4r8e3oEWF8hxTA, Accessed 12 May 2022

17. Stokoe, W.C.: Sign language structure. Ann. Rev. Anthropol. **9**(1), 365–390 (1980)

18. Vaswani, A., et al.: Attention is all you need. Adv. Neural Inf. Process. Syst. **30** (2017)

19. Von Agris, U., Knorr, M., Kraiss, K.F.: The significance of facial features for automatic sign language recognition. In: 2008 8th IEEE International Conference on Automatic Face & Gesture Recognition, pp. 1–6. IEEE (2008)

20. Zhou, H., Zhou, W., Zhou, Y., Li, H.: Spatial-temporal multi-cue network for continuous sign language recognition. In: Proceedings of the AAAI Conference on Artificial Intelligence, vol. 34, pp. 13009–13016 (2020)

TSPNet-HF: A Hand/Face TSPNet Method for Sign Language Translation

Péricles B.C. Miranda[1(✉)], Vitor Casadei[2], Emely Silva[3], Jayne Silva[2], Manoel Alves[2], Marianna Severo[2], and João Paulo Freitas[2]

[1] Universidade Federal Rural de Pernambuco, Recife, Brazil
`pericles.miranda@ufrpe.br`
[2] CESAR, Recife, Brazil
[3] Universidade Estadual de Campinas, Campinas, Brazil

Abstract. Sign Language is the language that the Deaf adopted to communicate. However, most hearing people do not know how to communicate in sign language, creating natural barriers between these groups. Aiming to reduce such barriers, Sign Language Translation (SLT) interprets sign video sequences into spoken language sentences. A recent SLT model, called TSPNet, explored sign videos' temporal and contextual semantic structures to learn more discriminative features. Although the TSPNet has reached promising results on the RWTH-PHOENIX-Weather 2014T (PHOENIX14T), the model only considers hand signs, ignoring facial and body information. Facial expressions reflect the extent of signs and play a relevant role in translation. The current work proposes the TSPNet-HandFace (TSPNet-HF), which considers hand and facial features in the SLT process. The proposal has two novel components: facial feature extraction and hand/face feature aggregation to combine hand and facial features. It was assessed on the PHOENIX14T in BLEU and ROUGE, and it was compared to TSPNet and CNN2dRNN. The results showed that the TSPNet-HF overcame the competing methods in all the translation metrics, showing that the inclusion of facial features positively impacts the SLT process.

Keywords: Sign language translation · Continuous Sign Language Recognition · Deep learning

1 Introduction

Sign Language Translation (SLT) is a task that produces spoken language translation for given continuously signing videos. As sign languages and natural languages differ linguistically, a signed sentence does not necessarily syntactically align with its respective spoken language translation [12]. SLT methods learn the embedding space of sign sentence videos and natural languages and their mappings. This process is not trivial and leads to a difficult sequential learning problem [2]. Sign language videos are composed of minimal units called sign gestures, which preserve semantics in the videos. Nonetheless, identifying

A. C. Bicharra Garcia et al. (Eds.): IBERAMIA 2022, LNAI 13788, pp. 305–316, 2022.
https://doi.org/10.1007/978-3-031-22419-5_26

boundaries between sign gestures is not trivial because the videos may have motion blurs, fine-grained gestural details, and transitions between the signs. Therefore, the output of SLT models usually represents sign visual features in a frame-wise manner to avoid explicitly segmenting the videos into isolated signs [2]. However, this manner captures only spatial appearance features, ignoring the temporal dependencies between sign gestures. This negligence may harm the translation because temporal information helps discriminate different signs with similar poses, avoiding ambiguity in translation. A recent work [9] proposed a bootstrapping method called TSPNet, that explored sign videos' temporal and contextual semantic structures to learn more discriminative features. The TSPNet learns from videos and their natural language sentences (e.g. subtitles) to produce translation, avoiding glosses. The TSPNet is composed of different procedures designed to reduce the necessity for accurate video segmentation, enhance semantical consistency of sign segments, and correct semantic ambiguity using non-local video context [9]. The TSPNet was assessed in the RWTH-PHOENIX-Weather 2014T (PHOENIX14T) dataset [2], largely used for SLT tasks, and compared to traditional SLT models. Although the TSPNet had reached promising ROUGE and BLEU scores overcoming the competing methods, it considers only hand signs in the translation process, ignoring facial expressions. Facial expressions presents an important role in the interpretation and translation of sign gestures, and neglect them may lead to mistranslation [14]. As a result, the TSPNet presented crucial limitations related to mistranslation and misinterpretation of videos.

This work proposes the TSPNet-HandFace (TSPNet-HF), an extended version of the TSPNet, which considers both hand signs and facial expressions in the translation process. Two novel procedures were included in the original pipeline: a facial expression feature extraction and a hand/face feature aggregation to combine hand and facial features. We implemented and tested five aggregation strategies. The main contribution of this work is to show the quantitative and qualitative impact that the addition of facial expressions has on the translation process, improving the SLT model's performance. The TSPNet-HF was assessed in the PHOENIX14T dataset and compared to the original TSPNet and the CNN2dRNN [2]. Evaluations were made for all methods in terms of ROUGE and BLEU with different grams. Our results showed that the TSPNet-HF, using the aggregation *average pooling*, overcame statistically all competing methods in all translation metrics. The qualitative analysis revealed that the TSPNet-HF avoided the mistranslation and misinterpretation problem faced by the original TSPNet, in most of the dataset's videos.

2   Related Works

In the last three decades, the computer vision community has studied sign languages [14], with the primary objective of translating SL videos to spoken language sentences and vice-versa, to facilitate the Deaf's daily lives. Most existent models are sorted under solution scenarios, representations and methods.

Conceptual video-based SLT methods were raised around the 2000s and employed a language model to generate spoken sentences from recognizing isolated signs [2]. However, studies for end-to-end SLT from a video (continuous data) were not performed until recently. A relevant point that reinforces the importance of treating SLT as a continuous data problem is that the association between sign sentences and their respective translations is non-monotonic because they have different ordering [2]. Besides, sign glosses and linguistic constructs do not necessarily have a one-to-one alignment with their respective spoken language. These characteristics have made the application of Continuous Sign Language Recognition (CSLR) methods unsuitable because they assume that sign language videos and annotations have the same temporal order [12].

Among all advancements to perform end-to-end SLT models, the proposition of transformer networks is considered one of the most important ones [15]. The adoption of Transformers in the SLT pipeline presented a relevant impact on the translation performance compared to legacy attention-based encoder-decoder methods. These improvements in Neural Machine Translation (NMT) methods motivated the development of other SLT methods [3,9], which contributed even more to the SLT process. Currently, these works can be divided in two categories: *two-staged* and *bootstrapping* methods. Two-staged SLT models demand videos with additional annotations, where each sign in the video has a respective word label in its occurring order. These extra annotations are called *gloss*, and they are used to help two-staged models learn to recognize gestures and structure the recognition results into spoken language sentences. These methods, also named *sign2gloss2text*, are considered the state-of-the-art in SLT tasks. The great drawback of these methods is the availability of datasets with a representative number of gloss annotations. Besides, curating and annotating sign language videos is a complex and arduous task. Then, these methods have been used in very specific scenarios [9]. On the other hand, the *bootstrapping* methods use only the spoken sentences available (subtitle), avoiding glosses [9]. The advantages are i) lighter networks in terms of components and parameters, and ii) the creation of datasets for *bootstrapping* methods is simpler because there is no gloss.

A recent work, developed by [9], proposed a *bootstrapping* method for the SLT task and assessed it in the PHOENIX14T dataset. The TSPNet obtained competitive results overcoming traditional *bootstrapping* SLT models in the literature. In addition, TSPNet's architecture is simple, presenting fewer components and parameters than other SLT methods, such as [3]. However, TSPNet' SLT process uses only hand signs, neglecting non-manual features. This work proposes a novel TSPNet architecture where not only hand signs are considered but also facial expressions.

3   TSPNet

The original TSPNet has an encoder-decoder architecture. The encoder is responsible for learning discriminative sign video representations using the hierarchical semantic structure among video segments. The encoder's result is provided to a Transformer decoder to produce the translation. Given a video input, the TSPNet has a specific sign videos segment representation, whose goal is to learn the temporal and spatial semantics of sign gestures. This novel representation tries to reduce the effect of inaccurate sign video segmentation by exploiting the semantic consistency among segments of distinct granularities in a hierarchy. Then, the authors implemented a sliding window approach to produce video segments (one pivot segment and its neighboring segments) with multiple window widths (multiple scales). In addition, to deal with the issue of ambiguity in gesture semantics, the TSPNet's authors focused on key observations: local consistency and context-dependency. Regarding the first topic, sign language videos are composed of gestures that evolve continuously, implying that video semantics change consistently. Hence, temporally close segments probably share consistent semantics. Regarding the context-dependency topic, sign gestures with high similarity may be translated to different words depending on the context [12]. Thus, non-local video information is vital for solving semantic ambiguity in single gestures, mainly when there is noise in the video segmentation. Guided by these remarks, the TSPNet's authors developed a hierarchical feature learning method that employs a local temporal structure to improve the semantic consistency and a non-local video context to decrease semantic ambiguity.

Figure 1 shows the TSPNet's workflow, which produces spoken language from sign language videos. Given a sign video, the TSPNet first generates its multi-scale segment representation and extracts features using the I3D networks $G_{I3D}(\cdot)$. The I3D networks were pre-trained on Kinetics [4], and then they were finetuned on two word-level sign language recognition (WSLR) datasets [8] in American Sign Language (ASL) to adapt to sign gesture videos. It is important to highlight that both WSLR datasets are composed of hand gestures, then the I3D focuses the feature extraction on the signers' hands. A *Shared Positional Embedding layer* (SPE) was created to identify the segments' positions in a sequence. Next, the authors developed a procedure to aggregate features in each local neighborhood for learning semantically consistent representations. Then, the TSPNet attempts to resolve the ambiguity of local gestures by collecting all the aggregated features and utilizing them to give non-local video context. Besides, the TSPNet also has a joint learning layer responsible for consolidating the feature learning by simultaneously using local and non-local information. Lastly, all information is provided to the decoder to generate the translation.

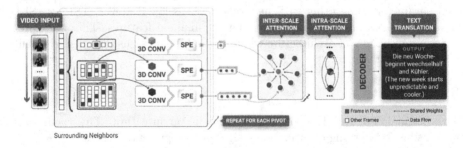

Fig. 1. TSPNet's workflow: sign language videos to spoken language. Source: [9]

Although the TSPNet's results are competitive when compared to consolidated methods in literature, its translation process considers only hand signs, not including facial expressions. Because of this, some semantic and synthatic errors are found in the translation, generating some misunderstandings. More details about this drawback can be found at [9].

4 Proposal

This work proposes the TSPNet-HF, an extended version of the original TSPNet, which, in addition to considering hand signs information, also includes facial expression in the translation process. Figure 2 presents the TSPNet-HF's workflow that has four main steps. The first step comes from the original TSPNet, where pre-trained I3D networks (in hand sign gestures) are used to extract characteristics of the frames with different widths. The second step is similar to the first, but in this case, I3D networks are used to extract characteristics of faces cropped from the video frames. This step was included in the process to produce feature vectors with facial information. The steps one and two produces feature vectors of hands and faces, respectively. Then, this work implemented different aggregation strategies to combine them (step 3). Finally, the forth step is equivalent to the TSPNet original, but now it receives a resultant vector with the combination of hands and faces information, in which both are considered in the translation process. Next, we detail steps two and three, which are the novelty of the TSPNet-HF's workflow.

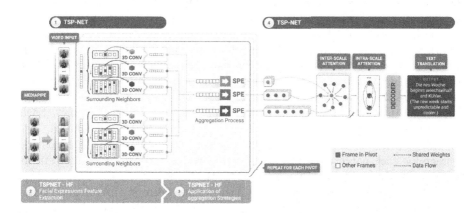

Fig. 2. TSPNet-HF's workflow: sign language videos to spoken language considering hand and face information. Source: Authors.

4.1 Facial Expressions Feature Extraction

As mentioned previously, the first step is responsible to extract features from hand signs. For this, I3D networks pre-trained on hand signs are applied in the entire video frames (one for each window). The output of this process is a set of feature vectors of shape 1×1024. Inspired in this process, we created a parallel procedure for the extraction of features from facial expressions (step two). In this procedure, every frame of the video had the face cropped using the Mediapipe tool[1]. The sequence of faces were organized in different windows (8, 12 and 16 sizes, as well as performed in the hand signs' procedure), and provided to I3D networks (not pre-trained). Herein, the I3D networks are used to extract characteristics from the sequenced facial expressions. The output of this procedure is a set of feature vectors of shape 1×1024 composed of facial information.

4.2 Aggregation Strategies

Each sign language video has 3 associated files (stride 8, 12 and 16), each of which consists of a list of feature vectors. The size of each list is equal to the number of video frames, divided by span, which is 2. A video of 50 frames generates a list of feature vectors of size 25. Each vector has the size (1×1024). As the span is equal to 2, the hand feature files have equivalent files in the face features, i.e. both have the same size. For example, if the video has 80 frames, the file of the hands features is size 40, and the file of the face features is also size 40.

The first and second steps, presented in Fig. 2, produce feature vectors of hand signs and facial expressions, respectively. How to combine hands and faces feature vectors so that both information can be used to improve the translation process? For this, we adopted different aggregation strategies (step 3) to combine the feature vectors and maintain the vectors' shape 1×1024. Each aggregation strategy operates over two vectors, one with hand signs features and the other with facial expressions features. Let H and F be two lists (files) referring to the same video (H being the features of Hands and F the features of Face), and let i be a number belonging to the natural numbers, for each equivalent pair of tensors $M[i]$ and $F[i]$, an operation was performed where $M[i]$ and $F[i]$ were used as the input for merging the features. The variable i was used so that it was guaranteed the alignment of the groups of hand and face frames. Each aggregation strategy returns a resultant vector with all features combined.

The aggregation strategies used here are: 1) Average Pooling: returns the average for each patch (2×2) of the feature vectors; 2) Max Pooling: returns the maximum value for each patch (2×2) of the feature vectors; 3) Sum: returns the sum of values for each patch (1×2) of the feature vectors; 4) Average: returns the average for each patch (1×2) of the feature vectors; 5) Weighted Average: returns the weighted average for each patch (1×2) of the feature vectors. Weight 2 for hands and 1 for faces.

As the translation process itself is already expensive computationally, we decided to use simple aggregation strategies to avoid additional computational cost. The combined feature vectors are then provided to the forth step, which produces spoken language considering hand signs and facial expressions.

---

[1] https://google.github.io/mediapipe/.

5 Experimental Methodology

This section presents the experimental methodology to assess the proposal. Details about the dataset, metrics, competing methods and implementation setup is presented next.

5.1 Dataset

The TSPNet-HF is assessed on PHOENIX14T dataset [2]. According to our knowledge, the PHOENIX14T is the only publicly and most used large-scale SLT dataset adoted for training and inference. Nine different signers produced all videos in GSL. Besides, each videos' translation in German is available. The PHOENIX14T protocol partitions the data into 7096, 519, 642 videos for training, validation, and testing sets, respectively. The PHOENIX14T includes a rich vocabulary of more than 3k German words. This amount of sentences and words are very relevant to increase the chances of reaching good results in the SLT task [2].

5.2 Metrics

This experiment considers BLEU and ROUGE-L metrics to assess the competing methods. These metrics are typically adopted to evaluate algorithms in machine translation tasks [5]. BLEU-n calculates the translation precision up to n-grams, summarizing the precision scores of 1, 2 and 3-grams. On the other hand, ROUGE-L calculates the $F_1 - score$ considering the longest common subsequences between ground-truth translations and predictions. Generally, these translation metrics return scores significantly lower than 100 because there are different valid translations of the same meaning in spoken language.

5.3 Competing Methods

Five instances of the TSPNet-HF were created: TSPNet-HF$_{avg\_pool}$, TSPNet-HF$_{max\_pool}$, TSPNet-HF$_{sum}$, TSPNet-HF$_{avg}$ and TSPNet-HF$_{weighted\_avg}$. Each, where each instance adopts a different aggregation method to combine hand and face feature vectors. All these TSPNet-HF variants were compared to: (i) Conv2dRNN, which uses AlexNet for feature extraction and performs a GRU-based encoder-decoder architecture for sequence modeling [2]. It also uses multiple attention strategies on top of recurrent units [10]; (ii) TSPNet-Joint (the best-performance TSPNet from [9]). This last is named here TSPNet-H, because it only considered hand signs for translation.

5.4 Implementation and Setup

The proposed TSPNet-HF followed the same implementation of the TSPNet-H. Both used FAIRSEQ framework [11] in PYTORCH[2]. Regarding the width of the sign gestures, since the signs last, on average, around half a second

---

[2] https://pytorch.org/.

($\sim$12 frames) [1]. The possible segment width scales were set to 8, 12 and 16 frame segments. Besides, we used a stride of 2 frames, in each width scale, to lessen the feature sequence lengths while preserving the most semantic information. The SENTENCEPIECE (using German subword embedding [6]) [7] was adopted to represent texts in the feature space. It is based on character units to deal with low-frequency words. The TSPNet-HF and TSPNet-H were set with Adam optimizer with a cross-entropy loss as in [15]. It is important to mention that all networks used an initial learning rate of 10^{-4} and a weight decay of 10^{-4}. We trained all networks for 200 epochs, sufficient for the convergence of all models. Finally, each method was executed ten times to produce average results.

6   Results and Discussion

This section presents the results achieved by the TSPNet-HF and the competing methods. The quantitative analysis presents the average results achieved by each approach in terms of n-BLEU and ROUGE. The qualitative analysis shows examples of translations produced by the TSPNet-HF and the original version (TSPNet-H). Our goal with this analysis is to present scenarios in which the addition of facial expressions contributed to the translation process.

6.1   Quantitative Analysis

Table 1 presents the average and standard deviation values of each approach. The first two approaches are the Conv2dRNN [2] using [10] and the TSPNet from [9], named here TSPNet-H. Both of them adopted only hand signs for the translation process. The other methods are variants of the proposed TSPNet-HF, each one adopting a specific feature aggregation strategy to combine hand and face information. As it can be seen, the TSPNet-HF$_{avg\_pool}$ overcame all the other methods in every translation metric. The results were evaluated in terms of statistical significance to conduct a fair comparison among all methods.

Table 1. Comparative performance analysis between the proposed technique and the competing methods in the Phoenix-Weather dataset.

Method	Hand	Face	ROUGE	BLEU-1	BLEU-2	BLEU-3
Conv2dRNN [2] + [10]	Yes	No	31.80 ($\pm 10^{-3}$)	32.24 ($\pm 10^{-3}$)	19.03 ($\pm 10^{-3}$)	12.83 ($\pm 10^{-3}$)
TSPNet-H [9]	Yes	No	34.96 ($\pm 10^{-3}$)	36.10 ($\pm 10^{-3}$)	23.12 ($\pm 10^{-3}$)	16.88 ($\pm 10^{-3}$)
TSPNet-HF$_{avg\_pool}$	Yes	Yes	**35.58** ($\pm 10^{-3}$)	**36.85** ($\pm 10^{-3}$)	**23.90** ($\pm 10^{-3}$)	**17.42** ($\pm 10^{-3}$)
TSPNet-HF$_{max\_pool}$	Yes	Yes	34.30 ($\pm 10^{-3}$)	35.61 ($\pm 10^{-3}$)	22.70 ($\pm 10^{-3}$)	16.39 ($\pm 10^{-3}$)
TSPNet-HF$_{sum}$	Yes	Yes	34.85 ($\pm 10^{-3}$)	35.92 ($\pm 10^{-3}$)	23.10 ($\pm 10^{-3}$)	16.70 ($\pm 10^{-3}$)
TSPNet-HF$_{avg}$	Yes	Yes	35.53 ($\pm 10^{-3}$)	35.99 ($\pm 10^{-3}$)	23.43 ($\pm 10^{-3}$)	17.14 ($\pm 10^{-3}$)
TSPNet-HF$_{weighted\_avg}$	Yes	Yes	34.00 ($\pm 10^{-3}$)	35.00 ($\pm 10^{-3}$)	20.00 ($\pm 10^{-3}$)	16.00 ($\pm 10^{-3}$)

The null hypothesis is: *there is no statistical difference among the results.* As the results do not follow a normal distribution, we used the *Friedman Aligned*

Ranks [16], a non-parametric hypothesis test, with a significance level of $\alpha = 0.05$. The null hypothesis was rejected with a *p*-value $= 1.17 \times 10^{-4}$, meaning that at least one result differs statistically from the others. To complement the previous hypothesis test, we employed the *Nemenyi post hoc* [13] test to identify groups of results that indicate statistical similarity. Two methods are statistically different if their corresponding average ranks differ by the Critical Difference (CD) value. Figure 3 shows the comparing results utilizing the CD diagram. This diagram is a graphical representation that ranks the methods and groups those statistically similar methods considering the CD value. As it can be seen in Fig. 3-ROUGE, the TSPNet-HF$_{avg\_pool}$ achieved the best ROUGE results, tying statistically only with the TSPNet-HF$_{avg}$ (second place) and overcoming all other methods. The TSPNet-H and the Conv2dRNN stayed in the fifth and seventh place in the rank.

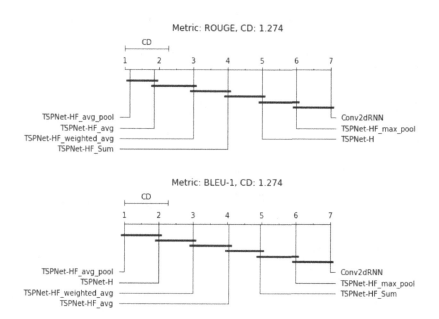

Fig. 3. Critical difference diagrams for translations metrics.

Regarding the BLEU-1 metric (Fig. 3-BLEU-1), the TSPNet-HF$_{avg\_pool}$ is the best ranked tied with the TSPNet-H. The other TSPNet-HF variants occupied the rank from the third to the sixth positions. The CNN2dRNN remained in the last position. Due to the lack of space, we did not include CD diagrams for BLEU-2 and BLEU-3, but next we present a discussion of the results in these metrics. Regarding BLEU-2, the TSPNet-HF$_{avg\_pool}$ overcame all other methods statistically. The second and third best methods, tied statistically, were the TSPNet-HF$_{weighted\_avg}$ and TSPNet-HF$_{avg}$. The TSPNet-H and Conv2dRNN stayed in the fourth and last positions, respectively. Regarding BLEU-3, the three best-ranked are the same. The TSPNet-HF$_{avg\_pool}$ is the best method, tying with the TSPNet-HF$_{avg}$ (second place). The TSPNet-HF$_{avg}$ reached the third-best position in the rank. In BLEU-3, the TSPNet-H reached the fourth position. In summary, the TSPNet-HF$_{avg\_pool}$ was statistically superior quantitatively than the other methods, overcoming them in all translation metrics.

6.2 Qualitative Analysis

The current section presents a qualitative analysis of the results. Table 2 shows three translation examples produced by the TSPNet-HF$_{avg\_pool}$, the best method from the quantitative analysis, and the TSPNet-H. Correctly translated 1-grams are highlighted in blue, and semantically correct translation in red. In the first example, the TSPNet-HF$_{avg\_pool}$ produces an accurate translation, hitting a 100% of the sentence. On the other hand, although the TSPNet-H produced a sentence semantically similar, it failed to generate the original spoken text.

Table 2. Comparison of some example translation results of the TSPNet-HF$_{avg\_pool}$ and the TSPNet-H (original version).

Ground Truth:	jetzt wünsche ich ihnen noch einen schönen abend (now i wish you a nice evening)
TSPNet-H:	guten abend liebe zuschauer (good evening dear viewers)
TSPNet-HF$_{avg\_pool}$:	jetzt wünsche ich ihnen noch einen schönen abend (now i wish you a nice evening)
Ground Truth:	morgen temperaturen von achtzehn grad an der küste bis sechsundzwanzig im breisgau (tomorrow temperatures from eighteen degrees on the coast to twenty-six in breisgau)
TSPNet-H:	morgen siebzehn grad am alpenrand bis achtundzwanzig grad am oberrhein (tomorrow seventeen degrees on the edge of the alps to twenty-eight degrees on the upper rhine)
TSPNet-HF$_{avg\_pool}$:	morgen reichen die temperaturen von achtzehn grad an der nordsee bis achtundzwanzig grad im breisgau (tomorrow the temperatures will range from eighteen degrees on the north sea to twenty-eight degrees in breisgau)
Ground Truth:	am mittwoch zieht von der nordsee regen heran der sich am donnerstag ausbreitet (on wednesday rain pulls in from the north sea which spreads on thursday)
TSPNet-H:	am mittwoch schneit es gebietsweise am mittwoch in der osthälfte auch am mittwoch ist es meist bewölkt oder neblig trüb (on wednesday it snows in some areas on wednesday in the eastern half also on wednesday it is mostly cloudy or foggy)
TSPNet-HF$_{avg\_pool}$:	am mittwoch fällt in der nordhälfte gebietsweise regen der breitet sich langsam ostwärts aus (on wednesday rain falls in the northern half of the area and it slowly spreads eastwards)

In the second example, the TSPNet-HF$_{avg\_pool}$ produced an accurate translation; however, it made two mistakes: it changed "coast" to "north sea," and it missed the value of the second temperature. An important point to highlight is that our approach is capable to identify the city name "Breisgau". The "Breisgau" sign is composed of hand gestures and facial expressions. As our approach includes both information in its pipeline, recognizing these signs became less complex. On the other hand, the TSPNet-H failed in the translation, producing a spoken sentence with critical mistakes. Finally, we bring the third example to highlight, even more, the importance of facial expressions in the translation process. In [9], the TSPNet-H faced some problems in differentiating similar signs with different meanings (e.g., rain, shower, and snow). These specific signs have similar hand gestures, but their main differences come from facial expressions. As the TSPNet-H does not consider facial expressions in its translation process, these signs are mistranslated most of the time. As seen in the fourth example, the TSPNet-H produced a sentence with the word "snow" instead of "rain". This mistranslation is produced for many other examples of the Phoenix-Weather dataset. On the other hand, as the TSPNet-HF$_{avg\_pool}$ includes facial expressions in its process, the mistranslation is avoided in most of the examples.

7 Conclusion

This paper proposes the TSPNet-HF, an algorithm for video sign language translation, an improved version of the well-known TSPNet. Unlike the TSPNet, the proposal considers both hand and facial features in the SLT process. For this, we integrated a procedure to extract facial expressions from video frames, and implemented aggregation strategies to combine the hand and face feature vectors. The TSPNet-HF was compared to the original TSPNet and CNN2dRNN in the PHOENIX14T dataset. All methods were assessed in three translation metrics: ROUGE and BLEU-1, BLEU-2 and BLEU-3. Quantitatively, the TSPNet-HF using the average pooling aggregation strategy overcame the TSPNet and CNN2dRNN in all metrics statistically. The best result that the original TSPNet reached was the second position in the BLEU-1 metric. We also performed a qualitative analysis to investigate how facial expressions impacted the spoken language generation. The results showed that the TSPNet-HF avoided the original TSPNet's mistranslation problem in the case of signs whose differences are in the facial expression (e.g., rain, shower, and snow). In future work, we intend to use an I3D model pre-trained in facial expressions (e.g., action units) to improve the extraction of facial information from video frames. Besides, we also plan to assess the current approach in different SLT datasets.

Acknowledgment. Lenovo partially funded this research as part of its R&D investment under Brazil's Informatics Law. The authors want to acknowledge the support of Lenovo R&D and CESAR Labs.

References

1. Buehler, P., Zisserman, A., Everingham, M.: Learning sign language by watching tv (using weakly aligned subtitles). In: 2009 IEEE Conference on Computer Vision and Pattern Recognition, pp. 2961–2968. IEEE (2009)
2. Camgoz, N.C., Hadfield, S., Koller, O., Ney, H., Bowden, R.: Neural sign language translation. In: Proceedings of the IEEE Conference on Computer Vision and Pattern Recognition, pp. 7784–7793 (2018)
3. Camgoz, N.C., Koller, O., Hadfield, S., Bowden, R.: Sign language transformers: joint end-to-end sign language recognition and translation. In: Proceedings of the IEEE/CVF Conference on Computer Vision and Pattern Recognition, pp. 10023–10033 (2020)
4. Carreira, J., Zisserman, A.: Quo vadis, action recognition? a new model and the kinetics dataset. In: proceedings of the IEEE Conference on Computer Vision and Pattern Recognition, pp. 6299–6308 (2017)
5. Farooq, U., Rahim, M.S.M., Sabir, N., Hussain, A., Abid, A.: Advances in machine translation for sign language: approaches, limitations, and challenges. In: Neural Computing and Applications, pp. 1–43 (2021)
6. Heinzerling, B., Strube, M.: Bpemb: tokenization-free pre-trained subword embeddings in 275 languages. arXiv preprint arXiv:1710.02187 (2017)
7. Kudo, T., Richardson, J.: Sentencepiece: a simple and language independent subword tokenizer and detokenizer for neural text processing. arXiv preprint arXiv:1808.06226 (2018)
8. Li, D., Rodriguez, C., Yu, X., Li, H.: Word-level deep sign language recognition from video: a new large-scale dataset and methods comparison. In: Proceedings of the IEEE/CVF Winter Conference on Applications of Computer Vision, pp. 1459–1469 (2020)
9. Li, D., et al.: Tspnet: hierarchical feature learning via temporal semantic pyramid for sign language translation. arXiv preprint arXiv:2010.05468 (2020)
10. Luong, M.T., Pham, H., Manning, C.D.: Effective approaches to attention-based neural machine translation. arXiv preprint arXiv:1508.04025 (2015)
11. Ott, M., et al.: fairseq: a fast, extensible toolkit for sequence modeling. arXiv preprint arXiv:1904.01038 (2019)
12. Pfau, R., Salzmann, M., Steinbach, M.: The syntax of sign language agreement: common ingredients, but unusual recipe. Glossa J. Gener. Linguist. 3(1) (2018)
13. Pohlert, T.: The pairwise multiple comparison of mean ranks package (pmcmr). R Package 27(2019), 9 (2014)
14. da Silva, E.P., Costa, P.D.P., Kumada, K.M.O., De Martino, J.M., Florentino, G.A.: Recognition of affective and grammatical facial expressions: a study for Brazilian sign language. In: Bartoli, A., Fusiello, A. (eds.) ECCV 2020. LNCS, vol. 12536, pp. 218–236. Springer, Cham (2020). https://doi.org/10.1007/978-3-030-66096-3_16
15. Vaswani, A., et al.: Attention is all you need. In: Advances in Neural Information Processing Systems, pp. 5998–6008 (2017)
16. Zimmerman, D.W., Zumbo, B.D.: Relative power of the wilcoxon test, the friedman test, and repeated-measures anova on ranks. J. Exp. Educ. 62(1), 75–86 (1993)

Phonetic Speech Segmentation of Audiobooks by Using Adapted LSTM-Based Acoustic Models

Zdeněk Hanzlíček$^{(\boxtimes)}$ ⓘ and Jindřich Matoušek ⓘ

NTIS - New Technology for the Information Society, Faculty of Applied Sciences,
University of West Bohemia, Univerzitní 22, 306 14 Plzeň, Czech Republic
{zhanzlic,jmatouse}@ntis.zcu.cz
https://www.ntis.zcu.cz/en

Abstract. This paper describes experiments on phonetic speech segmentation of audiobooks by using LSTM neural networks. The segmentation procedure includes an iterative adaptation of an initial speaker-independent model. The experimental data involves 5 audiobooks recorded by various renowned Czech speakers. About 20 min long portions of each audiobook were precisely manually segmented by phonetic experts. We focused mainly on the optimal setting of the iterative segmentation procedure and explored the effect of the most relevant parameters on the resulting segmentation accuracy.

Keywords: Audiobooks · Speech segmentation · LSTM

1 Introduction

The general aim of phonetic speech segmentation is the determination of phone boundaries in a speech recording. Accurate speech segmentation is important for many applications in the field of speech processing. The main segmentation task has several variations. This paper is concerned with the task when only an orthographic transcription is available and the segmentation process also involves the selection of correct phonetic transcriptions and the insertion of pauses between words.

The first methods for automatic speech segmentation utilized a table of acoustic-phonetic rules [20]. Afterwards, statistical segmentation methods based on the use of hidden Markov models HMMs [1,16] were predominant for a long time. A significant method competitive to HMMs utilized the dynamic time warping technique [10,14].

In the last decade, neural networks have played an important role in almost all speech processing applications [7]. One of the most widely used type of neural network are long short-term memory (LSTM) recurrent neural networks [8] which are suitable for modeling time series data [2]; therefore, they are convenient for speech processing applications [3] as well.

A. C. Bicharra Garcia et al. (Eds.): IBERAMIA 2022, LNAI 13788, pp. 317–327, 2022.
https://doi.org/10.1007/978-3-031-22419-5_27

This paper presents experiments on speech segmentation of Czech audiobooks by using LSTM neural networks. The segmentation procedure includes an iterative adaptation of an initial speaker-independent model. The segmentation performance was evaluated by comparison with a reference phonetic segmentation created by phonetic experts.

This paper is structured as follows: Sect. 2 describes speech data used in this research. Section 3 deals with the definition of the phonetic alphabet. Section 4 gives an overview of the neural network architecture and the segmentation process. Experiments and results are presented in Sect. 5. Conclusions and plans for the future work are given in Sect. 6.

2   Data Description

In this research, we used 2 different data sets:

1. Multilingual data used for training the default speaker-independent model are described in Table 1. The benefit of using multiple languages is a more general and robust resulting model. This data was mostly recorded by professional voice talents for the purposes of speech synthesis [18] and was supplemented by phonetic segmentation created by a HMM-based system with various additional correction procedures [17].
2. Audiobook data used for segmentation experiments are described in Table 2. All audiobooks were read by various renowned Czech speakers. About 20 min long part of each audiobook was available; this part was supplemented with its accurate text transcription and phonetic segmentation created manually by phonetic experts. Audiobook data was split into files with a duration limited to 1 min. Contrary to the speech-synthesis data recorded in a plain neutral style, the audio-book data is more expressive; the intonation, tempo, and style change often, reflecting the content. Data also contains frequent direct speech where the speaker can also variously change his/her voice characteristics to imitate the individual characters.

Table 1. Training data used for the default speaker independent model.

Language	Speakers	Utterances	Duration [hrs]
Czech	19	105k	127
American English	3	59k	52
British English	4	51k	35
German	2	33k	30
Slovak	1	12k	16
Russian	3	54k	41
Total	34	313k	301

Table 2. Description of audiobook data; speech and silence duration corresponds to the reference segmentation created by phonetic experts.

Audiobook	Total duration [min:sec]		Audio files	
	Speech	Silence	Number	Duration [sec]
AB1	16:11	4:33	23	40.8–59.3
AB2	15:19	5:31	23	39.8–62.1
AB3	16:46	4:43	25	27.1–59.7
AB4	15:50	6:37	25	43.8–59.5
AB5	15:26	7:19	26	31.6–58.1

3 Phonetic Alphabet

In our research, we used the International Phonetic Alphabet (IPA) [11], which allows a straight and consistent combination of different languages. The complete list of phones contained in our multilingual training data, including diacritic combinations, is presented in Table 3. Besides, we used 3 additional units: short interword pause, long pause (initial and terminal pauses of individual recordings), and loud breath. The complete phonetic set contains 139 units (136 phones + 3 additional units).

Many previous works on speech segmentation [6,9] demonstrated that using a large phonetic alphabet is often disadvantageous, especially for multilingual modeling; the overall robustness of the segmentation process can be increased by clustering the similar phones into phonetic groups and using a reduced phonetic alphabet.

In our previous study on cross-lingual modelling and segmentation [5], we proposed a substantially reduced alphabet, where all clustered phones were required to occur in all languages and the default multilingual phonetic alphabet was therefore reduced from 127 to 26 phones. In this work, we defined a "richer" alphabet with 39 clustered phones. The larger alphabet helped us to avoid the situation when several phonetic variants have the same reduced phonetic form and the default full phonetic form cannot be unambiguously selected, for example, distinguishing between an affricate and separated phones ($\widehat{ts}$ and t + s). The alphabet reduction process can be described by the following steps:

1. Diphthongs are split into particular single phones, e.g. ɛɪ → ɛ + j, ɛə → ɛ + ə, aʊ → a + ʊ, ɔʏ → ɔ + j, etc. Vowels ɪ and ʏ as less syllabic phones within diphthongs are changed to the consonant **j**.
2. Palatalized consonants (except for **dʲ**, **nʲ** and **tʲ** – see Table 4) are considered as the default consonant followed by the phone **j**, e.g. **pʲ** → p + j, **bʲ** → b + j, etc. When the palatalized consonant is followed by **j** or ɪ, the extra phone **j** is not added.[1]

---

[1] The palatalization phenomenon is more complex, yet this simplification is acceptable for this task.

3. Nasalized vowels (present in French loanwords in the German language) are changed to a vowel followed by phone ŋ, e.g. ã → a + ŋ, ɛ̃ → ɛ + ŋ, õ: → o: + ŋ, etc. When the nasalized vowel is followed by ŋ or n, the extra phone ŋ is not added.

Table 3. The complete list of phones: A = American English, B = British English, C = Czech, G = German, S = Slovak, R = Russian. The list is sorted by the IPA number in a line-wise manner.

IPA	Languages	IPA	Languages	IPA	Languages	IPA	Languages
p	A,B,C,G,R,S	pf	G	pʲ	R	b	A,B,C,G,R,S
bʲ	R	t	A,B,C,G,R,S	ts	C,G,R,S	tʃ	A,B,C,G,S
t̪	A	tʲ	R	d	A,B,C,G,R,S	dz	C,S
dʒ	A,B,C,G,S	dʲ	R	c	C,S	ɟ	C
ɟ̟	S	k	A,B,C,G,R,S	kʲ	R	g	A,B,C,G,R,S
gʲ	R	ʔ	A,C,G,S	m	A,B,C,G,R,S	mʲ	R
m̪	A,C	ɱ	B,C,G,S	n	A,B,C,G,R,S	n̪	S
nʲ	R	n̺	A	ɲ	C,S	ɲ̟	S
ŋ	A,B,C,G,S	ŋ:	G	r	C,G,R,S	r̝̊	C
rʲ	R	r̝	C	r̩	C,S	r̩:	S
ɾ	A	f	A,B,C,G,R,S	fʲ	R	v	A,B,C,G,R,S
vʲ	R	θ	A,B	ð	A,B	s	A,B,C,G,R,S
sʲ	R	z	A,B,C,G,R,S	zʲ	R	ʃ	A,B,C,G,S
ʒ	A,B,C,G,S	ʂ	R	ʐ	R	ç	G
x	C,G,R,S	xʲ	R	ɣ	C,S	ʁ	G
h	A,B,G	ɦ	C,S	ʋ	S	ɻ	A,B
j	A,B,C,G,R,S	l	A,B,C,G,S	lʲ	R	l̩	A,C,S
l̩:	S	ʎ	S	w	A,B	ɕ:	R
ʑ:	R	ɫ	R	tɕ	R	i	G,R
iː	A,B,C,G,S	ĭ	R	e	G,R	eɪ	A,B
eː	G	ě	R	ɛ	A,B,C,G,R,S	ɛʊ	C
ɛə	B	ɛ̃	G	ɛ̃:	G	ɛː	C,G,S
a	C,G,R,S	aɪ	A,B,G	aʊ	A,B,C	ã	G
ã:	G	aː	C,G,S	ɑ	A	ɑː	B
ɔ	G,S	ɔɪ	A,B	ɔɣ	G	ɔː	A,B
o	C,G,R	oʊ	A,C	õ:	G	oː	C,G,S
u	G,R,S	uː	A,B,C,G,S	y	G	yː	G
ø	G	øː	G	œ	G	œː	G
ɒ	A,B	ʌ	A,B	ɨ	A,R	ʉ	A,R
ɪ	A,B,C,G,R,S	ɪɛ	S	ɪa	S	ɪu	S
ɪə	B	i̯	S	ʏ	G	ʊ	A,B,C,G,R
ʊɔ	S	ʊə	B	u̯	S	ə	A,B,G,R
əʊ	B	ɚ	A	ɵ	R	ɐ	G,R
æ	A,B,G,R,S	ɜ	A	ɝ	A	ɜː	B

4. Removing all remaining IPA diacritical marks, e.g. vowel duration (aː → a, ɔː → ɔ, etc.) and consonant syllabicity (r̩ → r, l̩ → l, etc.)
5. Acoustically similar phones are clustered together. The resulting phonetic alphabet is presented in Table 4.

Table 4. Definition of the reduced alphabet. Each phonetic group contains a characteristic (highlighted) phone.

Single phones	p b t ts dz dʒ k g ʔ f j v s θ z ð æ					
Phonetic groups	**c** tʲ	**ɟ** dʲ	**m** ɱ	**n** ŋ	**ɲ** nʲ	**r** ɾ ʁ ɹ
	ʃ ʂ ɕ	**ʒ** ʐ ʑ	**x** ç h	**ɦ** ɣ	**l** ʎ ɫ	**i** ɨ ɪ
	y ʏ	**e** ɛ	**ø** œ	**a** ɑ ʌ ɐ	**o** ɔ	**ɒ** ɵ
	u ʉ ʊ	**ə** ɜ ɚ ɝ	**tʃ** tɕ	**w** ʊ		

4 System Description

For acoustic modeling, speech was represented by 13-dimensional MFCC vectors. The audio waveform was downsampled from the default sampling rate to 16 kHz and preprocessed by a preemphasis filter. The frame shift was 5 ms, i.e. the resolution of the resulting segmentation was 5 ms as well. The acoustic features were normalized.

The network for acoustic modeling consisted of 2 bidirectional LSTM layers followed by linear projection and softmax activation. Each bidirectional LSTM layer contained 2×512 cells[2]. During training, cross entropy loss was minimized and the standard ADAM optimizer was utilized [13].

The trained network performs frame-wise phonetic classification [4]; for a given utterance, it outputs so-called phonetic posteriorgram, which is a matrix in which rows correspond to phones and columns to speech frames. According to the phonetic transcription provided, the corresponding lines of the posteriorgram are composed to a new matrix in which the optimal alignment between frames and phones (i.e., the phonetic segmentation) can be determined by the dynamic programming [12].

Since each word can have more than one phonetic transcription, the alignment procedure is then performed concurrently through matrices corresponding to particular transcription variants; therefore, the resulting frames-to-phones assignment also includes the selection of the best-matching transcription variant. Although the most suitable phonetic transcription of each word is selected, the resulting segmentation may be not always fully correct. The misalignment has two basic reasons.

---

[2] Both forward and backward layers consisted of 512 LSTM cells, i.e. each bidirectional layer contained 1024 cells.

- The actual pronunciation does not fully correspond to the selected phonetic variant; either the annotation/transcription is not fully correct, or the word is mispronounced.
- The acoustic model may be insufficiently robust to work properly on a given voice.

We adopt a simple procedure for the detection of incorrectly aligned phones, referred to as *invalid phones*, similar as proposed in [5]. When more than 50% frames assigned to a phone do not have the highest posterior probability (value in the phonetic posteriogram) for the given phone, the segmentation is supposed to be inaccurate and the phone is labelled invalid.

The best segmentation performance would be naturally achieved when the segmented voice is included in the training data. Regarding the difference between training and segmented data, the acoustic model can fit the segmented data improperly. Using speaker adaptation methods can improve the acoustic model performance [15].

In our previous work on cross-lingual speech segmentation [5], we proposed a simple iterative procedure updating the default speaker-independent model by rotation of two steps: data segmentation and model retraining. The main idea is depicted in Fig. 1. The adapted model will certainly not achieve the same performance as if it were trained from a correct reference segmentation, but the experiments proved that the resulting segmentation is gradually improved during several initial iterations.

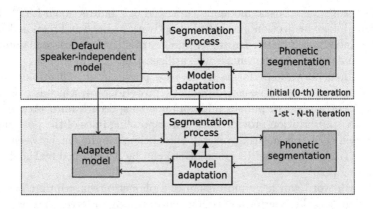

Fig. 1. The basic scheme of the iterative segmentation procedure.

5   Experiments

For evaluation of segmentation accuracy, we used the following metrics:

- Mean segmentation error, MSE [ms] – the average distance between matching phone boundaries

- Segmentation accuracy, SA [%] – the percentage of boundaries placed in the allowed interval (usually 20 or 10 ms) around the corresponding reference boundaries
- Relative number of invalid phones, IP [%] – explained in the previous section

MSE and SA are calculated for boundaries between phones that are identical in both reference and evaluated segmentation; phones that cannot be decidedly assigned, are not included; the similarity or interchangeability of particular phones was not taken into account (Table 5).

Table 5. Segmentation results achieved by the default speaker-independent model.

Speaker	MSE [ms]	SA [%]		IP [%]
		10 ms	20 ms	
AB1	11.7	67.0	85.9	17.0
AB2	10.7	71.9	87.5	15.2
AB3	17.9	57.6	76.6	38.1
AB4	10.7	68.3	86.8	14.8
AB5	15.8	57.8	79.0	28.3

5.1 Experiment I: The Effect of Detection and Excluding the Invalid Phones

During segmentation, some phones and speech frames may be incorrectly aligned. In Sect. 4, we described a simple procedure for detection of such phones (so-called invalid phones). Including such phones in the training data could have 2 potential contradictory effects:

1. Positive effect: The adapted model will be more robust and the segmentation accuracy will improve.
2. Negative effect: The adapted model will be less accurate and the segmentation accuracy will deteriorate.

To explore which effect predominates, we performed a comparison of 3 different settings:

1. All invalid phones were ordinarily included in the training data.
2. All invalid phones were excluded from the training data.
3. Invalid phones were excluded along with the immediate neighboring phones, since the incorrectly aligned phones can also affect their vicinity, e.g. by shifting the mutual boundary.

The segmentation results are presented in Fig. 2. In all cases, the best results were obtained for the third setting, i.e. excluding invalid phones with their immediate neighbours. The segmentation accuracy improved during ca. 4 initial iterations, which is in agreement with results presented in [5]. Afterwards, the iterative procedure diverged, whereas more stable course was reported in [5]. The reason is probably a different type and amount of speech data in both studies. Generally, we can conclude that selecting segmentation from the 3rd iteration is probably the optimal decision for any type of speech data; the corresponding results are presented in Table 6.

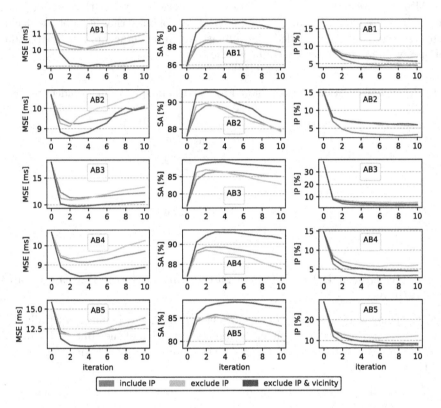

Fig. 2. The effect of including/excluding invalid phones (IP) from training data during the iterative segmentation procedure.

5.2   Experiment II: The Strategy of Model Adaptation

We compared 2 basic versions of the iterative segmentation procedure:

1. *Sequential Adaptation* - The model is gradually adapted, i.e. the model updated in the previous step is used in each iteration. This strategy corresponds to the procedure proposed in [5] and to the basic scheme in Fig. 1.
2. *Default Model Adaptation* - The default speaker-independent model is adapted in each iteration. This modification was primarily introduced to prevent the gradual deterioration of the acoustic model during for a higher number of iterations.

Table 6. Improvement of the segmentation results after 3 iterations when excluding the invalid and immediate neighboring phones from the training data.

Speaker	MSE [ms]	SA [%]		IP [%]
		20 ms	10 ms	
AB1	11.7 → 9.1	85.9 → 90.6	67.0 → 72.5	17.0 → 6.7
AB2	10.7 → 8.7	87.5 → 90.8	71.9 → 75.7	15.2 → 6.8
AB3	17.9 → 9.7	76.6 → 89.2	57.2 → 71.1	38.1 → 5.2
AB4	10.7 → 8.4	86.8 → 91.2	68.3 → 74.7	14.8 → 5.3
AB5	15.8 → 10.3	79.0 → 88.0	57.8 → 69.1	28.3 → 10.3

Besides, we compared various numbers of epochs in each iteration. The results are presented in Fig. 3. In most cases, the best segmentation accuracy was achieved around the 3rd–4th iteration. Both adaptation strategies produce similar results, none of them is definitely superior. The optimal number of epochs

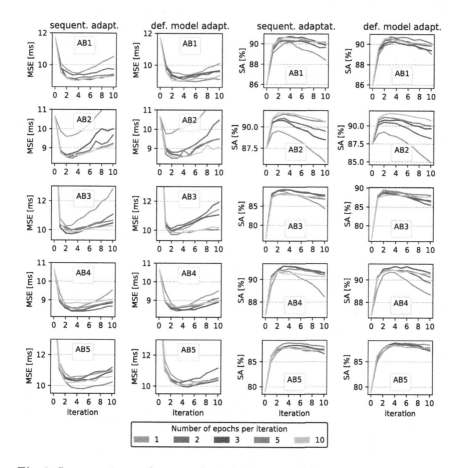

Fig. 3. Segmentation performance for various number of epochs in each iteration.

in each iteration cannot be clearly selected as well. Generally, the best results were obtained for 3–5 epochs in each iteration. A smaller number of epochs seems to be insufficient and more epochs do not guarantee a better segmentation accuracy.

Since the previous experiment used sequential adaptation with 3 epochs per iteration, which were selected as the recommended setting, the results presented in Table 6 are representative for this experiment as well.

6   Conclusion

This paper presented an initial research on speech segmentation of audiobooks using LSTM neural networks. The default speaker-independent model was trained on Czech, English, German, Slovak and Russian speech data mostly recorded by professional voice talents for the purposes of speech synthesis [18]. Segmentation experiments were performed on about 20 min long parts of 5 audiobooks recorded by various renowned Czech speakers. Segmentation accuracy was evaluated by comparison with a reference segmentation manually created by phonetic experts.

Many experiments were performed exploring various experimental conditions. The best segmentation performance was achieved after 3 iterations of the adaptive procedure with 3–5 epochs in each iteration, excluding the invalid phones along with the neighboring phones. The resulting segmentation accuracy was about 88–91% and the mean segmentation error was 8.4–10.3ms, which are comparable results to similar experiments performed on high-quality large speech corpora created for speech synthesis applications.

In our future work, we will analyze the segmentation accuracy and errors in more detail. We intend to explore the reason for the accuracy divergence after 3–4 improving initial steps of the iterative segmentation procedure. Our experiments shown, that it is not possible to select one generally optimal setting for all voices. An alternative way to increase segmentation performance is a felicitous combination of several independent segmentation systems [19,21].

Acknowledgement. This research was supported by the Czech Science Foundation (GA CR), project No. GA22-27800S. Computational resources were supplied by the project "e-Infrastruktura CZ" (e-INFRA CZ LM2018140) supported by the Ministry of Education, Youth and Sports of the Czech Republic.

References

1. Brugnara, F., Falavigna, D., Omologo, M.: Automatic segmentation and labeling of speech based on Hidden Markov Models. Speech Commun. **12**, 357–370 (1993)
2. Gers, F.A., Schraudolph, N.N., Schmidhuber, J.: Learning precise timing with LSTM recurrent networks. J. Mach. Learn. Res. **3**, 115–143 (2003)
3. Graves, A.: Supervised Sequence Labelling with Recurrent Neural Networks, Studies in Computational Intelligence, vol. 385. Springer-Verlag, Heidelberg (2012). https://doi.org/10.1007/978-3-642-24797-2_2

4. Graves, A., Schmidhuber, J.: Framewise phoneme classification with bidirectional LSTM and other neural network architectures. Neural Netw. **18**, 602–610 (2005)

5. Hanzlíček, Z., Vít, J.: LSTM-based speech segmentation trained on different foreign languages. In: Sojka, P., Kopeček, I., Pala, K., Horák, A. (eds.) TSD 2020. LNCS (LNAI), vol. 12284, pp. 456–464. Springer, Cham (2020). https://doi.org/10.1007/978-3-030-58323-1_49

6. Haubold, A., Kender, J.R.: Alignment of speech to highly imperfect text transcriptions. In: Proceeding of ICME 2007, pp. 224–227 (2007)

7. Hinton, G., et al.: Deep neural networks for acoustic modeling in speech recognition: the shared views of four research groups. IEEE Signal Process. Maga. **29**(6), 82–97 (2012)

8. Hochreiter, S., Schmidhuber, J.: Long short-term memory. Neural Comput. **9**, 1735–1780 (1997)

9. Hoffmann, S., Pfister, B.: Text-to-speech alignment of long recordings using universal phone models. In: Proceedings of Interspeech 2013, pp. 1520–1524 (2013)

10. Hunt, M.J.: Time alignment of natural speech to synthetic speech. In: Proceedings of ICASSP 1984, pp. 251–254 (1984)

11. International Phonetic Association: Handbook of the International Phonetic Association: A Guide to the Use of the International Phonetic Alphabet. Cambridge University Press, Cambridge (1999)

12. Kelley, M.C., Tucker, B.V.: A comparison of input types to a deep neural network-based forced aligner. In: Proceedings of Interspeech 2018, pp. 1205–1209 (2018)

13. Kingma, D.P., Ba, J.L.: Adam: a method for stochastic optimization. In: Proceedings of ICLR 2015 (2015)

14. Lennig, M.: Automatic alignment of natural speech with a corresponding transcription. Speech Commun. **2**, 190–192 (1983)

15. Liu, C., Wang, Y., Kumar, K., Gong, Y.: Investigations on speaker adaptation of LSTM RNN models for speech recognition. In: Proceedings of ICASSP 2016 (2016)

16. Ljolje, A., Riley, M.D.: Automatic segmentation and labeling of speech. In: Proceedings of ICASSP 1991, pp. 473–476 (1991)

17. Matoušek, J., Romportl, J.: Automatic pitch-synchronous phonetic segmentation. In: Proceedings of Interspeech 2008, pp. 1626–1629 (2008)

18. Matoušek, J., Tihelka, D., Romportl, J.: Building of a speech corpus optimised for unit selection TTS synthesis. In: Proceedings of LREC 2008 (2008)

19. Stolcke, A., Ryant, N., Mitra, V., Yuan, J., Wang, W., Liberman, M.: Highly accurate phonetic segmentation using boundary correction models and system fusion. In: Proceedings of ICASSP 2014 (2014)

20. Wagner, M.: Automatic labelling of continuous speech with a given phonetic transcription using dynamic programming algorithms. In: Proceedings of ICASSP 1981, pp. 1156–1159 (1981)

21. Wong, J.H.M., et al.: Ensemble combination between different time segmentations. In: Proceedings of ICASSP 2021, pp. 6768–6772 (2021)

Robotics and Computer Vision

Depth Estimation from a Single Image Using Line Segments only

José G. Nava Zavala$^{(\boxtimes)}$ and Jose Martinez-Carranza

Instituto Nacional de Astrofisica, Optica y Electronica, Luis Enrique Erro No. 1, Sta. Ma. Tonantzintla, Puebla, Mexico
{jose.nava,carranza}@inaoep.mx

Abstract. We present a method for depth estimation from a single image using an intermediate representation in the form of line segments. Rather than regressing depth from a chromatic image in RGB format, we explore the use an image containing line segments extracted from the original chromatic image using the Line Segment Detector (LSD), arguing that this image, even when sparse in visual data, still contains information to infer a depth image. Our proposed approach has been tested on the NYU-depth dataset for indoor scenes and on simulated images created with Airsim, seeking to assess the performance of our method with synthetic images. Our experiments show promising results confirming that it is possible to estimate a depth image from a single image containing line segments only.

Keywords: Depth estimation · Monocular camera · Line segments

1 Introduction

Since the pioneering work of Saxena [18], depth estimation from a single image has received wide attention in the last years, primarily due to deep learning techniques, which have contributed to the development of methods based on neural networks with impressive results [24]. However, several of these methods involve supervised training, thus requiring large amounts of labelled image data. One of the first datasets providing chromatic and depth images was that of KITTI [8], containing images captured from outdoor scenarios and intended to serve as a benchmark for robotic algorithms, e.g., visual odometry, visual SLAM and object detection algorithms. More specialised datasets have been made public such as the NYU depth dataset for indoor scenarios [19], and more recently, much larger datasets [17,20], whose particularity is that of providing thousand of synthetic images generated from virtual environments representing a wide set of scenarios, e.g., urban, neighbourhoods, woods, mountains, fields, farms, etc.

But, while the number of datasets could grow, facilitating much more varied examples to leverage the training, one could wonder whether all the examples are necessary, this is, whether they enhance or bias the model. These are typical questions in machine learning, taking relevance in the context of depth estimation

A. C. Bicharra Garcia et al. (Eds.): IBERAMIA 2022, LNAI 13788, pp. 331–341, 2022.
https://doi.org/10.1007/978-3-031-22419-5_28

from a single image where a depth image may correspond to several images representing the same scene. For instance, suppose that a camera is placed in the same corner of a room, capturing images in the morning, midday, and evening. There is no doubt that the corresponding dept image would be the same for all these images, assuming the objects in the room are still, and no dynamic objects pass by. However, these chromatic images could vary drastically upon illumination changes induced by natural and artificial light.

Motivated by the above, in this work, we explore the use of an intermediate representation that could be used as a prototype representing essential information of several image instances of the same scene, whose variability is produced by illumination changes. To this end, we propose to use an image containing line segments detected with the well-known Line Segment Detector (LSD) [9]. This proposal is inspired by the fact that humans are good at extracting complex information from sketches drawn on paper, such as recognising faces, objects and 3D structure. This has been explored in [3] for face recognition from sketches using deep learning, and the same concept can be extended to object and scene recognition [22].

Therefore, our results indicate that an image containing line segments of a scene only, even when sparse, could still serve as a sort of sketch from which depth could also be learned. To address this, we propose to train DenseDepth [1], a state-of-the-art method for depth estimation from single images, with images containing line segments labelled with their corresponding depth images, aiming at evaluating the quality of the depth estimation when using only line segments as opposed to using the full chromatic image, illustrated in Fig. 2 and sample results in Fig. 1.

To present our work, the rest of the paper is organised as follows: Sect. 2 discusses relevant related work; Sect. 3 describes our methodology; Sect. 4 presents our experiments and initial results, which we considered promising; finally, Sect. 5 summarises our conclusions.

2   Related Work

2.1   Ground Truth

There exist multiple methods for depth estimation, the state-of-the-art uses specialized hardware based methods such as LIDAR, Time of Flight or enhanced stereo-pair cameras such as Kinect and Kinect V2 or Intel Realsense. Hardware based methods are generally used directly, like LIDAR for autonomous cars, or, with the desire to remove this hardware barrier, to generate datasets such as the well known NYU-Depth and NYU-DepthV2 [19] and KITTI [7].

2.2   Monocular Methods

Monocular depth estimation is an ill-posed problem. Ambiguities can arise, scale ambiguities, translucent or reflective materials such as glass can have images where

Fig. 1. Comparison among images with chromatic input and our method. Input images are shown in the first row and result LSD images used are in the second row. Depth Ground Truth is shown in the third row, missing values in white. Depth image estimated with the network trained with RGB images are shown in the fourth row. Depth images estimated with the same network but trained with line segments (LSD) are shown in the fifth row.

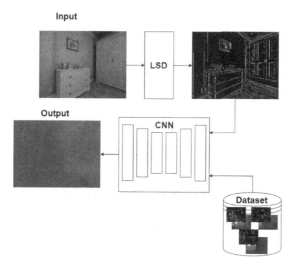

Fig. 2. Diagram Overview of the proposed method. First, we convert images to the intermediate representation and train a CNN. The output is the depth estimation for a Line segment image of the scene

the geometry of a scene cannot be derived. This problem has been approached by many by using a CNN [1,2,5,6,10,13,16,21] with a steady increase in performance as Vision tasks methods are applied and deeper networks are implemented, and changes in the network architecture like the use skip connections, have demonstrated that a monocular camera can have good results for this problem. Other methods include the use of supervised learning with MRF [18] with single image, and using this information to enhance stereo methods.

2.3   Line Segments and Edges

The use of Line Segments have been used to reconstruct 3D models from a series of images [14] to reconstruct 3d models from a series of photos, identifying the planes delimited by the line segments and fusing the planes to recreate the model.

The use of lines for pose, Yu et al. [23] uses Line Segment detection matches 3D lines to captured 2D lines, this correspondence between captured 2d image from the camera and the 3D model allows for the estimation of pose.

Complex reconstruction from edges have recently shown complex image translation by reconstructing human faces translating from a sketch [15], Li et al. generated photo-realistic photos of faces from edges from synthetically distorted sketches or human drawn sketches.

LSD has been used in [11], an unsupervised method for depth estimation, but in contrast to us, this unsupervised method uses two adjacent images in an image training dataset to estimate relative motion and thus be able to reconstruct a relative 3D model of the views. The authors modified DepthNet [4] to learn the coefficients of 3D planar structures given the relative 3D model and in this context, segmented planes and lines are used, in the loss function, to evaluate the linear consistency of the projected 3D planes on the images. Thus, lines are used as part of the evaluation mechanism instead of being used as visual data to infer depth.

3   Methodology

We propose using Line Segments from a monocular chromatic camera as an intermediate representation to train a neural network that can estimate a corresponding depth map from a single image in a scene. Human architecture consists of planes with texture-less surfaces (e.g., walls, the floor) whose edges represent structural features that can be represented by line segments. For the neural network, we propose using state-of-the-art networks that produce high-quality depth maps with RGB images and process the input with the proposed representation.

3.1   Intermediary Representation

For the LSD algorithm, we use OpenCV implementation, which follows the implementation of Grompone et al. [9]. First, the RGB image is loaded in grayscale to apply LSD algorithm, the network used for training still requires three-channel images, we used three different parameters for LSD initialization to find the Line Segments and each result is saved to a single image with each channel representing the different thresholds as illustrated in Fig. 3.

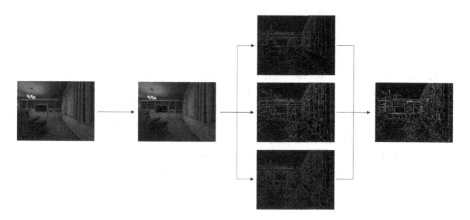

Fig. 3. Example of the generation of RGB-D data to train DenseDepth using LSD. To generate a chromatic image that could be read by the DenseDepth script, LSD parameters were varied for each channel.

We empirically found having three different parameters for LSD algorithm performed better than having the same parameters for the three channels. We found these values using a small (10) set of RGB images, selected values that gave the most line segments, and trained the network; the final values were those of the best performance reached in evaluation (654 samples). The best values reached where: length threshold = 8, distance threshold = 1.8, first canny threshold = 50.0, second canny threshold = 75.0, and aperture size for the Sobel operator = 3 as the parameters of one of the LSD, 10, 1.8, 25.0, 25.0, 3 as the parameters of another of the LSD and 8, 1.8, 25.0, 50.0, 3 as the parameters of one of the last LSD for each value respectably as reasonable values for the detectors.

3.2 Datasets

The datasets used are NYU Depth Dataset V2 and a synthetic dataset captured from a simulator. A copy of the NYUv2 dataset was saved, replacing the chromatic images and converting all scene chromatic images to the proposed intermediary representation.

For the synthetic dataset capture we use Airsim, a simulator that allows an RGB-D free-floating camera, and a script with position a rotation to capture the chromatic and corresponding depth image. The camera was positioned on a point in the room and captured 512 images while rotating 0.7° to capture around that point in the room, then moved to a different point and the process repeated to capture different points in all the rooms in the map.

Airsim uses the game engine Unreal Engine which allows ray-tracing technologies such as lighting, reflection, and ambient occlusion and these were enabled to achieve more realistic images. Both chromatic and depth images were saved with a resolution of 640×480.

3.3 Neural Network

For the neural network, we decided to use DenseDepth [1], a high-quality depth estimation from a monocular chromatic camera. DenseDepth uses transfer learning, using the DenseNet-169 network pre-trained on ImageNet for transfer learning for the encoder; other than changing the input data to train the network, no other changes have been made; we trained using the same architecture and start point as provided in [1] published code. This allows us to compare the results for the same image directly; by converting to the LSD intermediary representation for training and prediction, we can compare the results for the same input image.

3.4 Evaluation

For the quantitative evaluation to compare against state-of-the-art we propose the use of the same six standard metrics used in those works [1,5,6,12]

average relative error (rel):

$$\text{rel} = \frac{1}{n} \sum_{p}^{n} \frac{|y_p - \hat{y}_p|}{y}; \tag{1}$$

root mean squared error (rms)

$$\text{rms} = sqrt\frac{1}{n} \sum_{p}^{n} (y_p - \hat{y}_p)^2); \tag{2}$$

average $\log_{10}$ error (log10)

$$\log 10 = \frac{1}{n} \sum_{p}^{n} |\log_{10}(y_p) - \log_{10}(\hat{y}_p)| \tag{3}$$

threshold accuracy for $thr = 1.25, 1.25^2, 1.25^3$

$$\% \text{ of } y_p \mid \max(\frac{y_p}{\hat{y}_p}, \frac{\hat{y}_p}{y_p}) = \delta < thr \tag{4}$$

where y_p is a pixel in depth image y, $\hat{y}_p$ is a pixel in the predicted depth image $\hat{y}$, and n is the total number of pixels for each depth image.

4   Experiments   .

For this work, we focused on assessing our approach on images representing indoor scenarios. We use the popular NYU-Depth dataset and a simulated dataset created with Airsim from which high quality depth images can be generated, but also to assess the performance of our approach in synthetic images.

Since no changes were made to the DenseDepth network, the training times took the same time as a normal RGB input, around 24 h, the LSD step for all data was completed in 15 min for all data around 51K images. The training was performed on a computer with NVIDIA RTX 3090, AMD Ryzen 5950X, with 32GB of RAM.

For the simulated dataset, we used Airsim and a map from the Unreal Engine marketplace with minor changes, 129 different coordinates where taken and with a script for each scene 512 frames with chromatic and depth were saved as PNG, depth images where saved in range of 0–10 m (Kinect ranges between around 15cm to 10m) and 16 bits per pixel. 103 of the scenes were used for training, 17 for testing and 9 from a different room were used for evaluation.

The results obtained with the NYU-Depth dataset are summarised in Table 1, which show that our method performs very closely to state-of-the-art methods

Fig. 4. Comparison among images with higher and lower brightness. Input images are shown in the first row and result LSD images used are in the second row. Depth Ground Truth is shown in the third row, missing values in white. Depth image estimated with the network trained with RGB images are shown in the fourth row. Depth images estimated with the same network but trained with line segments (LSD) are shown in the fifth row. Note that our method is more consistent under these bright changes.

Table 1. Comparisons of different methods on the NYU Depth v2 dataset. The reported numbers are from the corresponding original papers. The best results are in bold, the higher the better.

Method	δ_1 ↑	δ_2 ↑	δ_3 ↑	rel↓	rms↓	log_{10} ↓
Eigen et al. [5]	0.769	0.950	0.988	0.158	0.641	–
Laina et al. [13]	0.811	0.953	0.988	0.127	0.573	0.055
MS-CRF [21]	0.811	0.954	0.987	0.121	0.586	0.052
Hao et al. [10]	0.841	0.966	0.991	0.127	0.555	0.053
Fu et al. [6]	0.828	0.965	0.992	**0.115**	0.509	**0.051**
Alhashim et al. [1]	**0.846**	**0.974**	**0.994**	0.123	**0.465**	0.053
Ours	0.766	0.945	0.984	0.162	0.675	0.068

Table 2. Comparison of generalization between the NYU-Depth V2 trained model and the synthetic dataset and with our proposed intermediate representation. The best results are bolded.

Input Image	Training	Evaluation	δ_1 ↑	δ_2 ↑	δ_3 ↑	rel↓	rms↓	log_{10} ↓
RGB	NYU	Synthetic	0.425	0.824	**0.957**	**0.293**	**1.331**	0.128
	Synthetic	NYU	0.253	0.524	0.752	0.668	1.561	0.203
LSD	NYU	Synthetic	**0.528**	**0.854**	0.949	0.304	1.348	**0.116**
	Synthetic	NYU	0.265	0.502	0.684	0.861	1.755	0.233

Table 3. Comparison of error between results on DenseDepth without and with our method

Method	Error mean	Error standard deviation	Error variance
Alhashim et al. [1]	−0.0786	0.5610	0.3147
Ours	−0.1218	0.6647	0.4419

with a difference of less than a tenth in the relative error, which could has no significance statistically speaking, The test statistic T equals 1.2701 with results obtained in Table 3, p-value equals 0.2043. This shows that even when sparse, line segments of a single image still provide enough information to estimate a depth image. When comparing with the simulated dataset a similar performance is observed for our method. For qualitative comparison, Fig. 4 shows some examples under illumination variations showing that our method exhibits a more consistent behaviour than when using RGB images for depth estimation. To evaluate the performance of our approach when evaluated with images whose visual texture may differ from that of the images used for training, we decided to train the DenseDepth network using images from NYU-Depth and then test it on synthetic images and vice-versa. This experiment was carried out using chromatic images as input for the training and also for images containing line segments. The results of this cross-over evaluation are shown in Table 2. RGB is used to indicate the use of chromatic images as input and LSD for images containing line segments. The results coincide with those obtained in Table 1 in that the difference is less than a tenth for the relative error.

5   Conclusion

We have presented initial results of a method for depth estimation from a single image using an intermediate representation in the form of a single image containing line segments extracted from the original chromatic image, arguing that even when these line segments contain sparse information, a depth image can still be estimated using a Convolutional Neural Network. This idea is motivated by the fact that supervised methods addressing this problem may require several chromatic images associated with the same depth image in order to become robust to illumination changes.

In contrast, inspired by methods performing neural inference from sketches such as face, object or scene recognition, we propose to use an image containing line segments as a form of sketch image. We have used the popular and robust LSD detector to extract line segments, carrying out experiments using the NYU dataset and a simulated dataset to asses the performance our approach with synthetic images. In both cases, our approach performs very closely to state-of-the-art methods with a difference of less than a tenth in the relative error, which could has no significance statistically speaking. In qualitative terms, depth images regressed from line segments are similar to those regressed with RGB images. Furthermore, our last set of experiments illustrate that depth estimation with line segments exhibit a more consistent performance under illumination variations than when using RGB images.

We considered these results encouraging and promising, and we will continue working with this approach for our future work testing with other networks and in outdoors scenes.

References

1. Alhashim, I., Wonka, P.: High quality monocular depth estimation via transfer learning. CoRR abs/1812.11941 (2018). https://arxiv.org/abs/1812.11941
2. Cao, Y., Wu, Z., Shen, C.: Estimating depth from monocular images as classification using deep fully convolutional residual networks. IEEE Trans. Circuits Syst. Video Technol. **28**(11), 3174–3182 (2018). https://doi.org/10.1109/tcsvt.2017.2740321
3. Chen, S.Y., Su, W., Gao, L., Xia, S., Fu, H.: Deep generation of face images from sketches. arXiv preprint arXiv:2006.01047 (2020)
4. CS Kumar, A., Bhandarkar, S.M., Prasad, M.: Depthnet: a recurrent neural network architecture for monocular depth prediction. In: Proceedings of the IEEE Conference on Computer Vision and Pattern Recognition Workshops, pp. 283–291 (2018)

5. Eigen, D., Puhrsch, C., Fergus, R.: Depth map prediction from a single image using a multi-scale deep network. CoRR abs/1406.2283 (2014). https://arxiv.org/abs/1406.2283
6. Fu, H., Gong, M., Wang, C., Batmanghelich, K., Tao, D.: Deep ordinal regression network for monocular depth estimation. CoRR abs/1806.02446 (2018). arxiv.org/abs/1806.02446
7. Garcia-Garcia, A., Orts-Escolano, S., Oprea, S., Villena-Martinez, V., Rodríguez, J.G.: A review on deep learning techniques applied to semantic segmentation. CoRR abs/1704.06857 (2017). https://arxiv.org/abs/1704.06857
8. Geiger, A., Lenz, P., Urtasun, R.: Are we ready for autonomous driving? the kitti vision benchmark suite. In: 2012 IEEE Conference on Computer Vision and Pattern Recognition, pp. 3354–3361. IEEE (2012)
9. Grompone von Gioi, R., Jakubowicz, J., Morel, J.M., Randall, G.: Lsd: a line segment detector. Image Processing On Line 2, 35–55 (2012). https://doi.org/10.5201/ipol.2012.gjmr-lsd
10. Hao, Z., Li, Y., You, S., Lu, F.: Detail preserving depth estimation from a single image using attention guided networks. CoRR abs/1809.00646 (2018). https://arxiv.org/abs/1809.00646
11. Jiang, H., Ding, L., Hu, J., Huang, R.: Plnet: Plane and line priors for unsupervised indoor depth estimation. CoRR abs/2110.05839 (2021). https://arxiv.org/abs/2110.05839
12. Ladicky, L., Shi, J., Pollefeys, M.: Pulling things out of perspective. In: 2014 IEEE Conference on Computer Vision and Pattern Recognition (2014). https://doi.org/10.1109/cvpr.2014.19
13. Laina, I., Rupprecht, C., Belagiannis, V., Tombari, F., Navab, N.: Deeper depth prediction with fully convolutional residual networks. CoRR abs/1606.00373 (2016). https://arxiv.org/abs/1606.00373
14. Langlois, P., Boulch, A., Marlet, R.: Surface reconstruction from 3d line segments. CoRR abs/1911.00451 (2019). https://arxiv.org/abs/1911.00451
15. Li, Y., Chen, X., Yang, B., Chen, Z., Cheng, Z., Zha, Z.J.: Deepfacepencil: creating face images from freehand sketches. In: Proceedings of the 28th ACM International Conference on Multimedia, MM 2020. ACM, New York (2020). https://doi.org/10.1145/3394171.3413684. https://doi.acm.org/10.1145/3394171.3413684
16. Liu, F., Shen, C., Lin, G.: Deep convolutional neural fields for depth estimation from a single image. CoRR abs/1411.6387 (2014), https://arxiv.org/abs/1411.6387
17. Lopez-Campos, R., Martinez-Carranza, J.: Espada: Extended synthetic and photogrammetric aerial-image dataset. IEEE Robot. Autom. Lett. 1, October 2021. https://doi.org/10.1109/LRA.2021.3101879
18. Saxena, A., Chung, S.H., Ng, A.Y.: 3-d depth reconstruction from a single still image. Int. J. Comput. Vision 76(1), 53–69 (2007). https://doi.org/10.1007/s11263-007-0071-y
19. Silberman, N., Hoiem, D., Kohli, P., Fergus, R.: Indoor segmentation and support inference from RGBD images. In: Fitzgibbon, A., Lazebnik, S., Perona, P., Sato, Y., Schmid, C. (eds.) ECCV 2012. LNCS, vol. 7576, pp. 746–760. Springer, Heidelberg (2012). https://doi.org/10.1007/978-3-642-33715-4_54
20. Wang, W., et al.: Tartanair: a dataset to push the limits of visual SLAM. CoRR abs/2003.14338 (2020). https://arxiv.org/abs/2003.14338
21. Xu, D., Ricci, E., Ouyang, W., Wang, X., Sebe, N.: Multi-scale continuous crfs as sequential deep networks for monocular depth estimation. CoRR abs/1704.02157 (2017). https://arxiv.org/abs/1704.02157

22. Xu, P., Hospedales, T.M., Yin, Q., Song, Y.Z., Xiang, T., Wang, L.: Deep learning for free-hand sketch: A survey. IEEE Transactions on Pattern Analysis and Machine Intelligence (2022)
23. Yu, H., Zhen, W., Yang, W., Scherer, S.: Line-based 2-d-3-d registration and camera localization in structured environments. IEEE Trans. Instrum. Measur. **69**(11), 8962–8972 (2020). https://doi.org/10.1109/tim.2020.2999137
24. Zhao, C., Sun, Q., Zhang, C., Tang, Y., Qian, F.: Monocular depth estimation based on deep learning: an overview. Sci. China Technol. Sci., 1–16 (2020)

Deep Learning Semantic Segmentation of Feet Using Infrared Thermal Images

Rafael Mejia-Zuluaga[1], Juan Carlos Aguirre-Arango[1(✉)],
Diego Collazos-Huertas[1], Jessica Daza-Castillo[2], Néstor Valencia-Marulanda[2],
Mauricio Calderón-Marulanda[2], Óscar Aguirre-Ospina[2],
Andrés Alvarez-Meza[1], and Germán Castellanos-Dominguez[1]

[1] Signal Processing and Recognition Group, Universidad Nacional de Colombia,
Manizales, Colombia
{rmejiaz,jucaguirrear,dfcollazosh,amalvarezme,
cgcastellanosd}@unal.edu.co
[2] SES Hospital Universitario de Caldas, Manizales, Colombia
yeje92@gmail.com, njvalencia@hotmail.com,
mauricio.calderon@ucaldas.edu.co, odaguirre@ses.com.co

Abstract. Regional neuraxial analgesia is a safe method for pain relief during labor, but its effectiveness must be assessed carefully. As a non-invasive technique, thermal imaging is gaining increasing acclaim as an objective way to quantify blood flow redistribution-related warm modifications. Hence, thermal measurements are acquired under controlled conditions at different timestamps to determine the anesthesia depth by characterizing earlier thermal changes. However, the procedures during labor are limited by two main factors: a relatively small sample size is possible and thermal images cannot be acquired with both feet in the same position. This work implements an automatic semantic segmentation approach using five state-of-the-art deep learning architectures and an artifact removal algorithm based on morphological operators to deal with this problem. The obtained results are evaluated on two databases (acquired at the Universidad Nacional de Colombia sede Manizales and the SES Hospital Universitario de Caldas): controlled and uncontrolled environments for thermal data acquisition. Obtained results indicate that U-Mobilenetv2 approach outperforms the rest of the compared models.

Keywords: U-Net · Semantic segmentation · Thermal images · Epidural anesthesia · Medical image

1 Introduction

Regional neuraxial analgesia is one of the safest methods for pain relief in the process of labor [5]. However, a quick and accurate assessment of its effectiveness is crucial for the optimal use of health care resources and patient wellness. Then, thermographic skin images can be used to measure the body temperature, which allows contrasting the cold sensation in predicting the efficacy and distribution of epidural anesthesia [6]. In this sense, thermal imaging poses as an objective

© The Author(s), under exclusive license to Springer Nature Switzerland AG 2022
A. C. Bicharra Garcia et al. (Eds.): IBERAMIA 2022, LNAI 13788, pp. 342–352, 2022.
https://doi.org/10.1007/978-3-031-22419-5_29

and non-invasive solution, allowing the characterization of warm modifications after the catheter placement due to blood flow redistribution [8]. Still, a suitable assessment requires temperature measurements from the patient's feet soles taken at different times after the catheter placement to characterize the earlier thermal modifications.

In this regard, the development of semantic segmentation approaches based on Deep Learning (DL) is a research field of interest [9]. In particular, biomedical applications have been proposed in the literature using infrared thermography in conjunction with deep learning techniques to perform decision-making toward improving the existing medical care [18]. In the concrete case of feet segmentation, DL approaches are reported using Fully Convolutional Networks, U-Net, and SegNet architectures, like in [1]. Nevertheless, this evaluation is frequently performed by combining thermal with other data sources, as in [3]. For the current proposal, however, only thermal images are available. Therefore, this work is developed similarly to the one suggested in [4] to prevent diabetic foot ulcers by identifying hyperthermia but placing the patient's feet in the optimal position. Unfortunately, this scenario is unlikely in an obstetric environment, with low-quality images for wrong placement of feet.

This work implements an automatic semantic segmentation approach using five state-of-the-art DL architectures and an artifact removal algorithm based on morphological operators. The obtained results are evaluated on two databases for feet segmentation: uncontrolled environment and controlled conditions of thermal data acquisition. The current proposal aims to assist in the assessment of the epidural anesthesia of obstetric patients in the *SES Hospital Universitario de Caldas*, improving the quality of the service provided by the hospital and the patient wellness. Attained results demonstrate that.

The remainder of this paper is organized as follows: Sect. 2 presents the methods. Section 3 shows the experimental set-up. Next, Sect. 4 describes the results and discussion. Finally, Sect. 5 reports the conclusions.

2   Methods

2.1   Thermography Database

In order to validate the DL models under test, a private thermography database was obtained during the application of epidural anesthesia prior to the active labor of women who had previously signed informed consent. Because of the complexity related to procedures during labor, two main restrictions on data acquisition arise: a relatively small sample is achievable, and acquiring images with both feet in the same position is not always possible. Therefore, the following timeline was developed by the clinicians of the *SES Hospital Universitario de Caldas* at Manizales: the first thermal picture was taken at the moment of catheter placement, the second one was made a minute later, and so on every five minutes up to complete a total of seven images acquired from each patient, as illustrated in Fig. 1b.

Nevertheless, the acquisition was conducted in an uncontrolled environment for the first part, capturing 196 thermal images from 22 women by a FLIR A320 infrared camera with a resolution of 640×480. Note that more than seven pictures were obtained for some patients. This camera had a spectral range of $7.5-13\,\mu m$ and thermal sensitivity of less than 0.05. As a result, this initial partition of pictures resulted in low-quality images with high variability due to the irregular positioning of the thermal camera. In the second part, the data collection was performed in a more controlled environment. A FLIR E95 thermal camera with enhanced sensitivity of less than $0.03°$ and more flexibility was used to enable better positioning of the infrared camera, resulting in 128 images with improved data quality. It is worth mentioning that high-quality images were chosen from the first partition to add to the second part, obtaining 166 images.

Both partitions were labeled for semantic segmentation by three researchers using a web-based tool for image labeling - namely, CVAT Computer Vision Annotation Tool, free available at[1].

2.2   Foot Segmentation

The main goal of semantic segmentation is to generate a pixel-wise class prediction from images. In this case, thermal images' segmentation aims to distinguish between feet and background. Thus, from a provided image $\boldsymbol{I} \in \mathbb{R}^{R \times C}$, an 2D mask $\hat{\boldsymbol{M}} \in \{0,1\}^{R \times C}$ is computed to encode the one-hot membership representation of each pixel according to the target class. To this end, a DL model is fed by an input-output set of labeled images $\{\boldsymbol{I}_n, \boldsymbol{M}_n : n \in N\}$.

For semantic segmentation, DL architectures exploit the spatial relationships locally by including a set of convolutional layers $\{\boldsymbol{W}_l \in \mathbb{R}^{P_l \times P_l \times D_l} : l \in L\}$, where P_l, D_l, and L denote the l-th layer size (termed kernel size), the number of filters at each layer, and the number of. Next, to estimate the mask $\hat{\boldsymbol{M}} \in [0,1]^{R \times C}$, the following cross-entropy-based optimization problem arises:

$$\{\mathbf{W}_l\}_{l=1}^{L} = \arg\min_{\mathbf{W}_l} -\mathbf{1}^\top \left(\mathbf{M} \odot \log(\hat{\mathbf{M}}) + (1 - \mathbf{M}) \odot \log(1 - \hat{\mathbf{M}}) \right) \mathbf{1}, \quad (1)$$

where $\odot$ stands for element-wise product and $\hat{\mathbf{M}} = (\varphi_L \circ \cdots \circ \varphi_1)(\mathbf{X})$ comprises the DL function composition. Besides, each φ_l holds the l-th convolutional layer. The optimization problem in Eq. (1) can be solved using backpropagation and automatic differantiation [7].

2.3   DL-Based Semantic Segmentation

To date, the following DL architectures have been reported for semantic segmentation:

---

[1] https://cvat.org/.

– Fully Convolutional Networks (FCN) [13] adapt contemporary classifications networks (AlexNet [12], and GoogLeNet [17]) into fully convolutional networks. Both DLs can take inputs of arbitrary sizes and produce correspondingly sized outputs with efficient inference and learning.
– The U-Net architecture relies on the use of data augmentation and allows training of small databases [15]. The architecture consists of two symmetric paths: a contracting one (or encoder) using convolutional layers followed by max-pooling to capture context and an expanding one (or decoder) using up-convolution layers to enable precise localization. Since the U-Net encoder extracts high-level characteristics of the image content, it can be replaced by other more elaborate networks (like VGG [16] or MobileNet [10]).
– SegNet has a similar structure as U-Net (i.e., encoder-decoder architecture), though the decoder upsamples the low-resolution feature maps differently [2]. To this end, it uses pooling indices computed in the max-pooling step of the corresponding encoder to perform non-linear upsampling.

3   Experimental Set-Up

3.1   Preprocessing and Data Augmentation

Figure 1a presents the block diagram used to evaluate the proposed approach for semantic segmentation of feet. Validation appraises the following steps: thermal data acquisition, preprocessing that includes data augmentation and image scaling, DL-based prediction, and artifact removal of the prediction. For this purpose, we test the following NN architectures: U-Net, U-Net with Mobilenetv2 encoder (U-Mobilenetv2), U-Net with VGG16 encoder (U-VGG16), Fully Convolutional Networks, and SegNet. Two training scenarios are considered during the learning step: using a partition of data acquired under an uncontrolled environment and employing data collected under more regulated conditions.

Although the thermal images had been captured at a resolution of 640×480, all dataset is scaled down to 224×224 using bilinear interpolation to supply the minimal size required for all tested NN architectures.

Some examples of the performed preprocessing are further displayed in Fig. 2: the top row depicts the input images, the middle row displays the computed segmentation masks, which are labeled using the CVAT application (or ground truth), and the bottom shows the segmented thermal images obtained after multiplying both previous rows. The examples illustrate images having different acquisition data: low-quality (left column) for which the feet are barely visible on the image; acceptable quality (central) where the feet soles are partially visible, and high-quality (right) where both feet soles are visible.

Moreover, to improve the models' robustness and reduce overfitting, a data augmentation scheme includes the random rotation, zoom, shearing, cropping, flip, and shifting procedures. These operations artificially increase the number of training examples, in our case to 500 images using the training dataset, and simulate different conditions such as diverse patient positioning. Besides, data augmentation enables the simulation of distinct configurations for the thermal camera.

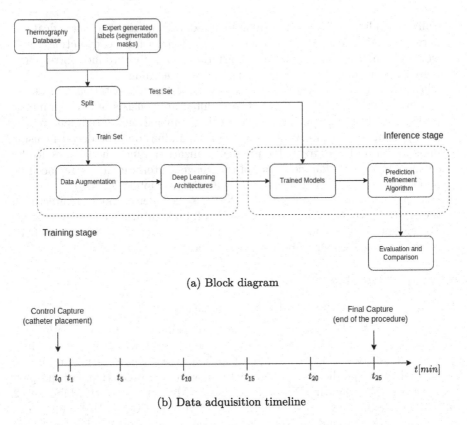

(a) Block diagram

(b) Data adquisition timeline

Fig. 1. Experimental setup. a) Deep-learning validation framework of the proposed approach for semantic segmentation of feet using thermal images; b) Timeline developed for thermal image acquicistion.

3.2   Artifact Removal

Afterward, artifact removal is carried out. For this purpose, the estimated segmentation masks outputted by the models can be further improved using morphological operations and filters to remove false positives and negatives. Although this procedure is highly manual and potentially not robust, it can successfully correct some models' predictions. Figure 3 illustrates the performed artifact removal: the left image depicts the ground truth; the central image presents the output segmentation mask, in which labeled with red are the artifacts to be removed and with green the portions that are conserved, according to the number of pixels of the structure. The right image shows the output image after removing the artifacts. As a result, the image after performing artifact removal is more similar to the ground truth than the original prediction.

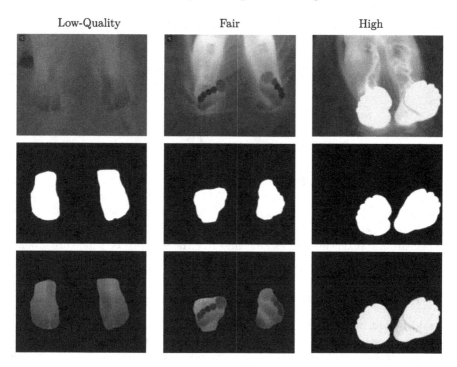

Fig. 2. Example images segmented after performing the preprocessing procedure using different acquisition quality data.

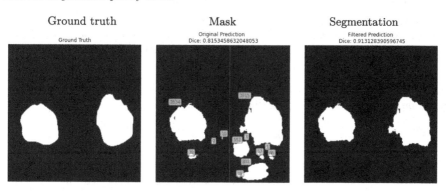

Fig. 3. Images segmented after performing artifact removal.

4 Results and Discussion

4.1 Comparing Metrics of Performance

For validation purposes, the following metrics are used to measure and compare the performance results obtained by the compared DL models:

$$\text{dice similarity: } \rho_{DSC}[\%] = 100\frac{2|M \cap \hat{M}|}{|M| + |\hat{M}|} = 100\frac{2TP}{2TP + FP + FN}$$

$$\text{intersection over union: } \rho_{IoU}[\%] = 100\frac{|M \cap \hat{M}|}{|M \cup \hat{M}|} = 100\frac{TP}{FN + FP + TP}$$

$$\text{sensitivity: } \rho_{sen}[\%] = 100\frac{TP}{TP + FN}$$

$$\text{specificity: } \rho_{spec}[\%] = 100\frac{TN}{TN + FP}$$

$$\text{precision: } \rho_{prec}[\%] = 100\frac{TP}{TP + FP}$$

where TP (True Positives) measures the pixels correctly labeled as feet, TN (True Negatives) the ones corrected labeled as background, FP (False Positives) the background pixels labeled as foot, and FN the foot pixels, which are labeled as background.

4.2   Performed Results of DL Semantic Segmentation

As said before, we validate the DL semantic segmentation method under two environment scenarios for capturing thermal data from lower limbs: controlled and uncontrolled. The training data set includes 148 images and 18 images for testing regarding the former partition. In turn, the second group consists of 163 images for training and 33 for testing.

Table 1 displays the results obtained for the compared DL segmentation architectures, which are divided into two approaches: i) Those that do not include transfer learning (i.e., U-Net, FCN, and SegNet) and those enabling it (U-VGG16 and U-MobilNetv2). The former approaches, reported upon the dashed line,

Table 1. Metrics of performance estimated for the uncontrolled environment data partition. Notation * stands for Artifact Removal.

Model	ρ_{DSC}	ρ_{IoU}	ρ_{spec}	ρ_{sen}	ρ_{prec}
U-Net	86.34 ± 6.21	76.46 ± 9.11	98.52 ± 1.16	83.84 ± 8.74	89.99 ± 8.59
U-Net*	86.15 ± 7.08	76.29 ± 10.07	98.58 ± 1.12	83.42 ± 9.94	90.23 ± 8.52
FCN	88.05 ± 4.4	78.9 ± 6.43	96.67 ± 2.96	82.89 ± 4.44	93.98 ± 5.06
FCN*	87.98 ± 4.38	78.79 ± 6.43	96.85 ± 2.83	82.58 ± 4.68	94.27 ± 4.89
SegNet	90.26 ± 5.32	82.66 ± 8.53	98.39 ± 1.08	**91.13 ± 7.05**	90.03 ± 7.49
SegNet*	90.28 ± 5.33	82.69 ± 8.56	98.4 ± 1.08	91.13 ± 7.05	90.07 ± 7.49
U-VGG16	90.22 ± 8.17	83.0 ± 11.13	98.57 ± 1.26	90.15 ± 9.53	90.76 ± 8.91
U-VGG16*	90.67 ± 7.43	83.63 ± 10.35	98.74 ± 1.04	90.15 ± 9.54	91.73 ± 7.32
U-Mobilenetv2	90.46 ± 4.77	82.92 ± 7.75	98.53 ± 1.25	90.53 ± 6.32	90.95 ± 6.91
U-Mobilenetv2*	**90.98 ± 4.51**	**83.75 ± 7.39**	**98.75 ± 1.0**	90.38 ± 6.52	**92.09 ± 5.99**
Average	89.066	80.788	98.136	**87.708**	91.142
Average*	**89.212**	**81.03**	**98.264**	87.532	**91.678**

achieve high performance, at least for U-Net and FCN. However, the SegNet architecture achieved performance values similarly to transfer learning-based architectures. As seen, the U-Mobilenetv2 outperforms all compared DL methods of feed segmenting in almost all metrics.

One more aspect of consideration is the inclusion of artifact removal. The last two rows show the metric estimates obtained without removal (penultimate row) and after (bottom row noted by *), indicating no statistical difference between the validation scenarios.

Table 2 shows the results obtained for the collected data under the controlled environment. The results are separated into two parts: the upper table displays the estimated metrics without data augmentation, while the lower table holds the metrics after including data augmentation. In the first case, the quality of

Table 2. Metrics of performance estimated for the controlled environment data partition. Upper table: performance obtained without data augmentation. Lower table: performance obtained with data augmentation. Notation * stands for Artifact Removal.

Model	ρ_{DSC}	ρ_{IoU}	ρ_{spec}	ρ_{sen}	ρ_{prec}
Without data augmentation					
U-Net	93.14 ± 1.42	87.19 ± 2.48	97.11 ± 1.15	91.26 ± 2.61	95.16 ± 1.79
U-Net *	93.16 ± 1.43	87.23 ± 2.5	97.15 ± 1.14	91.24 ± 2.6	95.22 ± 1.8
FCN	87.26 ± 3.03	77.52 ± 4.66	89.35 ± 3.15	90.69 ± 3.56	84.18 ± 3.78
FCN *	87.73 ± 3.14	78.28 ± 4.86	90.23 ± 3.15	90.47 ± 3.62	85.27 ± 3.9
SegNet	92.7 ± 1.83	86.44 ± 3.16	98.79 ± 0.57	88.13 ± 3.03	97.82 ± 1.06
SegNet *	92.62 ± 1.88	86.3 ± 3.24	98.8 ± 0.57	87.98 ± 3.12	97.83 ± 1.05
U-VGG16	95.87 ± 2.26	92.16 ± 3.97	98.05 ± 2.11	**94.98 ± 1.67**	96.82 ± 3.26
U-VGG16 *	95.89 ± 2.25	**92.2 ± 3.94**	98.09 ± 2.11	94.96 ± 1.68	96.89 ± 3.25
U-Mobilenetv2	95.21 ± 1.63	90.89 ± 2.89	98.05 ± 0.94	93.77 ± 2.1	96.71 ± 1.72
U-Mobilenetv2 *	**95.9 ± 1.76**	92.17 ± 3.15	**99.07 ± 0.84**	93.58 ± 2.32	**98.36 ± 1.64**
Average	92.836	86.84	96.27	**91.766**	94.138
Average*	**93.06**	**87.236**	**96.668**	91.646	**94.714**
Including data augmentation					
U-Net	76.46 ± 4.39	62.09 ± 5.57	91.05 ± 3.33	70.9 ± 4.37	83.16 ± 5.64
U-Net *	75.99 ± 5.93	61.62 ± 7.39	95.2 ± 3.17	66.16 ± 6.38	89.65 ± 6.28
FCN	90.56 ± 1.01	82.76 ± 1.68	96.35 ± 0.82	87.61 ± 1.63	93.74 ± 1.16
FCN *	90.97 ± 0.89	83.45 ± 1.49	96.98 ± 0.59	87.51 ± 1.69	94.75 ± 0.76
SegNet	97.45 ± 1.46	95.07 ± 2.7	98.26 ± 1.53	**97.7 ± 1.3**	97.24 ± 2.31
SegNet *	97.45 ± 1.48	95.07 ± 2.73	98.28 ± 1.52	97.66 ± 1.33	97.27 ± 2.3
U-VGG16	98.09 ± 0.39	96.26 ± 0.75	99.44 ± 0.18	97.15 ± 0.69	99.07 ± 0.3
U-VGG16 *	**98.10 ± 0.40**	**96.28 ± 0.77**	**99.45 ± 0.18**	97.15 ± 0.69	**99.09 ± 0.3**
U-Mobilenetv2	96.52 ± 1.26	93.31 ± 2.33	98.18 ± 1.04	96.03 ± 1.72	97.05 ± 1.64
U-Mobilenetv2*	97.12 ± 1.07	94.41 ± 1.99	98.98 ± 0.49	95.97 ± 1.84	98.31 ± 0.75
Average	91.816	85.898	96.656	**89.878**	94.052
Averag, yielding *	**91.926**	**86.166**	**97.778**	88.89	**95.814**

Table 3. Number of model parameters.

Model	Parameters
U-Net	27,897,345
FCN	17,972,697
SegNet	29,459,461
U-VGG16	17,003,714
U-Mobilenetv2	**6,497,505**

feet segmentation improves proportionally in each DL architecture. Note that U-Mobilenetv2 outperforms once again.

Next, the lower table presents the results of including data augmentation, also improving the quality of feet segmentation. However, that is not the case for the tested U-Net architecture, which has a notable reduction in performance. This situation may be explained because the simpler learning model can not generalize if fed by enlarged datasets with higher variability. This concern also affects the FCN learner but to a lesser extent. In contrast, the remaining DL architectures benefit from data augmentation, yielding the most elevated quality of feet segmentation. Moreover, although the U-VGG16 algorithm reaches the best mean performance values, there is no statistical difference between them.

Lastly, it is worth noting that artifact removal barely influences the DL performance of segmentation regardless of the validating scenario tested.

5   Concluding Remarks

Here, we develop a DL model for the segmentation of feet using infrared thermal images. Semantic segmentation is validated within several DL architectures using a thermography database obtained during the application of epidural anesthesia, showing promising results. From the performed evaluation, the following aspects are worth mentioning:

Data Acquisition Protocol. A timeline devised by the clinicians is evaluated that deals with restrictions derived from the complexity of procedures during labor. For the implementation of transfer learning approaches, supplementary thermal data in similar foot positions outside the maternity ward are collected on the same timeline.

Another aspect influencing the segmentation performance is the quality of data acquisition. Two data acquisition scenarios are evaluated: uncontrolled environment (as a common situation present in procedures during labor) and controlled conditions. As expected, the former scenario results in an inferior quality of feet segmentation, being the reduction more evident in the simpler algorithms of DL (Like U-Net and FCN).

Semantic Deep Learners. Semantic segmentation approaches are employed to provide pixel-level predictions for the location of different objects in a given scene. Here, a couple of DL architectures are tested for semantic segmentation: learners without transfer learning and DL architectures with it. The latter DL architectures allow a better foot segmentation in cases of nonregulated conditions of thermal images. An additional finding concerns the inability of learners without transfer learning to benefit from the evaluated data augmentation, possibly caused by their lack of generalization capability. Thus, the performance of U-Net decreases by more than 10 points when including this training strategy. It is worth noting that SegNet achieves a performance close to the algorithms with transfer learning but at a higher computational cost, as shown in Table 3. By comparison, a more complex architecture like Movilenetv2 shortens the tuning process by more than five times.

Artifact Removal. Regarding this strategy for improving the input representations, the obtained results show its weak impact on the considered performance metrics. This situation can be better appreciated in the best-performing algorithms (SegNet, U-VGG16, and U-Mobilentev2), for which the impact of artifact removal reduces to zero.

As future work, the authors plan to evaluate more effective transfer learning methods [14], increase the data collection, and explore shallow architectures based on kernel methods, i.e., Random Fourier Features [11], to enhance the system performance.

Acknowledgements. Under grants provided by the Minciencias's project: "Herramienta de apoyo a la predicción de los efectos de anestéticos locales vía neuroaxial epidural a partir de termografía por infrarrojo" code 111984468021.

References

1. Arteaga-Marrero, N., Hernández, A., Villa, E., González-Pérez, S., Luque, C., Ruiz-Alzola, J.: Segmentation approaches for diabetic foot disorders. Sensors **21**(3) (2021). https://doi.org/10.3390/s21030934
2. Badrinarayanan, V., Kendall, A., Cipolla, R.: Segnet: a deep convolutional encoder-decoder architecture for image segmentation. IEEE Trans. Pattern Anal. Mach. Intell. **39**(12), 2481–2495 (2017). https://doi.org/10.1109/TPAMI.2016.2644615
3. Bouallal, D., Bougrine, A., Douzi, H., Harba, R., Canals, R., Vilcahuaman, L., Arbanil, H.: Segmentation of plantar foot thermal images: application to diabetic foot diagnosis. In: 2020 International Conference on Systems, Signals and Image Processing (IWSSIP), pp. 116–121 (2020). https://doi.org/10.1109/IWSSIP48289.2020.9145167
4. Bougrine, A., Harba, R., Canals, R., Lédée, R., Jabloun, M.: On the segmentation of plantar foot thermal images with deep learning. In: 2019 27th European Signal Processing Conference (EUSIPCO), pp. 1–5 (2019)
5. Brown, D.T., Wildsmith, J.A.W., Covino, B.G., Scott, D.B.: Effect of Baricity on Spinal Anaesthesia with Amethocaine. BJA: British J. Anaesthesia **52**(6), 589–596 (1980). https://doi.org/10.1093/bja/52.6.589
6. Bruins, A.A., Kistemaker, K.R.J., Boom, A., Klaessens, J., Verdaasdonk, R., Boer, C.: Thermographic skin temperature measurement compared with cold sensation in predicting the efficacy and distribution of epidural anesthesia. J. Clin. Monit. Comput. **32**, 335–341 (2018)

7. Géron, A.: Hands-on machine learning with Scikit-Learn, Keras, and TensorFlow: Concepts, tools, and techniques to build intelligent systems. O'Reilly Media, Inc. (2019)
8. Haren, F., Kadic, L., Driessen, J.: Skin temperature measured by infrared thermography after ultrasound-guided blockade of the sciatic nerve. Acta anaesthesiologica Scandinavica **57**, August 2013. https://doi.org/10.1111/aas.12170
9. He, Y., et al.: Infrared machine vision and infrared thermography with deep learning: a review. Infrared Phys. Technol. **116**, 103754 (2021). https://doi.org/10.1016/j.infrared.2021.103754. https://www.sciencedirect.com/science/article/pii/S1350449521001262
10. Howard, A.G., et al.: Mobilenets: efficient convolutional neural networks for mobile vision applications. CoRR abs/1704.04861 (2017). http://arxiv.org/abs/1704.04861
11. Jimenez-Castaño, C.A., Álvarez-Meza, A.M., Aguirre-Ospina, O.D., Cárdenas-Peña, D.A., Orozco-Gutiérrez, Á.A.: Random fourier features-based deep learning improvement with class activation interpretability for nerve structure segmentation. Sensors **21**(22), 7741 (2021)
12. Krizhevsky, A., Sutskever, I., Hinton, G.E.: Imagenet classification with deep convolutional neural networks. In: Pereira, F., Burges, C.J.C., Bottou, L., Weinberger, K.Q. (eds.) Advances in Neural Information Processing Systems, vol. 25. Curran Associates, Inc. (2012). https://proceedings.neurips.cc/paper/2012/file/c399862d3b9d6b76c8436e924a68c45b-Paper.pdf
13. Long, J., Shelhamer, E., Darrell, T.: Fully convolutional networks for semantic segmentation (2015)
14. Morid, M.A., Borjali, A., Del Fiol, G.: A scoping review of transfer learning research on medical image analysis using imagenet. Comput. Biol. Med. **128**, 104115 (2021)
15. Ronneberger, O., Fischer, P., Brox, T.: U-net: Convolutional networks for biomedical image segmentation. CoRR **abs/1505.04597** (2015). http://arxiv.org/abs/1505.04597
16. Simonyan, K., Zisserman, A.: Very deep convolutional networks for large-scale image recognition (2015)
17. Szegedy, C., et al.: Going deeper with convolutions. CoRR abs/1409.4842 (2014). http://arxiv.org/abs/1409.4842
18. Vardasca, R., Magalhaes, C., Mendes, J.: Biomedical applications of infrared thermal imaging: current state of machine learning classification. Proceedings **27**(1) (2019). https://doi.org/10.3390/proceedings2019027046. https://www.mdpi.com/2504-3900/27/1/46

Where Are the Gates: Discovering Effective Waypoints for Autonomous Drone Racing

Leticia Oyuki Rojas-Perez[(✉)] and Jose Martinez-Carranza[(✉)]

Department of Computer Science at Instituto Nacional de Astrofísica, Óptica y Electrónica (INAOE), Puebla, Mexico
{oyukirojas,carranza}@inaoep.mx

Abstract. We present a two-step approach for autonomous drone racing that does not require information about the race track (i.e., number of gates, their position and orientation), only the drone's position in the arena, which could be obtained with GPS, a motion capture system or visual SLAM. We use a neural pilot, trained to regress basic flight commands from camera images where a gate is observed, to enable a drone to navigate the unknown race track autonomously, although at a low speed. During this navigation stage, we use a Single Shot Detector to visually detect the gates. The latter is used to identify the drone's positions before and after crossing the gate. Once discovered, these positions are used as waypoints in a flight controller to perform a much faster flight to navigate throughout the race track. This approach resembles how human pilots train on an unknown race track, performing several laps to discover key areas where the drone must increase or decrease its speed to cross all the gates successfully. Our approach has been evaluated in the RotorS simulator, comparing it with the performance of a human pilot and obtaining very similar time results.

Keywords: Autonomous drone racing · Deep learning · Neural pilot

1 Introduction

Autonomous Drone Racing (ADR) poses the problem of developing an artificial pilot capable of flying on a race track autonomously and, as an ultimate goal, competing against a human pilot aiming to beat them [19]. The way human pilots tackle this challenge is intriguing. It has been shown that human pilots do not seek an optimal flight. Instead, the flight policy seems simple: fly as quickly as possible towards the next gate, and once closer to it, manoeuvre the drone such that the gate can be crossed without colliding with it. For the latter, the pilot pays attention to potential areas of collision at the gate and reduces the speed in the case the next gate requires a significant turn in the flight direction [22]; all of that by only observing live video transmitted from a camera on board the drone to head-mounted glasses worn by the pilot. To achieve outstanding

© The Author(s), under exclusive license to Springer Nature Switzerland AG 2022
A. C. Bicharra Garcia et al. (Eds.): IBERAMIA 2022, LNAI 13788, pp. 353–365, 2022.
https://doi.org/10.1007/978-3-031-22419-5_30

performance, pilots fly several laps on the race track to get familiar with it and to learn good places to increase/decrease speed.

Inspired by the learning stage performed by humans, some approaches have proposed to use Deep Learning (DL) to train a Convolutional Neural Networks (CNN) to be used as neural pilots that can regress flight commands from camera images [23]. DeepPilot is an approach where a CNN is taught basic flight commands given a set of consecutive images depicting a gate. Following a reactive approach, the flight commands aim to align the drone w.r.t. the gate's centre for further crossing. Although the model could pilot the drone to effectively navigate throughout a race track, the flight speed is not very fast. Nevertheless, a noticeable feature of DeepPilot is its ability to maintain a forward direction once the gate stops from being observed. This is known as the *blind spot* zone of the gate [2], which is the flight zone where the drone has to fly forwards only to cross the gate; otherwise, it could hit the side frame of the gate.

Therefore, in the effort toward developing an autonomous artificial pilot for ADR, in this work, we present a two-step approach that requires no information about the race track except for the drone's position in the arena. In a real scenario, it is reasonable to expect that positioning could be provided by GPS, a motion capture system [7] or visual SLAM [18]. In the first step, our approach employs two well-known methods used in ADR: 1) an artificial pilot that has been trained to fly towards a gate and cross it as trained to [23]; 2) a gate detector based on the Single Shot Detector [1]. These two CNN-based methods are easy and inexpensive to be trained compared to state-of-the-art DL methods requiring a large amount of data. Thus, we combine them to automatically discover what we call the enter and exit 3D points of the blind spot zone. This is, when the drone is ready to cross the gate and when the gate has been crossed, and it is safe for the drone to change direction. Thus, in this first step, the drone flies autonomously throughout the race tack using the neural pilot (DeepPilot) and the gate detector to discover the last drone's position before entering the blind spot and the next drone' position where the drone has safely exited the blind spot. For the second step, the discovered waypoints will be used by a flight controller seeking to fly at a faster speed.

The aforementioned strategy has been evaluated in the RotorS simulator implemented for Gazebo [8]. We compared it against the performance of a human pilot on the same race track with very similar time results in average. We show experiments on the flight performance when our approach performs one lap only to discover the enter/exit blind spot waypoints and compare it to refined waypoints obtained after completing several laps to get a group of enter/exit waypoints from which we use the average position.

This paper has been organised as follows to convey our approach: Sect. 2 discusses the related work; Sect. 3 describes in more detail our two-step approach; Sect. 4 presents our experimental framework; and finally, Sect. 5 outlines our conclusions and future work.

2   Related Work

Autonomous Drone Racing (ADR) increases the complexity of the environment's interpretation by an artificial agent as drones have to go as fast as possible in a small, confined environment without human intervention. Moreover, during the race, there are difficulties in perception (blurred images, occlusions, or partial views of the target to cross), control and trajectory tracking due to the vehicle dynamics and localisation without GPS. Competitions such as the IEEE IROS Autonomous Drone Racing (ADR) [19,20], the AlphaPilot [6], and Microsoft Game of Drones [17] motivate the development strategies for artificial pilots. Authors such as [1,9–11,19,24] used deep learning to mitigate susceptibility to lighting conditions and the overlapping between gates.

Other works use complementary information obtained from IMU [2,13–15], LiDAR sensors [9,10], ultrasonic sensors or optical flow [19], due to the latency from one frame to the next affects the vehicle's reaction. Even the visual information does not allow us to know the status of the drone, i.e. whether it has crossed the gate or is in the middle. [9,10] uses LiDAR sensors to detect the gate base and [2] uses IMU and Kalman Filter to estimate when the drone is out of the blind spot.

Another approach for ADR is flight path planning. For this, it is necessary to know the positions of the gates and a drone's position estimator. In robotics, the most common localisation systems are Simultaneous Localisation and Mapping (SLAM) and Visual Odometry [5]. Recently, position estimation methods based on the IMU [19], using either the Kalman Filter or the Extended Kalman Filter to obtain a position, are popular for embedded systems, such as [14]. Since the visual or inertial position estimation system accumulates error, the trajectory planning can be modified by the gate location [1,9,10], relative position [2–4,12,14], actions [19], velocity, or even the drone's direction [11,12].

In contrast to visual localisation or pose estimation methods, motion capture systems do not accumulate errors and provide positions up 500 Hz. This system allows executing agile flight at high speed. Authors in [7] present a solution using 36 VICON (motion capture system) and calculate the time-optimal trajectory. The authors give the positions of the gates (waypoints) and pre-calculate the optimal flight path concatenating the waypoints for a single lap multiple times.

However, authors in [21] state that in the solution presented by Foehn in [7], the drone is incapable of exploring and learning from experience to improve the visibility of gates by choosing an entry angle as human pilots. Instead, the drone can perform a flight only if it knows a sequence of waypoints.

3   Methodology

In this section, we present the components of our system, see Fig. 1. We begin by describing the first step composed of the Neural Pilot used to guide the drone on the race track and the gate detection network used to discover a set of waypoints. Next, we describe the second step, which consists of the flight controller that uses the discovered waypoints, whose position is taken from the drone's position observed with the Gazebo's global localisation system.

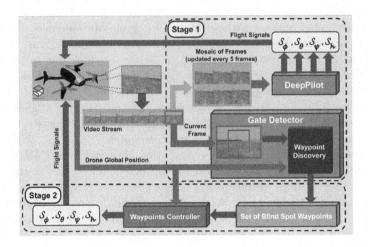

Fig. 1. Schematic view of our approach where we propose to use a gate detection to discover the waypoints while DeepPilot, a neural pilot, flies the drone autonomously on a race track. Our approach comprises two stages. The first stage consist of exploring the race track to discover 3D points related to the drone's position when it is about to enter the gate and after exiting the gate during crossing. This step generates a set of waypoints used by the controller in the second step to effectively fly the drone throughout the track without collisions. A video illustrating our approach can be seen at https://youtu.be/kbpiWEC-mhU

3.1   Neural Pilot

As a part of the navigation stage (1), see Fig. 1, we used DeepPilot to guide the drone on the race track autonomously. DeepPilot is a neural pilot that associates a set of images with a flight signal to produce translational and rotational motion as a tuple $(S_\phi, S_\theta, S_\psi, S_h)$, which corresponds to the signal values in roll, pitch, yaw and altitude [23]. The tuple acquired by DeepPilot is sent to the drone's internal control to navigate autonomously through the race track.

The computational model learned seven basic movements: right, left, up, down, right rotation, left rotation and forward displacement. Since there is ambiguity in the control signals between roll and yaw, the authors generated three specialised models, one for roll and pitch signals, one for yaw and one for altitude. Therefore their architecture is composed of three branches to obtain the flight signals in parallel. Each branch comprises four convolutional layers, three inception modules, one fully connected layer and one regressor.

3.2   Gate Detector

In this work, we use a reduced variant of the Single Shot Detector (SSD) network [16], called SSD7, to minimise the training and search time. This network is a fast multi-category object detector that combines predictions on multiple feature maps of different sizes, producing detections at high and low levels of the image by applying convolution filters.

The SSD7 network architecture is based on a VGG16 image classification network and an auxiliary structure composed of 7 convolution layers. The auxiliary structure obtains multi-scale feature maps and convolutional predictions to know the bounding box displacement and the box's position relative to the location of each feature map.

To detect the gate in the race track, the SSD7 was trained to identify the gates in different positions, and overlapping views as in [1, 2, 9, 10] where images from real environments (indoors and outdoors) and from the Gazebo simulator are used.

3.3   Waypoint Discovery

To discover the effective waypoints to cross a gate, we use the DeepPilot network [23] to guide a drone on a race track autonomously and a Single Shot Detector to visually identify the gates. Also, we use the global position of the drone provided by the simulator. Finally, to obtain the set of successful waypoints autonomously to complete the race track without collisions, we implement a state machine composed of four estates as detailed below, see Fig. 2.

In state one, there are two cases. First, if the detection of the gate is active, it continues in state 1. Additionally, the algorithm evaluates whether the area of the current detection is larger than the largest registered area, then the largest area updates its value by the area of the current detection. In the second case, if there is no detection of the gate, it transitions to state two and initialises cont1 to zero, which indicates the number of times the gate has been missed.

In-state two, we carried out three conditions. In the first condition, if there is no detection of the gate, there is no transition to another state and cont1 is incremented by one. In the second condition, if condition 1 is interrupted, i.e. if there is detection, the cont1 is reset to zero and transitions to state 1. Finally, in the third condition, if cont1 is equal to a number of frames (nf), the algorithm adds to a list drone's 3d coordinates to indicate the entrance to the blind spot, then transitions to state three and initialises cont2.

State three prevents a waypoint from being placed in the centre of the gate. For this, a given time is set for the drone to leave the blind spot because if a waypoint is placed in the centre of the gate, during the navigation using the waypoints, the drone could turn towards the next gate without first exiting the blind spot zone, then the risk of collision is bigger. Therefore cont2 increments by one unit each time it is in state three and can only transition to state four if there is a gate detection and if cont2 is greater than or equal to a nf.

In-state four, two actions are performed to remove a false positive, in case one has been detected, otherwise it adds the exit waypoint. For the former, the algorithm evaluates if the gate area is greater than 75% of the size of the largest recorded area; if this is true, then that waypoint is removed since once the drone exits the blind spot, the next gate to be observed should have an area smaller than the largest area recorded before entering the blind spot zone. Otherwise, the drone's 3D coordinates are added to the list to indicate the exit waypoint, and then it transitions to state one and resets the largest area to zero.

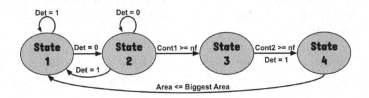

Fig. 2. State machine implemented in our proposed approach to discover the waypoints representing enter and exit drone positions in the blind spot zone. These waypoints are used by the waypoint controller to perform a much faster flight later on.

3.4   Waypoint Controller

We implemented a proportional integral controller for roll and height, a proportional controller for yaw, and pitch was set to a the maximum control signal 1 to move the drone forward as quick as possible. When the distance between the drone and the enter waypoint was less than 3 m, the controller switches to a proportional controller that reduced the speed from 1 up to 0.5. This helped the drone to gradually reduce the speed to enable it to reduce the motion inertia, thus enabling the drone to turn towards the exit waypoint. This was particularly useful in curve sections of the race track where the turn implied a large angle change in yaw. This controller uses the drone's current position obtained by the Gazebo simulator to calculate the errors between the current drone's position and the reference waypoint.

Assuming that the drone flies on a horizontal plane, we operate with vectors obtained from the translation $\mathbf{t}$ and rotation matrix $\mathbf{R}$ estimated with the Gazebo simulator. Using a unit vector $\mathbf{v} = [1, 0, 0]$, a heading vector is set as $\mathbf{h} = \mathbf{R}\mathbf{v}$. A departing waypoint $\mathbf{w}_s$ and the next waypoint $\mathbf{w}_g$ are used to define the direction vector $\mathbf{d} = \mathbf{w}_g - \mathbf{w}_s$, with its corresponding rotation matrix representation $\mathbf{R}_d = Rot(\mathbf{d})$, where $Rot(\cdot)$ is a function that calculates such matrix. Finally, we compute the drone's position relative to $\mathbf{w}_s$:

$$\mathbf{r} = \mathbf{R}_d^\top (\mathbf{t} - \mathbf{w}_s) \tag{1}$$

The control signals are calculated as follows:

$$s_\theta = \begin{cases} 1 : (\|\mathbf{d}\| - r_x) > 3 \\ 0.5 + 0.5 \frac{(\|\mathbf{d}\| - r_x)}{3} : (\|\mathbf{d}\| - r_x) \leq 3 \end{cases} \tag{2}$$

$$s_\phi = K_{p_\phi}(-r_y) + K_{i_\phi} \int (-r_y)dt \tag{3}$$

$$\mathbf{n} = \mathbf{d} \times \mathbf{h} \tag{4}$$

$$s_\psi = K_{p_\psi} sign(\mathbf{n}) acos\left(\frac{\mathbf{d} \cdot \mathbf{h}}{\|\mathbf{d}\|\|\mathbf{h}\|} \right) \tag{5}$$

$$s_h = K_{p_h}(w_{gz} - r_z) + K_{i_h} \int (w_{gz} - r_z)dt \tag{6}$$

where $sign$ is defined as:

$$sign(\mathbf{n}) = \begin{cases} 1 : n_z \geq 0 \\ -1 : n_z < 0 \end{cases} \tag{7}$$

and $\mathbf{w}_g = [w_{gx}, w_{gy}, w_{gz}]$, $\mathbf{r} = [r_x, r_y, r_z]$, $\mathbf{n} = [n_x, n_y, n_z]$.

Our controller received the pose estimation, published by the Gazebo simulator at $1000\,Hz$, see Fig. 1 and the set of enter/exit blind spot waypoints. Finally, the gains $K_{p_\phi}, K_{i_\phi}, K_{p_\psi}, K_{p_h}, K_{i_h}$ were tuned empirically aiming to avoid oscillations or excessive flight speed that would make the drone hit a gate.

4 Experiments

4.1 System Overview

We used the Alienware R5 laptop to carry out our experimental framework. This laptop has a corei7 processor, 32GB of RAM and an NVIDIA GTX 1070 graphics card. Also, it runs the Ubuntu 18.04 LTs operating system and the Robot Operating System (ROS) Melodic Morenia version. To run the SSD7 and DeepPilot networks, we used an NVIDIA GTX 1070 graphics card, Keras 2.3.1 and TensorFlow 1.15 frameworks. To emulate the Parrot Bebop 2 vehicle for the experimental framework, we used the Gazebo 9 simulator and the rotors_simulator package [8].

Our communication architecture is based on ROS. For the simulation environment, we use RotorS [8] and its corresponding package rotors_simulators that runs on top Gazebo, which provides a video stream from the onboard camera of the Bebop2 at 80 fps and simultaneously receives information to control the Bebop2. The second node, DeepPilot, acquires the video stream, generates an mosaic image (updated every five frames) and provides four flight signals $(S_\phi, S_\theta, S_\psi, S_h)$ at 25 fps. The third node, Gate Detector, receives the video stream from the Bebop2, applies the trained gate detection model, and provides the two waypoints discovered for each gate, labelling as enter and exit waypoints accordingly; this node runs at 80 fps. We used DeepPilot and Gate

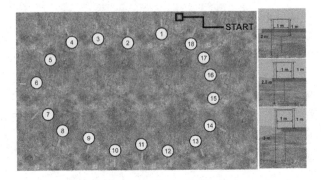

Fig. 3. Left figure show a top view of the race track, composed of 18 gates at different heights and orientations in the RotorS simulator. The race track extends over 68 m × 49 m and has gate spacing of 6 m to 15 m. The dimensions of gates are shown to the right. Each gate has a square frame of 1 m × 1 m × 0.05 m. Note that we highlighted the gates to indicate the order in which the gates have to be crossed.

Detector nodes simultaneously to obtain the blind spot waypoints while Deep-Pilot Guided the Drone throughout the race track.

The fourth node called Waypoint Controller uses the discovered waypoints and the global position of the drone (obtained from the Gazebo simulator) to fly the drone to cross all gates in the race track. Finally, a five-node Keyboard is used only to initiate or cancel the autonomous flight. It should be clear that before each experiment, no information is given to the system regarding the race track (i.e., number of gates or their position/orientation/height).

4.2   Race Track Description

We show our test race track in Fig. 3, which extends over 68 m × 49 m, and we set 18 gates in an ellipse shape to vary the orientations and height. Each gate has a square frame of 1 m × 1 m × 0.05 m, a reduce crossing space that adds up to the difficulty of racing in the track. Gates 1, 2, 3, 5, 9, 12 and 14 are 2m high, gates 4, 7, 8, 11, 13, 15, 16 and 18 are 2.5 m high, and gates 6, 10 and 17 are 3 m high as indicates Fig. 3, and the spacing between gates is 6 m to 15 m. Note that some gates in this track imply significant turns for the drone, which makes it difficult to fly at constant speed, such as it happens in real race tracks.

4.3   Experiment Constraints

We defined the following constraints for our experiments: 1) No prior information about the track is provided to the system. This means that there is no information about the size of the arena or the shape of the race track; 2) The number of gates, their size, height, orientation and positions are unknown to the system; 3) The drone only knows its global position in the arena provided by Gazebo; 4) During the discovery stage, the vehicle will not have access to

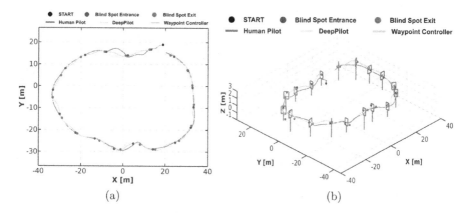

Fig. 4. Drone trajectories generated when the drone is flown by a human pilot (magenta), when flown by DeepPilot (yellow) and when flown by the waypoint controller (cyan). The discovered enter waypoint is depicted in blue and the discover exit waypoint is depicted in red. Top view of these trajectories are shown in (a). The side view in (b) helps to appreciate tha DeepPilot exhibits more oscillations, whereas the human pilot and the waypoint controller performs very similarly. A video illustrating our approach can be seen at https://youtu.be/kbpiWEC-mhU (Color figure online)

external feedback; 5) In the navigation stage (second step of our approach), the modules of step 1 are disabled, and only the controller for waypoint navigation is active.

It should also be stated that the human pilot in these experiments has a large experience in flying real drones, mostly quadrotors, in both indoor and outdoor scenarios. Although this pilot is not a professional drone racer, the pilot is well acquainted in flying the drone in the RotorS simulator and had the opportunity to fly dozens of times in the race track used in these experiments.

Figure 4(a) compares the trajectories performed by a human pilot, illustrated in magenta, and the performed by DeepPilot, highlighted in yellow. Also shown are the entry (blue) and exit (red) points discovered in the blind spot zone of each gate. These points were discovered with one lap. In the same figure, the human pilot seems to perform with a more stable trajectory. DeepPilot produces multiple oscillations, especially between gates. When navigating with the waypoint controller, it exhibits stable movements, very similar to that of the human pilot. Figure 4(b) helps to appreciate the variations in height for all these approaches, with DeepPilot having larger oscillations.

In Fig. 5, we show a top view of the race track, where the entry (cyan blue) and exit (yellow) points of the blind spot are found after 10 laps using DeepPilot and the gate detection with our approach. Also, the averaged waypoints of the enter and exit points of the blind spot are shown in blue and red. We also show the drone trajectories obtained with the waypoint controller for 10 runs. To the right, the square shows a zoom-in area to show the trajectories and, in this case,

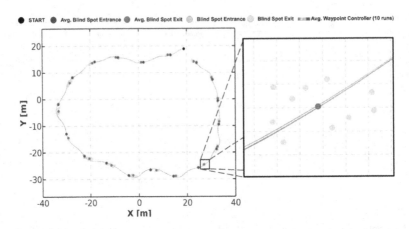

Fig. 5. Top view of ten drone trajectories generated by the waypoint controller using average of the discovered waypoints. The latter were discovered during 10 laps of our approach using DeepPilot and the gate detector. These discovered waypoints are depicted in cyan and yellow corresponding to the enter and exit waypoints of the blind spot zone. The average waypoints are indicated in blue and red respectively. A zoom-in view helps to appreciate discovered enter waypoints (yellow points) and the average waypoint (red point) plus the drone trajectories generated by the waypoint controller. Note that the trajectories are very similar. A video illustrating our approach can be seen at https://youtu.be/kbpiWEC-mhU (Color figure online)

Table 1. Average results for 10 runs testing the performance of a human pilot, the neural pilot (DeepPilot), the waypoint controller using the discovered waypoints in 1 lap with our proposed approach, and the waypoint controller using averaged waypoints discovered after 10 laps with our proposed approach.

	# Runs	Average			Best	Worst
		Trajectory [m]	Time [sec.]	Speed [m/s]	Time [sec.]	Time [sec.]
Human Pilot	10	188.19	62.64	3.0	62.4	64.8
DeepPilot	10	218.72	533.4	0.41	517.8	541.8
Waypoint Controller (Wpts 1 Lap)	10	188.81	74.18	2.54	74	75.8
Waypoint Controller (Avg. Wpts 10 Laps)	10	187.14	72.04	2.6	72	72.4

the different exit points found during the 10 laps of DeepPilot. The average position is shown in red. Note that for the average waypoints, the waypoint controller generates very similar drone trajectories.

To summarise our results, Table 1 shows the average results for 10 runs testing the performance of a human pilot, the neural pilot (DeepPilot), the waypoint controller using the discovered waypoints in 1 lap with our proposed approach, and the waypoint controller using averaged waypoints discovered after 10 laps with our proposed approach. Clearly, DeepPilot with the gate detector takes much more time than the human pilot (8.1 times more). However, once the waypoints have been discovered in 1 lap, the waypoint controller reduces this time significantly (7.8 times less). Furthermore, if more laps are allowed (10 laps) to discover more waypoints, and the average of these waypoints is used by the waypoint controller, then there is an additional reduction of 0.76 s, meaning that the waypoints get better as more laps are performed. In average, the human flies the drone at a speed of 2.88 m/s, whereas, the artificial pilot using the discovered waypoints reaches a speed of 2.80 m/s, a difference that could be deemed no significant. Still, the best time is produced by the human with less than 1.5 s of difference with respect the waypoint controller. We are aware that in reality, humans fly much faster than 2.88 m/s. However, for this work, we used the minimum and maximum control signal values in the range of $[-1, 1]$ allowed by the RotorS simulator and that coincide with the SDK of the Bebop drone, nevertheless, we consider these results encouraging.

5 Conclusion

Inspired by how human pilots train on an unknown race track to get familiar with it, we have presented an approach for Autonomous Drone Racing where an artificial pilot, trained to regress basic flight commands from camera images where a gate is observed, is used to navigate an unknown race track at low speed autonomously. No information about the gates (i.e., position, orientation, height, number of gates) is known beforehand, only the drone's position in the arena. We use the Single Shot Detector to enable the drone to visually detect the gates in the track and, in combination with the neural pilot, to automatically discover the enter and exit positions of what we call the *blind spot zone*, i.e., the zone where the gate is no longer visually observed during the crossing. Once discovered, these positions are used as only waypoints by a flight controller to perform a much faster flight. Our approach was assessed in the RotorS simulator and compared against the performance of a human pilot with almost the same flight time in average, although the human performed the best flight with 1.5 s less than the waypoint controller. We consider, however, this results encouraging and promising for our proposed approach.

For future work, we will test with more sophisticated flight controllers that use the discovered waypoints. We will also consider using the control signals calculated by the waypoint controller to update the dataset and retrain DeepPilot.

References

1. Cabrera-Ponce, A.A., Rojas-Perez, L.O., Carrasco-Ochoa, J.A., Martinez-Trinidad, J.F., Martinez-Carranza, J.: Gate detection for micro aerial vehicles using a single shot detector. IEEE Lat. Am. Trans. **17**(12), 2045–2052 (2019)

2. Cocoma-Ortega, J.A., Rojas-Perez, L.O., Cabrera-Ponce, A.A., Martinez-Carranza, J.: Overcoming the blind spot in cnn-based gate detection for autonomous drone racing. In: 2019 Workshop on Research, Education and Development of Unmanned Aerial Systems (RED UAS), pp. 253–259. IEEE (2019)
3. Cocoma-Ortega, J.A., Martinez-Carranza, J.: A cnn based drone localisation approach for autonomous drone racing. In: 11th International Micro Air Vehicle Competition and Conference (2019)
4. Cocoma-Ortega, J.A., Martínez-Carranza, J.: Towards high-speed localisation for autonomous drone racing. In: Mexican International Conference on Artificial Intelligence, pp. 740–751. Springer (2019)
5. Davison, A.J., Reid, I.D., Molton, N.D., Stasse, O.: Monoslam: real-time single camera slam. IEEE Trans. Pattern Anal. Mach. Intell. **29**(6), 1052–1067 (2007)
6. Foehn, P., et al.: Alphapilot: autonomous drone racing. arXiv preprint arXiv:2005.12813 (2020)
7. Foehn, P., Romero, A., Scaramuzza, D.: Time-optimal planning for quadrotor waypoint flight. Sci. Robot. **6**(56), eabh1221 (2021)
8. Furrer, F., Burri, M., Achtelik, M., Siegwart, R.: RotorS—a modular gazebo MAV simulator framework. In: Koubaa, A. (ed.) Robot Operating System (ROS). SCI, vol. 625, pp. 595–625. Springer, Cham (2016). https://doi.org/10.1007/978-3-319-26054-9_23
9. Jung, S., Hwang, S., Shin, H., Shim, D.H., et al.: Perception, guidance, and navigation for indoor autonomous drone racing using deep learning. IEEE Robot. Autom. Lett. **3**(3), 2539–2544 (2018)
10. Jung, S., Lee, H., Hwang, S., Shim, D.H.: Real time embedded system framework for autonomous drone racing using deep learning techniques. In: 2018 AIAA Information Systems-AIAA Infotech@ Aerospace, p. 2138 (2018)
11. Kaufmann, E., Gehrig, M., Foehn, P., Ranftl, R., Dosovitskiy, A., Koltun, V., Scaramuzza, D.: Beauty and the beast: Optimal methods meet learning for drone racing. In: 2019 International Conference on Robotics and Automation (ICRA), pp. 690–696. IEEE (2019)
12. Kaufmann, E., Loquercio, A., Ranftl, R., Dosovitskiy, A., Koltun, V., Scaramuzza, D., et al.: Deep drone racing: Learning agile flight in dynamic environments. In: Conference on Robot Learning, pp. 133–145. PMLR (2018)
13. Li, S., De Wagter, C., de Visser, C., Chu, Q., de Croon, G., et al.: In-flight model parameter and state estimation using gradient descent for high-speed flight. Int. J. Micro Air Vehicles **11**, 1756829319833685 (2019)
14. Li, S., van der Horst, E., Duernay, P., De Wagter, C., de Croon, G.C., et al.: Visual model-predictive localization for computationally efficient autonomous racing of a 72-g drone. J. Field Robot. (2020)
15. Li, S., Ozo, M.M., De Wagter, C., de Croon, G.C., et al.: Autonomous drone race: a computationally efficient vision-based navigation and control strategy. Robotics and Autonomous Systems, p. 103621 (2020)
16. Liu, W., et al.: SSD: single shot MultiBox detector. In: Leibe, B., Matas, J., Sebe, N., Welling, M. (eds.) ECCV 2016. LNCS, vol. 9905, pp. 21–37. Springer, Cham (2016). https://doi.org/10.1007/978-3-319-46448-0_2
17. Madaan, R., et al.: Airsim drone racing lab, pp. 177–191 (2020)
18. Martinez-Carranza, J., Rojas-Perez, L.O.: Warehouse inspection with an autonomous micro air vehicle. Unmanned Systems, pp. 1–14 (2022)
19. Moon, H., et al.: Challenges and implemented technologies used in autonomous drone racing. Intel. Serv. Robot. **12**(2), 137–148 (2019). https://doi.org/10.1007/s11370-018-00271-6

20. Moon, H., Sun, Y., Baltes, J., Kim, S.J., et al.: The iros 2016 competitions [competitions]. IEEE Robot. Autom. Mag. **24**(1), 20–29 (2017)
21. Pfeiffer, C., Scaramuzza, D.: Expertise affects drone racing performance. arXiv preprint arXiv:2109.07307 (2021)
22. Pfeiffer, C., Scaramuzza, D.: Human-piloted drone racing: visual processing and control. IEEE Robot. Autom. Lett. **6**(2), 3467–3474 (2021)
23. Rojas-Perez, L.O., Martinez-Carranza, J.: Deeppilot: a cnn for autonomous drone racing. Sensors **20**(16), 4524 (2020)
24. Sanket, N.J., Singh, C., Ganguly, K., Fermuller, C., Aloimonos, Y.: Gapflyt: active vision based minimalist structure-less gap detection for quadrotor flight. IEEE Robotics and Automation Letters (2018)

Probabilistic Logic Markov Decision Processes for Modeling Driving Behaviors in Self-driving Cars

Héctor Avilés[1] , Marco Negrete[2] , Rubén Machucho[1(✉)] ,
Karelly Rivera[1] , David Trejo[2] , and Héctor Vargas[3]

[1] Polytechnic University of Victoria, 87138 Cd. Victoria, Tamaulipas, Mexico
{havilesa,rmachuchoc,1930435}@upv.edu.mx
[2] Faculty of Engineering, National Autonomous University of Mexico,
04510 Mexico City, Mexico
{mnegretev,davidtrejofs}@comunidad.unam.mx
[3] Popular Autonomous University of Puebla, 72410 Puebla, Mexico
hectorsimon.vargas@upaep.mx

Abstract. Rule-based strategies and probability models are among the most successful techniques for selecting driving behaviors of self-driving cars. However, there is still the need to explore the combination of the flexibility and conceptual clarity of deterministic rules with probabilistic models to describe the uncertainty in the spacial relationships among the entities on the road. Therefore, in this paper we propose an action policy obtained from a probabilistic logical description of a Markov decision process (MDP) as a behavior selection scheme for a self-driving car to avoid collisions with other vehicles. We consider three behaviours: *keep distance*, *overtaking*, and *steady motion*. The state variables of the MDP signal the presence or absence of other vehicles in the surroundings of the ego car. Simple probabilistic logic rules characterize the probability distribution of the immediate future state of the autonomous car given the current state and action. The utility model of the MDP rewards the autonomous car when no car is ahead and it penalizes two types of crashes accordingly to their severity. We simulated our proposal in 16 possible scenarios. The results show the appropriateness of both, the overall system and the decision-making strategy to choose actions that prevents potential accidents of the self-driving car.

Keywords: Autonomous vehicles · Probabilistic logic · Markov decision processes · Computer vision

1 Introduction

The development of self-driving cars have received significant attention in the last decades due to the potential positive effects for the future of mobility. Some

This work was supported by UNAM-DGAPA under grant TA101222 and Consorcio de IA CIMAT-CONACYT.

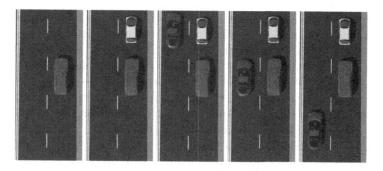

Fig. 1. Example of the modeled environments. Left: Because there is no other car around, the self-driving vehicle moves at a cruise speed. Second left: The self-driving detects another car and should pass it. Three rightmost: Vehicle detects a car in front but should not overtake and just keep distance.

of these benefits include the increase of safety or health care, and the reduction of economic costs [1–3]. An autonomous car has several components that strongly interact among themselves and can roughly be classified as perceptual and decision making subsystems. One of the most important elements of decision making is the behavior selection module that decides what reactive or short-term action (e.g., braking, accelerating, or stopping) is more convenient given the current environment.

Neural networks, rule-based models and probabilistic models are some of the most common representations for behavior selection. On the one hand, neural networks has a outstanding performance in learning from numerical examples and in classifying noisy or partial observations. Unfortunately, the internal knowledge representation of neural networks is not human-readable. This condition hinders the inspection of the decision-making process, that it is essential for self-driving cars. On the other hand, rule-based models provides understandable descriptions of the domain, usually in the form of predicate logic rules and facts which simplify the explanation of the decisions. Moreover, probability theory is the most widely used scheme for handling uncertainty and despite the propositional nature of the random variables, it relaxes the binary truth assumptions of first-order logic. Therefore, in this document, we propose an action policy obtained from a probabilistic logic representation of an MDP as a behavior selection strategy for a self-driving car to avoid accidents with other moving vehicles. The MDP is encoded in MDP-ProbLog [4] that is developed on top of ProbLog [5]. Our approach allows us: i) to express spatial relationships between the self-driving car and the other cars on the road using fairly simple probabilistic logic rules and facts, and ii) to assign negative utilities for different potential crash situations and positive rewards for desirable states and actions. In our setting the self-driving car and four vehicles travel on a one-way street with two lanes. The ego car considers three simple actions: *overtaking*, *keep distance* and *steady motion*. The self-driving starts its movement in the right lane. The perceptual system reports if there is another car in front of it in the right lane, a car ahead in the left lane, a car aside to the left, or a car behind it in the left lane (we call these positions North, North-West, West and South-West, respectively). Then, the self-driving car has to decide which action is more convenient given the presence or absence of the other vehicles. Figure 1 depicts 5 of the 16 possible settings and the action taken on each one. We performed 160 repetitions of the previous experiment. Our initial results shows a correct classification rate of 92.5% on a simulated environment when testing at low speeds and 88.1% at faster speeds.

The main contribution of this paper is twofold. First, in the best of our knowledge this is the first time a probabilistic logic approach is used to select behaviors in autonomous vehicles. Second, it is shown the suitability of the probabilistic logic representation to describe driving knowledge and to generate action policies that prevent potential collisions.

2   Related Work

In [6] Chen and et al. present a hierarchical controller for autonomous vehicle management, this controller is divided into two levels: high and low. The high level uses readings from vehicle sensors such as camera, gps, digital map, radar, etc. and has a finite state machine (FSM), with which it identifies the relative position of surrounding cars, it also predicts a possible collision. This information is passed to another FSM, which determines the behavior of the driver to avoid possible collisions with other vehicles. At the low level, they develop a pair of controllers: the first one for the lateral movement that the vehicle will develop and the second one for the longitudinal movement; guaranteeing with this the stability of the vehicle, during its advance and lane change.

In [7] Lu Huang et al. construct a 9-region grid, where the autonomous vehicle is placed in the center, surrounded by 8 possible neighboring vehicles or empty places. They also consider 3 classes, the obstacles, the road network and the driving scenario. In the first class, there are the static obstacles, such as trees, buildings, cones on the road, and also the dynamic obstacles, such as cyclists, pedestrians and animals. In the second class, highways, urban areas and parts of them, such as traffic lights, lane markings, etc., are considered. In the last class, scenarios include passing through a tunnel, a bridge or a road intersection. With this information they develop a Prolog program for decision making, and perform the autonomous driving of the vehicle.

In [8] A. Rizaldi and M. Althoff investigate the case of a possible collision between two autonomous vehicles, they consider that if an autonomous vehicle complies with the traffic rules established by international conventions, then it would be exempted from crash liability. To do this, they construct a set of high-order logic rules and use Isabelle/HOL software to test their results.

3 Theoretical Background

3.1 ProbLog Overview

ProbLog is a declarative language and a suite of algorithms that integrate independent ground probabilistic facts into the syntax of Prolog programs. ProbLog supports inductive definitions, cyclic or recursive rules, transitive relations, multiple non-mutually exclusive bodies with the same head, arbitrary atoms as evidence, marginal and joint probabilities given the evidence, and *annotated disjunctions* [9]. ProbLog follows *distribution semantics* proposed by T. Sato [10] for calculating the probability of a query or goal given a probabilistic logic program. A ProbLog program P_L is an ordered pair (F_p, R), where:

1. $F_p = \{f_1, f_2, ..., f_n\}$ is a finite set of independent probabilistic facts
2. $R = \{r_1, r_2, ..., r_m\}$ is a finite set of probabilistic normal rules

A probabilistic fact is a ground Prolog fact augmented with a probability value. They can be also understood as Boolean random variables. In a similar form, probabilistic rules are Prolog rules extended with probabilities. An important restriction is that no head of a rule in R is a member of F_p. The probability of a query q given a probabilistic logic program P_L is obtained by the next equation:

$$P(q|P_L) = \sum_{\forall L, L \models q} P(L|P_L) \tag{1}$$

where L is a ground logic program and the probability of q is equal to the sum of the probability of each ground logic program L that logically entails q. The probability $P(L|P_L)$ is calculated as follows:

$$P(L|P_L) = \prod_{f_i \in L} P(f_i) \prod_{f_i \notin L} (1 - P(f_i)), \tag{2}$$

that is, the product of the probability of every fact f_i assumed to be true or false under the interpretation (total choice) associated to L. For n Boolean facts, 2^n logic programs can be derived. Well-founded semantics [11] is adopted for determining whether a query is logically entailed by a ground logic program via the well-founded model of the logic program L.

3.2 Markov Decision Processes

Markov decision processes [12] are standard models for sequential decision-making. A (discrete time) MDP is a 4-tuple (S, A, P, R) in which:

a) $S = \{s_1, s_2, ..., s_n\}$ is a finite set of n states that are completely observable (i.e., the current state is known)
b) $A = \{a_1, a_2, ..., a_m\}$ is a finite set of m actions

c) $P(s'|a, s)$ is a transition function that models the uncertainty of reaching a future state $s' \in S$ given an action $a \in A$ executed in the current state $s \in S$

d) $R(s, a)$ is a reward function for each pair (s, a)

The reward function is used to compute the cumulative expected reward $V^\pi(s)$ of a state s, that corresponds to the immediate reward for the state-action pair, plus the average of the cumulative future rewards from the next states following a fixed policy π. The solution of the MDP is an optimal action policy π^* that maximizes the value $V^\pi(s)$ of each state s accordingly to the Bellman's equation:

$$V^\pi(s) = R(s, \pi(s) = a) + \gamma \sum_{s' \in S} V^\pi(s')P(s'|\pi(s) = a, s) \tag{3}$$

where $\gamma \in [0, 1)$ is a discount factor for infinite-horizon MDP. The optimal policy can be obtained by algorithms such as value iteration (iterative maximization of the value of each state over all possible actions until convergence) and policy iteration (the iterative improvement of policies from an arbitrary initial policy). As the number of states increases, a factorization of the transition function [13] using *state variables* is more convenient. The assignment of values to state variables defines the states of the MDP. This factored representation of states has the advantage of simplifying the transition model by focusing on the effect of the actions over subsets of the state variables.

3.3 MDP-ProbLog

MDP-ProbLog models infinite horizon factored MDPs. The tuple (A_t, F_p, R) is a n MDP-ProbLog program if:

i) A_t is a finite set of atomic formulas that is divided into:
  (1) S_F, a finite set of state (or fluent) variables
  (2) A, a finite set of action predicates
  (3) U, a set of utility predicates
ii) F_p is a finite set of auxiliary probabilistic facts
iii) R is a set of rules categorized as:
  (a) T_r, a finite set of transition rules
  (b) R_r, a finite set of reward rules

Special predicates are used to declare these elements. For example, state variables are identified with the unary predicate 'state_fluent', and actions are denoted with the predicate 'action' that receive as parameter the identifier of a state variable and action, respectively. The predicate 'utility' assigns a positive or negative reward to actions and states of the MDP. MDP-ProbLog uses the algorithm value iteration for solving the MDP.

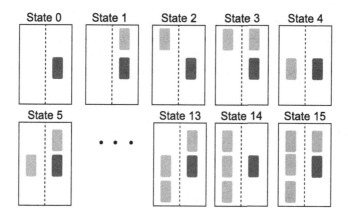

Fig. 2. MDP States to model the environment. The dark grey rectangle represents the self-driving car and the light grey rectangles are the nearby (obstacle) vehicles.

4 Methodology

4.1 Design of the Probabilistic Logic MDP

As stated above, our setting consists on a autonomous vehicle moving at a steady speed and every time it has to decide whether to continue at a constant speed, to overtake or to keep its distance accordingly to the existence of other cars nearby. Figure 2 shows some of the states considered to model the environment.

Herein after North, North-West, West and South-West will be referred to as N, NW, W and SW, respectively. States are factored by four Boolean state variables that we call *free_N*, *free_NW*, *free_W* and *free_SW*. These state variables represent regions around the car which can be free or occupied. In MDP-ProbLog, the syntax for the declaration of the state variables is simple. For example, state_fluent(free_N) is the definition of the variable free_N. The definition of the actions is straightforward too, e.g., action(keep_distance).

The transition model of the MDP-ProbLog requires to know the truth value of each state variable. These values are obtained by our visual perception system as discussed in Sect. 4.2. Understandable probabilistic logic rules were designed to model the transition to potential future states given the current values of the state variables and the actions. For example, for the overtaking action we have:

```
0.9::free_N(1)  :- free_NW(0), free_W(0), overtaking.
0.05::free_N(1) :- not(free_NW(0)), overtaking.
0.05::free_N(1) :- not(free_W(0)), overtaking.
```

Accordingly to first rule, free_N will be true in the next epoch with a probability of 0.9 if free_NW and free_W are true and the action overtaking is performed (i.e., there is enough space to overtake). The latter two rules indicate that there is a low probability of 0.05 of observing free_N as true if there is not free space to the northwest or to the west. Initially, we manually designed probabilistic logic rules for the factored transition function by considering all the state variables. This approach resulted in several probabilistic logic rules. However, we reduced the rules for each action by applying the resolution-like algorithm by hand on rules that share the same probability value. In this manner, we derived fewer and simpler rules that subsumes the original ones. Finally, the reward model is defined as:

```
utility(free_N(1), 5).
utility(rear_crash(1), -30).
utility(side_crash(1), -10).
utility(keep_distance, -10).
utility(overtaking, -1).

0.99::rear_crash(1) :- not(free_N(1)), steady_motion,
not(keep_distance).
0.95::rear_crash(1) :- not(free_NW(1)), overtaking.
0.95::side_crash(1) :- not(free_W(1)), overtaking.
```

In this case, we assign a positive reward whenever the self-driving car has free space in front of it, and different negative rewards for rear crashes and side swipe crashes (Fig. 3). These negatives rewards reflects a difference on the severity of the consequences of these types of accidents. The reward is also negative for overtaking, as we consider that this action involves some risk of accident, and for keeping the distance, because selecting that action delays the arrival to the destination.

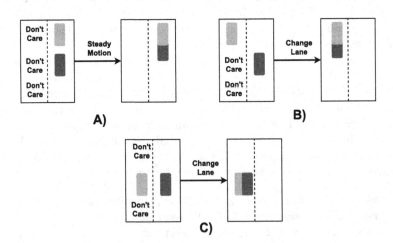

Fig. 3. Examples of states and actions that lead to the different types of crashes in our setting: A) and B) show rear crashes. C) shows a side swipe

4.2   Perceptual System

To simulate the environment we used the Webots simulator [14]. The simulated car is equipped with an RGB camera, used for lane segmentation and tracking, and a 3D-Lidar, used to detect and estimate position and speed of other vehicles. Noise was added to both sensors using the tools provided by Webots. The simulated car allows to set linear speed v and steering δ as control inputs.

Lane Segmentation and Tracking. The segmentation process is based on Canny edge detection and Hough Transform for finding lines. Since almost half of the image corresponds to the landscape above the horizon, and considering that due to perspective, lanes form a triangle in front of the car, we define a region of interest where lanes are most likely to be found, reducing the computation time. General process for lane segmentation can be summarized in the following steps:

1. Get the set of edges using a Canny edge detector
2. Set a triangular Region of Interest where lanes are most likely to be found
3. Use the Hough-Transform to find lines in the edges

Figure 4 show the general steps to detect lanes. Lane tracking is performed using a proportional control that combines distance and angle to the lane border. Consider the Fig. 5. Green lines represent the desired lines we should observe if the car is well centered in the lane. Cyan lines represent the lines actually observed.

Expressing lines in the normal form, (ρ_{ld}, θ_{ld}) and (ρ_{rd}, θ_{rd}) represent the desired line parameters for left and right lane borders respectively, and (ρ_l, θ_l) and (ρ_r, θ_r) represent the actual observed borders. Consider the kinematic model of the vehicle:

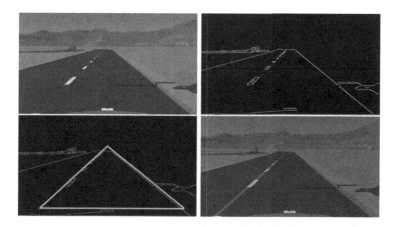

Fig. 4. Segmentation Process. From top-left to bottom-right: Original image, border detection with Canny, triangular region of interest and resulting segmented lines.

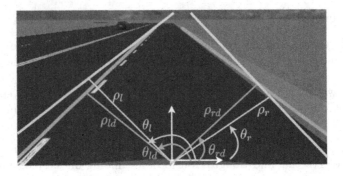

Fig. 5. Variables for lane tracking

$$\dot{x} = v\cos(\theta + \delta) \tag{4}$$
$$\dot{y} = v\sin(\theta + \delta) \tag{5}$$
$$\dot{\theta} = \frac{v}{l}\sin(\delta) \tag{6}$$

with $[x, y, \theta]$, the current configuration of the vehicle, l, the distance from rear to front tires and (v, δ) the linear speed (set by throttle and brake) and steering angle, respectively, taken as control inputs. We propose the following control laws to track lane:

$$v = C_b \tag{7}$$
$$\delta = K_\rho e_\rho + K_\theta e_\theta \tag{8}$$

with

$$e_\rho = ((\rho_{ld} - \rho_l + \rho_{rd} - \rho_r)/2)$$
$$e_\theta = ((\theta_{ld} - \theta_l + \theta_{rd} - \theta_r)/2)$$

Steering is calculated based on the error between desired and observed lines. Linear speed is set as a constant selected based on the current behavior: cruise motion, keep distance and overtaking.

Vehicles Speed Estimation. As mentioned before, the situation we tested consist on a decision system to choose whether to pass another car or not. Thus, we need to estimate other cars' position and velocity. To achieve this, we used the 3D Lidar sensor mounted on the top of the simulated vehicle. Vehicle position and speed estimation can be summarized in the following steps:

- Point cloud filtering by distance and height
- Clustering by K-means
- Velocity estimation by Kalman Filters

As a first step, we eliminate the points in the point-cloud that are part of the ground. Since the 3D Lidar detects its own car, we also filter points with coordinates inside a bounding box representing the car. We cluster the filtered cloud using K-means and each centroid is taken as a position measurement to estimate speed with a Kalman Filter.

5   Experiments and Results

Table 1 represents the policy obtained with the MDP previously described. The symbol **F** means there is a car to the N, NW, SW or W and **T** means that the corresponding position is free. Depending of what is detected using the lidar sensor, we choose one of the three behaviors.

Table 1. Policy obtained from the probabilistic logic MDP.

N	NW	SW	W	Action	N	NW	SW	W	Action
F	**F**	**F**	**F**	keep distance	**F**	**F**	**F**	**T**	keep distance
T	**F**	**F**	**F**	Steady motion	**T**	**F**	**F**	**T**	Steady motion
F	**T**	**F**	**F**	keep distance	**F**	**T**	**F**	**T**	Overtaking
T	**T**	**F**	**F**	Steady motion	**T**	**T**	**F**	**T**	Steady motion
F	**F**	**T**	**F**	keep distance	**F**	**F**	**T**	**T**	keep distance
T	**F**	**T**	**F**	Steady motion	**T**	**F**	**T**	**T**	Steady motion
F	**T**	**T**	**F**	keep distance	**F**	**T**	**T**	**T**	Overtaking
T	**T**	**T**	**F**	Steady motion	**T**	**T**	**T**	**T**	Steady motion

This policy was implemented using a finite state machine with the vehicles positions as input signals and the chosen behavior as output signals. In order to add uncertainty to the environment, we added Gaussian noise with zero-mean both to camera and lidar. Camera noise has a Std Dev of 5% of the maximum value. Lidar noise has a Std Dev of 0.1 m. In each test, speeds of surrounding cars were set randomly with uniform distribution in the interval (3.0, 4.0).

To evaluate our approach, we performed an experiment with 20 repetitions of each possible configuration listed in Table 1. 10 repetitions with slow random speeds (in the range [3.5,5] m/s) and another 10 repetitions with faster random speeds (in the range [5,10] m/s). The confusion matrix of the 160 repetitions for slower speeds, that includes the number of repetitions in which the system took correct and wrong actions is shown in Table 2. The confusion matrix for faster speeds is shown in Table 3. Figure 6 shows an example situation where the car chooses to overtake. As it is shown, in most cases, the car performed the expected action and only in the 7.5% of cases, for lower speeds, and 11.87% for faster speeds, the car selected a non-expected behavior. However, only in 4 cases the car crashed with one of the surrounding vehicles, when moving at lower speed, and only in 7 cases, when moving at faster speeds, most likely because an error in the position estimations.

Table 2. Performance of the decision system with speeds in [3.5, 5] m/s

Ideal action	Chosen action		
	Steady motion	Keep distance	Overtaking
Steady motion	80	0	0
Keep distance	0	57	3
Overtaking	0	9	11

Table 3. Performance of the decision system with speeds in [5, 10] m/s

Ideal action	Chosen action		
	Steady motion	Keep distance	Overtaking
Steady motion	80	0	0
Keep distance	0	44	16
Overtaking	0	3	17

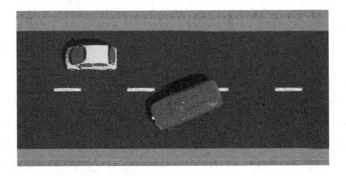

Fig. 6. Example of the overtaking behavior.

6   Conclusions and Future Work

We have presented a probabilistic logic method to generate the action policies of an autonomous vehicle in three behaviors: steady motion, keep distance and overtaking. We tested the method in a simulated environment using Webots and MDP-ProbLog. The test environment consisted of two lanes, with four possible vehicles surrounding the autonomous vehicle. The results obtained demonstrate that the vehicle can be driven safely and collision-free by making use of the generated policies. As future work, the system will be tested considering more lanes and other possible collision risk situations. Also, the decision system will be improved to take into account the speed of surrounding vehicles, and not only their relative positions. Finally, we propose to implement our proposal in AutoMiny Version 4, an open source 1:10 scale autonomous vehicle developed at the Free University of Berlin, and that is equipped with a small PC, a Lidar sensor, a RGB-D camera, an IMU, and actuators for driving and steering. Testing will be performed in a small car test track of 3 × 6 m.

Acknowledgments. The authors would like to thank Vincent Derkinderen (KU Leuven) and Thiago P. Bueno (University of São Paulo) for their comments and guidance and the two anonymous reviewers for their insightful suggestions.

References

1. Silva, O., Cordera, R., González-González, E., Nogués, S.: Environmental impacts of autonomous vehicles: a review of the scientific literature. Sci. Total Environ. **830**, 154615 (2022)
2. Howard, D., Dai, D.: Public perceptions of self-driving cars: the case of Berkeley, California. In: Transportation Research Board 93rd Annual Meeting, vol. 14, pp. 1–16 (2014)
3. Xu, W., Wei, J., Dolan, J.M., Zhao, H., Zha, H.: A real-time motion planner with trajectory optimization for autonomous vehicles. In: 2012 IEEE International Conference on Robotics and Automation, pp. 2061–2067. IEEE (2012)
4. Bueno, T.P., Mauá, D.D., De Barros, L.N., Cozman, F.G.: Markov decision processes specified by probabilistic logic programming: representation and solution. In: 2016 5th Brazilian Conference on Intelligent Systems (BRACIS), pp. 337–342. IEEE (2016)
5. Fierens, D., et al.: Inference and learning in probabilistic logic programs using weighted boolean formulas. Theory Pract. Logic Program. **15**(3), 358–401 (2015)
6. Chen, K., Yang, B., Pei, X., Guo, X.: Hierarchical control strategy towards safe driving of autonomous vehicles. J. Intell. Fuzzy Syst. **34**(4), 2197–2212 (2018)
7. Huang, L., Liang, H., Yu, B., Li, B., Zhu, H.: Ontology-based driving scene modeling, situation assessment and decision making for autonomous vehicles. In: 2019 4th Asia-Pacific Conference on Intelligent Robot Systems (ACIRS), pp. 57–62. IEEE (2019)
8. Rizaldi, A., Althoff, M.: Formalising traffic rules for accountability of autonomous vehicles. In: 2015 IEEE 18th International Conference on Intelligent Transportation Systems, pp. 1658–1665. IEEE (2015)
9. Vennekens, J., Verbaeten, S., Bruynooghe, M.: Logic programs with annotated disjunctions. In: Demoen, B., Lifschitz, V. (eds.) ICLP 2004. LNCS, vol. 3132, pp. 431–445. Springer, Heidelberg (2004). https://doi.org/10.1007/978-3-540-27775-0_30
10. Sato, T., Kameya, Y.: PRISM: a language for symbolic-statistical modeling. In: IJCAI, vol. 97, pp. 1330–1339. Citeseer (1997)
11. Van Gelder, A., Ross, K.A., Schlipf, J.S.: The well-founded semantics for general logic programs. J. ACM (JACM) **38**(3), 619–649 (1991)
12. Puterman, M.L.: Markov decision processes: discrete stochastic dynamic programming. John Wiley & Sons (2014)
13. Hoey, J., St-Aubin, R., Hu, A., Boutilier, C.: Spudd: stochastic planning using decision diagrams. In: Proceedings of International Conference on Uncertainty in Artificial Intelligence (UAI 1999) (1999)
14. Webots. http://www.cyberbotics.com. Open-source Mobile Robot Simulation Software

Simulation and Forecasting

Design of E. coli Growth Simulator Using Multi-agent System

Salvador E. Ayala-Raggi[1], Luís Efraín López-García[1],
Jesús Manuel Roa-Escalante[1(✉)], Lourdes Gabriela Soid-Raggi[2],
Aldrin Barreto-Flores[1], and José Francisco Portillo-Robledo[1]

[1] Facultad de Ciencias de la Electrónica, Benemérita Universidad Autónoma de Puebla, Puebla, Puebla, Mexico
salvador.raggi@correo.buap.mx , jesus.roae@alumno.buap.mx
[2] Facultad de Ciencias, Universidad Nacional Autónoma de México, Ciudad de México, Mexico

Abstract. This work proposes the design of a multi-agent system to simulate the growth of E. coli bacteria in a growth medium, based on the behavior of the bacteria in four main phases: lag phase, exponential phase, stationary phase and death phase. Using the multi-agent system, we simulated each bacterium within a culture medium, employing four main behaviors: moving, dividing, eating, and dying. Bacterial growth is related to the temperature and volume of the growth medium that the user enters before the simulation. The simulator was integrated as a web application that allows visualizing the movement and growth of bacteria, as well as graphing the population of bacteria with respect to elapsed time, using Python, JavaScript, HTML and CSS for the design of the website.

Keywords: Multi-agent system · Web simulator · Python · Html · JavaScript

1 Introduction

The bacterium E. coli (Escherichia coli) is a facultative microorganism that predominates in the intestinal flora of humans, which generally remains harmless in the intestine. However, when the host is weakened or immunosuppressed, it can cause infection [9], leading to illnesses such as diarrhoea, kidney failure and death [1]. Bacteria called anaerobes have the peculiarity of not being able to live in contact with oxygen when they are outside the human body. On the other hand, E. coli can survive until it finds another host [2]. There are several ways that humans can become infected with E. coli, including contact with carrier animals and drinking or swimming in infected water. E. coli bacteria can grow in multiple types of water, even with different temperature ranges. Thus, the growth of E. coli becomes more optimal when they are at high temperatures and slows down when they remain in environments with low temperatures [1].

A. C. Bicharra Garcia et al. (Eds.): IBERAMIA 2022, LNAI 13788, pp. 381–392, 2022.
https://doi.org/10.1007/978-3-031-22419-5_32

E. coli make decisions to control their movement and detect the properties of the growth medium, they have functions that allow them to focus and optimize their search for nutrients, as well as an optimal development temperature. This bacterium can avoid alkaline and acidic environments, also substances that affect its ability to move and replicate. In [7] the authors mention that E. coli populations show intelligent behavior, so they can be considered as a multi-agent system, whose objective is to move away from substances that affect their functions and search for food. In this way, they intentionally move in a group doing a synchronized race, looking for patterns of swarm behavior. The search result, decision making and movement are based on the detection of a nutrient gradient.

When E. coli grows, it elongates and then splits into two halves. If fed well and kept at 37 °C, E. coli bacteria will divide every 20 min [2]. In [14] the authors build a monitoring system for the growth of E. coli capable of displaying the growth curve. During the first phase, E. coli adapts to new environmental conditions and is called the lag phase. In the second phase the E. coli population shows exponential growth, at this stage the DNA replication activity is at its highest level. This exponential phase ends when the nutrients in the culture medium are depleted and the third stationary phase begins. During the stationary phase, the E. coli population stops replicating due to lack of nutrients and accumulation of metabolic waste. Agent-based modeling is a computational modeling that can show an emergent interaction between cells in E. coli population [4]. The aim of use a multi-agent system is simulate complex systems such as a population of bacterias. Every bacteria is an independent agent and can be considered as a decentralized system [13].

In this work we present a simulator based on a multi-agent system, which is used to model the growth of E. coli bacteria as a function of the analysis temperature. Using programming tools, a web page was developed, which uses a graphical interface to initialize the study parameters, giving the user options to take temperature values for the environment and the number of initial bacteria within a culture medium, so that when running the simulator shows the growth curve as a function of time, as well as the duplication and movement of the bacteria.

2   Related Work

Predictive growth models have been generated for E. coli in order to strengthen food quality control, giving rise to a synergy between the areas of microbiology, statistics, mathematics and computer science, so that microbial growth models are used to optimize food safety control [8]. In [12] different bacterial growth simulators are reviewed, these simulators are oriented to microbial behavior in food, and it is mentioned that the most representative database is ComBase, which consists of inactivation and microbial growth data. There are also tools such as Sym'Previus, MicroHibro and Baseline [12], which provide analysis on the kinetics of bacterial growth in vegetables and fruits, to understand the general trend of bacterial behavior against environmental factors, which is very useful for food processors. Speaking of other types of simulators based on multi-agent in other areas, in [3] they make a web application based on agents, which simulates the presence and movement of occupants in buildings considering spatial and temporal diversity, generating a new model stochastic to simulate meeting events based on real meeting data, in addition, they manage to improve the mathematical algorithms to solve the homogeneous Markov chain model for random motion, which allows to obtain better computing performance and simulate any number of spaces and occupants inside the building.

At [5], they use a probabilistic evacuation approach for automatic exit route planning during a fire, and they call their simulator Aamks, which delivers the risk value using different probability distributions and evacuation animations. Aamks is intended for use in buildings such as offices, factories, and shopping malls and is not intended for use on streets or stadiums. In [10], they make a pedestrian simulator, with which they can control and improve the detailed behaviors of the pedestrians and the changes in the conditions of the environments, this provides a simulation of the dynamics of the pedestrians to satisfy different requests. They add a new functionality to control the simulation conditions through a Ruby script within the simulation of the CrowdWalk simulator. The result of their contribution allows to change the environments and behaviors of the agents during the simulation, managing to carry out several simulations in a uniform way.

3   Bacteria Model

E. coli is representing as agents, each agent perceives its environment to make decisions. The environmental information conditions the agent behavior, every decision modifies its internal state. We must mention that it is not possible to have an analytical model that models the entire population since we are not modeling the interaction globally. Having this, the simulator reinforced with a multi-agent methodology, where it is first necessary to conceptualize the behavior of the agent in a simple way. With this in mind, Fig. 1 shows that the bacterium can move, divide, eat and die, and this bacterium lives in an environment that provides a substrate for the bacterium to feed on and therefore there is a limit on the amount of bacteria, the environment also has a temperature that modifies the specific growth rate of the bacteria.

4   Simulator Design

Mainly it is intended that this simulator can be used through a website, this greatly influences its development, that is why the design of this simulator is divided into two sections, the architecture and the algorithmic stage. In the first place, the architectural design of the simulator is based on the development of a web application that will be executed in a free hosting service, this generates some limitations, the most important one lies in the volatile memory provided by the server, being in this way that the properties and memory of both the agents and the simulator cannot be stored inside the server, they must be stored on the client side. The agent is created using the OOP paradigm, so all the code will be encapsulated in an object structure. In the Table 1 the properties of the agent are shown, these can be accessed by the simulator and other agents. In the Table 2 you can see the agent's memory registers, these registers are private, only the agent has access to them. In this way, the simulator uses HTML, CSS, JavaScript and Python, the latter as server control. Through the use of HTML, the structure of the page is generated, with JavaScript, the HTML elements are handled to generate the simulation, through JavaScript, through a POST request, the data is sent to the server, which processes it and returns a response.

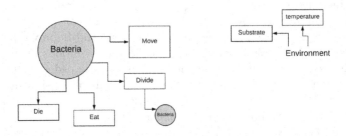

Fig. 1. General diagram of the behavior of the bacterium seen as an agent and its interaction with the environment, which is made up of an amount of substrate and a temperature value.

Table 1. Agent properties

Properties	Description	Data type
Self.X	Value at position X	Integer
Self.Y	Value at position Y	Integer
Self.State	Live state	Boolean
Self.StateDiv	Duplication flag	Boolean

Table 2. Agent memory

Memory	Description	Data type
Self.dupTick	Ticks needed to duplicate	Integer
Self.countTick	Tick counter	Integer
Self.Food	Food counter	Float

Figure 2 shows that the execution of the simulation is controlled by the start button. When this button is pressed, the bacteria are initialized. There is also a stop button, which controls whether the simulation continues to run or stops, each bacterium in BACTERIA is evaluated if the real bacterium divides, a new bacterium is added to the list and the next one is evaluated, when there are no more bacteria to evaluate, the bacteria are redrawn and the procedure is repeated. The simulation is controlled by a timer, this timer can be configured by the user in the interface, each execution of this timer is called *tick*.

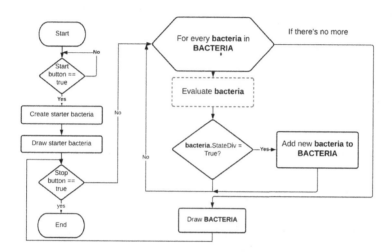

Fig. 2. Flowchart that describes the general operation of the simulator.

To test the bacterium, it is first necessary to check whether it lives or dies, if it dies, no action is taken and the test ends. On the other hand, if it lives, the bacterium moves, feeds and evaluates whether to divide or not. The Fig. 3 shows the flow diagram of this process.

To evaluate if bacteria lives or dies, the amount of food the agent has in his memory is checked. If this amount is less than the established parameter, a false value is assigned to the **Self.State** property, that is, the bacterium is considered dead, otherwise a true value is assigned to mark it as alive, in the Fig. 4 shows the flow diagram for this process. Figure 5 shows the flow diagram for the movement of the agent, which is done by creating a Δx and a Δy with a small random number, these are added with the **X** and **Y** property of the agent, so redrawing the bacterium generates movement.

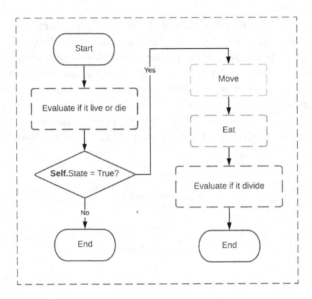

Fig. 3. Flow diagram corresponding to the process of evaluating bacteria.

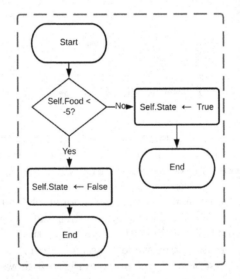

Fig. 4. Flowchart of the agent feeding process.

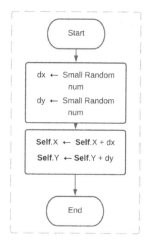

Fig. 5. Flowchart for the process of random movement of agents.

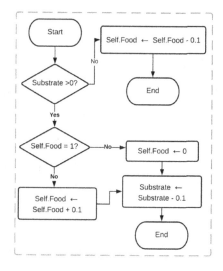

Fig. 6. Process flow chart to evaluate if the bacteria can live or should die in relation to the substrate.

There is the process by which the agent is fed, in which it is checked if the current substrate value is greater than zero, if it is not, the amount of agent feed is decreased and the process ends. If the value of the substrate is greater than zero, then the amount of food of the agent is evaluated, if this amount is equal to one, then the mentioned amount is reset to zero and the value of the substrate decreases. If this quantity is not equal to one, the value of the food increases and the value of the substrate decreases. In Fig. 6 the corresponding diagram illustrates the process. And finally, to perform the division, it is verified if the tick counter is equal to the number of ticks necessary for the duplication, if this is true the duplication flag is activated, otherwise it remains deactivated, in the Fig. 7 shows the corresponding diagram.

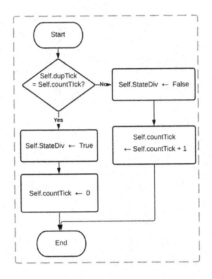

Fig. 7. Flowchart of the agent evaluation process before splitting.

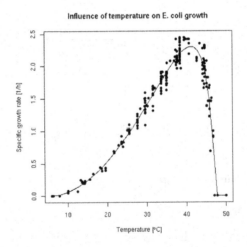

Fig. 8. Graph of relationship between temperature and specific growth factor for E. coli bacteria taken from [6].

Finally, to assign the number of clock ticks necessary for the duplication of the bacterium, the temperature is used, with this the specific growth factor is calculated following the curve presented in the Fig. 8. This shows the relationship between the specific growth factor with respect to a temperature variation [6], with this the doubling time is obtained and the number of ticks is calculated considering the simulation resolution introduced by the user this resolution in minutes.

5   Results

The simulator implementation was built through a website, the *Frontend* was built using **HTML, CSS, JavaScript**, the *Backend* used **Python 3.9 .7**. The complete project can be found on GitHub through the following link: https:// github.com/Freyenn/SimuladorBacterias

Figure 9 shows the interface of the web application, this web is hosted by a free service called *Heroku*, the simulator can be accessed through the following link: https://movbolitatest.herokuapp.com

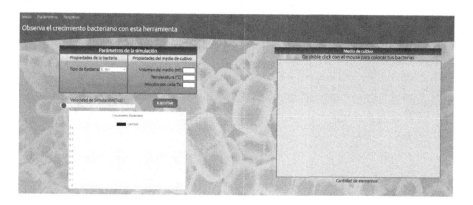

Fig. 9. Graphical interface of the growth simulator for E. coli bacteria.

In Fig. 10 a simulator test is carried out, in this test we set a temperature of 35 °C where the population of bacteria grows at high speed, on the right side of the web simulator you can see the number of bacteria in that specific time, on the left side there is a graph showing the total number of bacteria per entity.

Once the simulator was implemented, several tests were carried out to verify its operation. The simulator was tested by assigning the same amount of substrate, the same simulation resolution, and the temperature was modified to validate the change in the doubling factor of the bacteria with respect to changes in temperature. From this, the graphs shown in Fig. 11 were obtained,

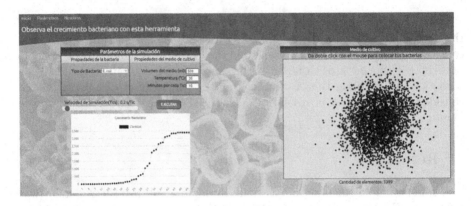

Fig. 10. Execution of the simulation and visualization of the results within the graphical interface.

where the growth phases of a culture can be observed, that is, the lag phase, the exponential phase, the stationary phase and the death phase, you can see a similar behavior seen in [11].

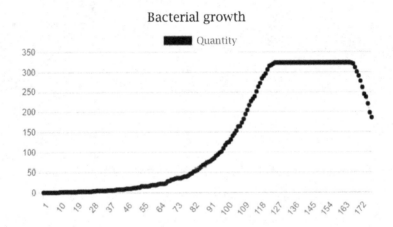

Fig. 11. Comparison of the bacterial growth obtained from the simulation and the growth reference.

6   Conclusions

During the modeling of the properties of the bacteria, challenges were found to generate mathematical relationships that could represent their behavior in the face of variations in temperature and volume of the culture medium, likewise, it should not be forgotten that each bacterium is governed by unique characteristics, given by their DNA.

Our main contribution is to provide the user with a functional simulator, using a web page to have open access to any interested party. The design of the Web page interface was developed to show, in a didactic way, the growth phases of the E. coli bacteria with respect to the initial temperature and volume of the culture medium, in addition, the user can place any number of bacteria before starting the simulation, it will give you an understandable animation about the behavior of bacteria within a culture medium, in addition to this, the population of bacteria is displayed in a graph at each tick of the simulation clock. At the end of the simulation, when the nutritional properties of the culture medium are exhausted, the graphical results show all the growth phases that resemble the real behavior subject to temperature conditions. Through the experimental results it can be observed that the simulator generates a behavior similar to the real behavior of the E. coli bacteria.

In the future, it is intended to develop a better growth behavior for E. coli, including new parameters for the simulation, such as the pH in the medium, the type of culture medium, as well as a movement of the bacteria controlled by the concentration of nutrients in the culture medium. We also want to continue adding new bacteria and the option to use a fully customized bacteria, through its behavior against humidity, pH and temperature.

References

1. Aziz, S.Z., Jamlos, M.F., Jamlos, M.A.: Escherichia coli detection in different types of water. In: 2014 IEEE Symposium on Wireless Technology and Applications (ISWTA), pp. 125–129 (2014). https://doi.org/10.1109/ISWTA.2014.6981170
2. Berg, H.C.: E. coli in Motion. Springer-Verlag, New York (2004). https://doi.org/10.1007/b97370
3. Chen, Y., Hong, T., Luo, X.: An agent-based stochastic occupancy simulator. Build. Simul. **11**(1), 37–49 (2017). https://doi.org/10.1007/s12273-017-0379-7
4. Dib, L.: Multi-agent systems simulating the physiological role of plasmic membrane. Comput. Biol. Med. **38**(6), 676–683 (2008). https://doi.org/10.1016/j.compbiomed.2008.03.005, https://www.sciencedirect.com/science/article/pii/S0010482508000486
5. Krasuski, A., Kreski, K.: A-EVAC: the evacuation simulator for stochastic environment, August 2018. https://doi.org/10.13140/RG.2.2.28854.60481
6. Lobry, J., Chessel, D.: Internal correspondence analysis of codon and amino-acid usage in thermophilic bacteria. J. Appl. Genet. **44**, 235–61 (2003)
7. Lyshevski, S.: Modeling, simulation, control and optimization paradigms for E.coli bacteria. In: 2003 Third IEEE Conference on Nanotechnology, 2003. IEEE-NANO 2003, vol. 2, pp. 690–693 (2003). https://doi.org/10.1109/NANO.2003.1231006
8. Mahdinia, E., Liu, S., Demirci, A., Puri, V.M.: Microbial growth models. In: Demirci, A., Feng, H., Krishnamurthy, K. (eds.) Food Safety Engineering. FES, pp. 357–398. Springer, Cham (2020). https://doi.org/10.1007/978-3-030-42660-6_14
9. Nataro, J.P., Kaper, J.B.: Diarrheagenic Escherichia coli. Clin. Microbiol. Rev. **11**(1), 142–201 (1998)
10. Noda, I., Yamashita, T.: Pedestrian simulator with flexible framework to enhance detailed behavior and environmental change. Artif. Life Robot. **22**(3), 308–315 (2017). https://doi.org/10.1007/s10015-017-0371-4

11. Quigley, T.: Monitoring the growth of E. coli with light scattering using the synergy TM 4 multi-mode microplate reader with hybrid technology (2008)
12. Koseki, S.: Ensuring fresh produce safety and quality by utilizing predictive growth models and predictive microbiology software tools. In: Pérez-Rodríguez, F., Skandamis, P., Valdramidis, V. (eds.) Quantitative Methods for Food Safety and Quality in the Vegetable Industry. FMFS, pp. 213–222. Springer, Cham (2018). https://doi.org/10.1007/978-3-319-68177-1_10
13. Wasik, S., Jackowiak, P., Figlerowicz, M., Blazewicz, J.: Multi-agent model of hepatitis c virus infection. Artif. Intell. Med. **60**(2), 123–131 (2014). https://doi.org/10.1016/j.artmed.2013.11.001, https://www.sciencedirect.com/science/article/pii/S093336571300153X
14. Yao, L., et al.: CMOS conductometric system for growth monitoring and sensing of bacteria. IEEE Trans. Biomed. Circuits Syst. **5**(3), 223–230 (2011). https://doi.org/10.1109/TBCAS.2010.2089794

Quantitative Models for Forecasting Demand for Perishable Products: A Systematic Review

Jonathan Vinicius Kaizer[1]([⊠]), Rodrigo Clemente Thom de Souza[1,2], and Linnyer Beatrys Ruiz Aylon[1]

[1] State University of Maringá (UEM), Maringá, PR, Brazil
{pg403761,lbruiz}@uem.br
[2] Advanced Campus in Jandaia do Sul, Federal University of Paraná (UFPR), Jandaia do Sul, PR, Brazil
thom@ufpr.br

Abstract. Demand forecasting impacts business profitability by assisting in decision-making regarding production and inventory levels to meet demand, without the occurrence of product shortages or waste due to excess. The supply chain of perishable products faces a lack of application, and it is common to observe significant losses in these sectors. To understand how demand forecasting for perishable products has been performed in the literature, this study aims to conduct a systematic review to identify the methods, the data required, and how these studies have been evaluated. Although we usually associate perishable products with food, any other product that cannot be stocked, or that becomes obsolete quickly or has a short shelf life can be considered perishable. This study has shown that the most recent approaches involve classical models such as ARIMA, and machine learning models such as Gradient Boosting, long short-term memory (LSTM) and support vector regression (SVR). In addition, these models are characterized by the use of time series supplemented with external data. To evaluate the models, statistical indicators, such as Mean Absolute Percentage Error (MAPE), are used to measure their results and calculate Root Mean Square Error (RMSE). The results presented in this systematic review allow new studies in this area to be developed based on the main approaches in the literature and are also an opportunity to identify new approaches for methods yet to be more systematically explored.

Keywords: Demand forecasting · Perishable products · Forecasting models · Systematic review

1 Introduction

The supply chain of perishable products is an activity that deals with many uncertainties. Demand variability causes the industry to incur significant losses when it is unable to flow products or when there is a shortage of products because of unexpected high demand. Retailers of perishable consumer goods face the challenge of forecasting demand accurately in order to manage their daily operations, since their products have a short shelf life and a high volatility of demand [1].

A. C. Bicharra Garcia et al. (Eds.): IBERAMIA 2022, LNAI 13788, pp. 393–404, 2022.
https://doi.org/10.1007/978-3-031-22419-5_33

The main studies involving perishable products involve optimization [3]; the use of simulation [4]; forecasts of price, yield, demand [5] or the forecast of other variables, such as the weight of broilers approached by Johansen et al. (2019) [6]; or even involve the use of Internet of Things (IoT) devices as an auxiliary tool in data collection and decision-making [7].

Although we usually associate perishable products with food, any other product that cannot be stored or has a short shelf life can be considered perishable, such as blood used for transfusions in hospitals, electronic products that quickly become obsolete, and fashionable clothing and footwear products that go out of style as the seasons progress.

Through computational techniques, it is possible to employ models that assist in forecasting demand for these products. But where to start? There is a plethora of techniques and variations found in the literature, especially with the advancement of machine learning. There is the need for an accurate forecast model with good efficiency and application potential, and that reflects the real demand, with lean stock, but with margin so that no shortage occurs.

The main question that this research sought to answer was: How have studies on demand forecasting for perishable products been developed in the existing literature? To understand how demand forecasting for perishable products has been performed in the literature, this study aims to conduct a systematic review to identify the methods, the required data, and the way in which the studies were evaluated, which may serve as a basis for the development of new works within this theme.

The rest of this work is organized as follows: Sect. 2 presents the methodology of this systematic review; in Sect. 3, the research results are presented; Sect. 4 discusses the results; and finally, Sect. 5 shows the conclusions of this research.

2   Systematic Literature Review

This section describes the search and selection process that was used to obtain quality articles for this systematic review.

2.1   Research Problem

The questions that sought to be answered were:

- What are the main data involved in forecasting demand for perishable products?
- What are the main quantitative models used in demand forecasting for perishable products?
- What indicators were used to evaluate the solutions?

The first research question seeks to understand what data is necessary for model experimentation, in order to guide new studies by directing what data can be collected.

The second research question seeks to understand the main quantitative models applied in the studies used for this review, so that this study can serve as a guide for the choice of the most appropriate models to be used in future studies within this theme.

The third research question aims to identify the methodology used in those works to assess the effectiveness of the quantitative models addressed. A well-defined evaluation method is very important to find out if the results are as expected or if it is necessary to make adjustments to the input data, changes or discard the model due to low efficiency.

2.2 Search String Definition

The databases consulted were IEEE, ACM, Web of Science and Science Direct using a search string with the following terms: "demand AND (forecast OR forecasting OR prediction OR predicting OR predict) AND (perishable OR fresh)".

2.3 Inclusion and Exclusion Criteria

The inclusion and exclusion criteria for articles were defined to assist in the selection of relevant papers that corresponded with the intended research objective and research questions.

The exclusion criteria for the defined papers were as follows:

- Studies with less than 4 pages.
- Studies not in English.
- Studies that do not approach demand forecasting as a proposal.
- Studies that do not approach perishable products.
- Studies that do not approach quantitative models.

The inclusion criteria for papers were:

- Studies that approach quantitative models in demand forecasting for perishable products.

2.4 Search Execution

The execution of the search and extraction of the studies occurred in the month of October 2021 and the search string was used to search for studies published between 2017 and 2021, a 5-year interval, in order to obtain the most recent papers.

A total of 308 articles were returned and cataloged in the "StArt" [11] tool to perform data evaluation, selection, and extraction, as can be seen in Fig. 1.

Database

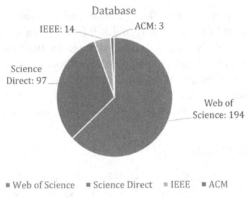

■ Web of Science   ■ Science Direct   ■ IEEE   ■ ACM

Fig. 1. Articles returned by database.

In the selection phase, from the 308 articles obtained, 60 were classified as duplicates, 211 were rejected, and 37 articles were selected after reading the title, abstract and keywords. The "StArt" [11] tool was important in this task because it has an automatic duplicate identification mechanism and a system that compares the similarity between the titles, abstracts, and keywords of the cataloged articles.

In the extraction phase, the articles selected in the previous step were read in full and the inclusion and exclusion criteria were applied again. Of the 37 articles, 20 were rejected for not meeting the inclusion criteria and 17 articles were selected for extraction of the information from the systematic review.

3   Results

This section will present the selection of the most recent studies on demand forecasting for perishable products and the main data found in these works, the objectives of the studies, the techniques used, and the indicators employed in the evaluation of the results.

In the first study, Chen et al. (2019) [1] develop a model considering sales and inventory data with data from external factors, such as weather and temperature. The approach involves grouping franchised stores that have similar patterns.

Cristensen et al. (2021) [12] propose a new indicator for assessing forecast accuracy by considering the availability, freshness and shelf life of fresh food. The proposal aims to be a more efficient alternative than the other commonly used statistical indicators by trying to reduce the variations of forecast deviations and reduce expected losses.

Cristensen et al. (2019) [13] also present the proposal of a new asymmetric valuation indicator, aiming to penalize deviations from the forecast that compromise on-time sales, available stock and future demand for some different types of meat. The results which were observed are in line with what was also proposed in the study [12].

Dellino et al. (2018) [8] approach prediction for highly perishable fresh produce, starting from preprocessing and normalization of the data and comparing three different models. They also approach automatic optimization of the parameters used to prevent a forecast from generating negative stocks, selling expired products, and assisting other important decisions that help plot an executable plan on top of the generated forecast.

Since the classic ARIMA model cannot use data combined with other variables, it is then necessary to resort to ARIMAX and other models to make comparisons.

Dellino et al. (2017) [14] also propose a decision-making system based on demand forecasting with parameters adjusted by heuristic optimization. In this work, data preprocessing to identify seasonality and remove noise is also approached. The study showed there is no one model that is adequate for all cases.

Hua et al. (2021) [15] take an approach to forecasting demand as a function of the discount offered, so that demand can be manipulated to meet the constraints of perishable products according to the price being set.

Huber and Stuckenschmidt (2021) [16] work with a bakery in which it is necessary to optimize the use of the oven to bake various products and prevent these from losing their freshness. A noteworthy point is that the authors made a comparison between hourly and daily forecasting to verify which would be more assertive. This shows us that the interval of each of the forecasts is important and is directly related to the shelf life of the product.

Huber et al. (2017) [17] also approach a bakery and make a daily forecast for 16 products sold for 18 months in six stores. The authors did a grouping of the products to then use ARIMA in these groups.

Khatibi et al. (2020) [18] approach the tourism industry, which is considered to be perishable, that is, it disappears if not used. The highlight of this study is in the use of data extracted from social media and travel websites combined with environmental data to forecast demand for tourists. The study proved to be very relevant, and it is clear that it is possible to work with data extracted from an alternative correlated database.

Another study involving tourism is Kulshrestha et al. (2020) [19] in which a combination of LSTM and Bayesian Network was used and showed superior results compared to other models. The results showed that the combination of techniques can be beneficial in some problems, however, the proposed model works as a black box (similar to other machine learning models) in which the prediction is accurate but cannot be explained easily.

Li, N. et al. (2021) [2] involves the demand for blood in hospitals using statistical modeling, machine learning, and operations research to control stock levels. Keeping excess stock can hinder availability in other hospitals, causing loss of stock that has a shelf life of a few weeks, as well as shortages that can result in loss of life.

Li, C. et al. (2021) [20] use Exponential Factorization Machine (EFM) for products with short life cycles, such as footwear. The appealing thing about this study is that the authors use the sales history of other shoes that have similar attributes, since the new shoes are new releases and had no sales history.

Priyadarshi et al. (2019) [9] approach demand forecasting for fresh food and conclude that the models and results cannot be generalized, but they may serve as a basis for other studies on perishable products. The authors state that the industry lacks more forecasting applications and that factors such as consumer preference, promotions, holidays, and seasonality affect demand.

Puchalsky et al. (2018) [21] work with two approaches: soybean price forecasting and soybean demand forecasting, using a neural network combined with optimization techniques to find the best parameters for the model. As in the study by Kulshrestha et al.

(2020) [19], optimization techniques were employed to search for the best parameters for each model, and this may contribute to more accurate and optimized results.

The work of Uzundumlu et al. (2018) [22] is a brief study that addresses the use of ARIMA in demand forecasting for fig in Turkey. This study serves as a basis for understanding the ease of applying ARIMA with time series data without other external variables involved.

Yang and Sutrisno (2018) [23] approach short-term demand forecasting for a bakery in China, using the number of sales from the first few hours to forecast the demand for the rest of the day. One of the great advantages of having a good amount of data is the facility of obtaining high accuracy results and easier models to build.

Finally, the study by Zhao and Setyawan (2020) [10] approaches demand forecasting for bread on a daily and weekly basis. The data presented involves time series that needed treatment to correct missing data, incorrect data, outliers and to normalize data. Clustering helps to decrease the dimensionality of the problem and facilitates the development and execution of the model, and is a suitable approach for when a significant amount of data is involved.

The following is a summarized compilation of the main information from the studies presented in relation to the research questions posed.

3.1 Main Data Found

In all the selected studies, the main source of data was time series which dealt with demand over time, such as daily sales. In addition, these time series were used in conjunction with other external data, for example, holiday dates, weather data, shelf life, production and inventory volume, prices, product attributes, cost of living and customer income, shipping information, data extracted from social networks, the discount rate offered, and others. Table 1 shows the objectives and data used in the selected studies.

Table 1. Main data found.

Studies	Objective	Data
[1]	Grouping of stores of perishable products and demand forecast for these groups	Time series of 100 products and 1500 stores, weather data, shelf life, production volume and stock
[12]	Proposal of new asymmetric indicator for daily forecast for meat	Time series of 17 products, data from promotions and campaigns
[13]	Proposal of new asymmetric indicator for daily forecast for meat	Time series of sales of meat categories, attributes such as shelf life and discount
[8]	Demand forecasting for fresh food (milk, cheese, salmon) with optimization of parameters	Time series of sales of 156 products from 19 stores (4 products were used in the study), plus exogenous variables such as shelf life

(*continued*)

Table 1. (*continued*)

Studies	Objective	Data
[14]	Demand forecasting for fresh products supply chain (milk, cheese, yogurt, salmon) with optimization of parameters	Time series and prices of sales of 156 products from 19 stores (4 products were used in the study)
[15]	Demand forecasting as a function of discounts for online retail electronics	Time series of sales of 11,000 products from 100 stores in 10 cities and discounts offered on these sales
[16]	Hourly demand forecast for a bakery	Time series of sales of 14 products from 9 stores over a period of 987 days, location, special dates, transactions and business hours information
[17]	Daily demand forecast for a bakery	Time series of sales of 16 products sold for 18 months in 6 stores, plus special dates
[18]	Tourism demand forecasting	Time series of 27 museums and 76 US national parks extracted from social media and travel websites, combined with environmental and weather data
[19]	Tourism demand forecasting	Time series with 75 records including country of origin, cost of living in the country of origin, and tourist income
[2]	Predicting demand for blood in hospitals	Time series with 369,481 transfusions in 60,141 patients from 2008 to 2018, plus patient information such as age, sex, blood type, diagnosis and 200 other variables
[20]	Demand forecasting for newly launched shoes that have no sales history	About 5000 sales records and 45 attributes related to the products sold
[9]	Demand forecasting for vegetables (tomato, potato and onion)	Time series of daily sales for three products for 22 weeks
[21]	Soybean demand forecast	Sales time series with 108 samples for each of the three available soybean groups
[22]	Fig demand forecast	Production time series for the last 26 years
[23]	Daily demand forecast for the products of a bakery	Time series with more than 10 million sales records from 53 franchised stores selling 2000 different products, plus supporting variables such as time, temperature, day of the week and weather condition
[10]	Daily and weekly demand forecast per bread	Time series with more than 49 thousand records, 18 thousand stores grouped into 9 groups and 5 of the main products most sold, in addition to variables such as day of the week, holidays, store code and consumption profile

3.2 Main Quantitative Models Used

The articles consulted employ more than one model to make comparisons and determine which one had the best results. The models that obtained the best results in each study are arranged in Table 2. The number of uses of each model does not coincide with the number of papers, since some studies found more than one suitable model for the problem presented.

Table 2. Models that obtained the best results.

Model	Uses	Class	Studies
ARIMA	3	Classic	[14, 17, 22]
ARIMAX	2	Classic	[14, 17]
SVR	2	Machine Learning	[9, 18]
LSTM	2	Machine Learning	[9, 16]
XGBoost and Gradient Boosting	2	Machine Learning	[2, 15]
Naïve Bayes	2	Classic	[12, 13]
Transfer Function	2	Classic	[8, 14]
Holt-Winter	1	Classic	[10]
Wavelet Neural Network (WNN)	1	Machine Learning	[21]
Bayesian Network	1	Machine Learning	[19]
Feed-Forward Neural Network (FFNN)	1	Machine Learning	[23]
Nonlinear Autoregressive Network with Exogenous Inputs (NARX)	1	Machine Learning	[1]
Exponential Factorization Machine (EFM)	1	Classic	[20]

3.3 Indicators Used to Evaluate the Solutions

The main indicators used in the studies can be seen in Table 3. The evaluation of the performance of the solution generated by the models was mostly done by statistical indicators. Some works have proposed new evaluation indicators derived from these statistical indicators, while other studies have used quantitative indicators as an evaluation parameter, such as waste minimization or profit maximization.

Table 3. Main indicators used.

Indicator	Uses	Studies
Mean Absolute Percentage Error (MAPE)	11	[2, 9, 10, 12, 13, 16–21]
Measure Root Mean Square Error (RMSE)	10	[2, 8, 9, 12–14, 16, 17, 19, 23]
Mean Absolute Error (MAE)	7	[8, 13, 14, 16, 19, 20, 23]
Mean Square Error (MSE)	6	[9, 10, 13, 14, 21, 23]
Median Absolute Deviation (MAD)	2	[9, 10]
Max Absolute Error (MaxAE)	2	[8, 14]

4   Discussion

The results show that recent studies have similar methodologies, with the particularity that some use classical models, while others use machine learning approaches in forecasting demand for perishable products.

14 studies directly approached the use of quantitative models, while other two studies focused on proposing or comparing the indicators of these quantitative models. A single study took an approach with a semi-parametric model in which it combined demand forecasting with the discounts offered, and in this way, it was possible to obtain an estimated demand according to the discount rate.

The studies selected predominantly involve food as perishable products, but we also obtained studies focusing on the demand for blood in hospitals, for footwear and fashion items, for tourism, and on the demand for electronic products. All these approaches are considered perishable products since they have a shelf life in which they deteriorate or become obsolete, or they cannot be stocked as in the case of tourism.

The main data of these works were provided by time series. Time series can be organized so that patterns can be found, such as trends and seasonality [10]. A time series is a stochastic process observed sequentially over time in which a mathematical model can be identified to describe it [8].

In addition to demand time series, many authors have approached the use of external data to add attributes to the information in these time series to improve predictability and establish a better relationship between demand and external variables. It is necessary to obtain detailed data that includes sales and inventory records, with environmental factors such as temperature, weather, and wind speed [1]. Exchange rates, fuel prices, climate change, crises and epidemics could cause deviations in forecasting demand for perishables, especially in tourism [18].

Khatibi et al. (2020) [18], for example, approach the extraction of social network data to be used as external variables in conjunction with tourism time series for more assertive demand forecasting. Tourism products are considered perishable, since a seat on a flight or an unoccupied hotel room represents loss of profitability relative to the time elapsed. In the case of using classical models, it is necessary to pre-process the data for smoothing and normalization, removing outliers if necessary. On the other hand, machine learning models can handle these discrepancies better, without the need for preprocessing, but this is not to say that it cannot be done.

When analyzing the most used models, it was noted that the classical models are the most covered in the studies, especially the ARIMA and ARIMAX. The ARIMA is the classical model most used in the literature, because it performs well, it is flexible and easy to implement, and it is normally used for comparison with more sophisticated models [14]. It is one of the models most widely used to understand and forecast seasonal demands and time series [9]. The main application of ARIMA is in short-term forecasting, and ARIMAX is an approach more suitable for seasonal data [18]. ARIMAX considers the time series together with other external variables [14].

When the focus is on machine learning, LSTM, SVR and Gradient Boosting (discussed as a variation called XGBoost) are the most widely applied models, although many studies approach other variations involving neural networks. The LSTM neural network is a specialized recurrent neural network (RNN) that can learn from long-term data and performs better on nonlinear models than the ARIMA and the backpropagation neural network. The Gradient Boosting (GBR and XGBoost) consists of using decision trees to make predictions and these techniques can be classified as Ensemble Learnings [9]. The XGBoost is a variation of Gradient Boosting that consists of using smaller models to build a more general model. SVR (Support Vector Regression) is derived from the widely used SVM (Support Vector Machine) for regression, which uses a linear regression on a nonlinear data set [18].

As for the evaluation of the model, studies generally use the same indicators to evaluate the results. These indicators are statistical and the most used to compare different models on the same dataset are RMSE, MAE, MSE and MaxAE. When having a different set of data for the same model, it is necessary to resort to relative indicators such as MAPE, being careful to avoid bases with missing data or very close to zero. Usually, the indicator of the quality of ARIMA forecast models is the Mean Square Error (MSE). The MAE is used when one wants to minimize the absolute deviation [14]. The RMSE is used to compare forecast errors of models using the same data [18]. The most used indicator in the studies selected is the Mean Absolute Percentage Error (MAPE).

This study can serve as a basis for the development of new works on demand forecasting for perishable products, helping to identify some of the most widely used quantitative computational models and the evaluation indicators used to measure the results, as well as contributing to the contextualization of the problem and the construction of the models.

5 Conclusion

Demand forecasting for perishable products is an approach that has been gaining space, but there are still few reports of its application in the literature. The supply chain of this sector needs to better forecast and control its operations to prevent losses due to obsolete or degraded products, or to prevent the lack of product availability from leading to the loss of the opportunity to meet the demand.

This study showed that the most recent approaches involve classical models, such as ARIMA, and machine learning models such as Gradient Boosting, LSTM, and SVR. In addition, they are characterized by the use of time series supplemented with external data. To evaluate these models, statistical indicators are used to measure the results, such as Mean Absolute Percentage Error (MAPE) and Measure Root Mean Square Error (RMSE).

The results presented in this systematic review allow new studies in this area to be developed based on the main approaches in the literature and are also an opportunity to identify new approaches for methods yet to be more systematically explored. It is recommended that new in-depth studies be developed on the models and indicators raised, since there is an infinite number of possibilities of application and complexity that have been approached only briefly in this paper, or else, that have not been approached because they are not part of the recent literature.

Acknowledgment. This research was funded by CNPq, Conselho Nacional de Desenvolvimento Científico e Tecnológico - CNPq grant - Brazil (311685/2017–0 and 163961/2021–2) and Fundação Araucária (17.633.124–0).

The authors would like to thank the Academic Writing Center (Centro de Escrita Acadêmica, CEA) of the State University of Maringá (UEM) for assistance with English language developmental editing.

References

1. Chen, C., Wang, Y., Huang, G., Xiong, H.: Hierarchical demand forecasting for factory production of perishable goods. In: 2019 IEEE International Conference on Big Data (Big Data). IEEE (2019). https://doi.org/10.1109/bigdata47090.2019.9006161
2. Li, N., Chiang, F., Down, D.G., Heddle, N.M.: A decision integration strategy for short-term demand forecasting and ordering for red blood cell components. Operations Res. Health Care **29**, 100290 (2021). https://doi.org/10.1016/j.orhc.2021.100290
3. Teimoury, E., Nedaei, H., Ansari, S., Sabbaghi, M.: A multi-objective analysis for import quota policy making in a perishable fruit and vegetable supply chain: a system dynamics approach. Comput. Electron. Agric. **93**, 37–45 (2013). https://doi.org/10.1016/j.compag.2013.01.010
4. Lin, X., Negenborn, R.R., Lodewijks, G.: Predictive quality-aware control for scheduling of potato starch production. Comput. Electron. Agric. **150**, 266–278 (2018). https://doi.org/10.1016/j.compag.2018.04.020
5. Motevalli-Taher, F., Paydar, M.M., Emami, S.: Wheat sustainable supply chain network design with forecasted demand by simulation. Comput. Electron. Agriculture **178**, 105763 (2020). https://doi.org/10.1016/j.compag.2020.105763
6. Johansen, S.V., Bendtsen, J.D., Jensen, M.R., Mogensen, J.: Broiler weight forecasting using dynamic neural network models with input variable selection. Comput. Electron. Agric. **159**, 97–109 (2019). https://doi.org/10.1016/j.compag.2018.12.014
7. Santa, J., Zamora-Izquierdo, M.A., Jara, A.J., Gómez-Skarmeta, A.F.: Telematic platform for integral management of agricultural/perishable goods in terrestrial logistics. Comput. Electron. Agric. **80**, 31–40 (2012). https://doi.org/10.1016/j.compag.2011.10.010
8. Dellino, G., Laudadio, T., Mari, R., Mastronardi, N., Meloni, C.: Microforecasting methods for fresh food supply chain management: a computational study. Math. Comput. Simul. **147**, 100–120 (2018). https://doi.org/10.1016/j.matcom.2017.12.006
9. Priyadarshi, R., Panigrahi, A., Routroy, S., Garg, G.K.: Demand forecasting at retail stage for selected vegetables: a performance analysis. J. Model. Manag. **14**, 1042–1063 (2019). https://doi.org/10.1108/jm2-11-2018-0192
10. Zhào, M.A., Setyawan, B.: Sales forecasting for fresh foods: a study in Indonesian FMCG. In: 2020 International Conference on Information Science and Communications Technologies (ICISCT). IEEE (2020). https://doi.org/10.1109/icisct50599.2020.9351484
11. Start (state of the art through systematic review). http://lapes.dc.ufscar.br/tools/start_tool, 2021. Accessed 1 October 2021
12. Christensen, F.M.M., Solheim-Bojer, C., Dukovska-Popovska, I., Steger-Jensen, K.: Developing new forecasting accuracy measure considering product's shelf life: Effect on availability and waste. J. Clean. Prod. **288**, 125594 (2021). https://doi.org/10.1016/j.jclepro.2020.125594
13. Christensen, F.M.M., Dukovska-Popovska, I., Bojer, C.S., Steger-Jensen, K.: Asymmetrical evaluation of forecasting models through fresh food product characteristics. In: Ameri, F., Stecke, K.E., von Cieminski, G., Kiritsis, D. (eds.) APMS 2019. IAICT, vol. 566, pp. 155–163. Springer, Cham (2019). https://doi.org/10.1007/978-3-030-30000-5_21
14. Dellino, G., Laudadio, T., Mari, R., Mastronardi, N., Meloni, C.: A reliable decision support system for fresh food supply chain management. Int. J. Prod. Res. **56**, 1458–1485 (2017). https://doi.org/10.1080/00207543.2017.1367106

15. Hua, J., Yan, L., Xu, H., Yang, C.: Markdowns in e-commerce fresh retail: a counterfactual prediction and multi-period optimization approach. In: Proceedings of the 27th ACM SIGKDD Conference on Knowledge Discovery & Data Mining. ACM (2021). doi:https://doi.org/10.1145/3447548.3467083

16. Huber, J., Stuckenschmidt, H.: Intraday shelf replenishment decision support for perishable goods. Int. J. Prod. Econ. **231**, 107828 (2021). https://doi.org/10.1016/j.ijpe.2020.107828

17. Huber, J., Gossmann, A., Stuckenschmidt, H.: Cluster-based hierarchical demand forecasting for perishable goods. Expert Syst. Appl. **76**, 140–151 (2017). https://doi.org/10.1016/j.eswa.2017.01.022

18. Khatibi, A., Belém, F., da Silva, A.P.C., Almeida, J.M., Gonçalves, M.A.: Fine-grained tourism prediction: Impact of social and environmental features. Inf. Process. Manage. **57**, 102057 (2020). https://doi.org/10.1016/j.ipm.2019.102057

19. Kulshrestha, A., Krishnaswamy, V., Sharma, M.: Bayesian BILSTM approach fortourism demand forecasting. Ann. Tour. Res. **83**, 102925 (2020). https://doi.org/10.1016/j.annals.2020.102925

20. Li, C., Cheang, B., Luo, Z., Lim, A.: An exponential factorization machine with percentage error minimization to retail sales forecasting. ACM Trans. Knowl. Discov. Data **15**, 1–32 (2021). https://doi.org/10.1145/3426238

21. Puchalsky, W., Ribeiro, G.T., da Veiga, C.P., Freire, R.Z.: dos Santos Coelho. Agribusiness time series forecasting using wavelet neural networks and metaheuristic optimization: an analysis of the soybean sack price and perishable products demand. Int. J. Prod. Econ. **203**, 174–189 (2018). https://doi.org/10.1016/j.ijpe.2018.06.010

22. Uzundumlu, A., Oksuz, M., Kurtoglu, S.: Future of fig production in Turkey. J. Tekirdag Agric. Faculty **15**, 138–146 (2018)

23. Yang, C.-L., Sutrisno, H.: Short-term sales forecast of perishable goods for franchise business. In: 2018 10th International Conference on Knowledge and Smart Technology (KST). IEEE (2018). https://doi.org/10.1109/kst.2018.8426091

Short Papers

FeetGUI: A Python-based Computer Vision Tool to Support Anesthesia Assessment Procedures Using Infrared Thermography

B. Lotero-Londoño[1], M. Loaiza-Arias[1], M. Tobon-Henao[1(✉)],
D. Collazos-Huertas[1], J. Daza-Castillo[2], N. Valencia-Marulanda[2],
M. Calderón-Marulanda[2], O. Aguirre-Ospina[2], A. Alvarez-Meza[1],
and G. Castellanos-Dominguez[1]

[1] Signal Processing and Recognition Group, Universidad Nacional de Colombia,
Manizales, Colombia
{bloterol,mloaizaa,mtobonh,dfcollazosh,amalvarezme,
cgcastellanosd}@unal.edu.co
[2] SES Hospital Universitario de Caldas, Manizales, Colombia
yeje92@gmail.com;njvalencia@hotmail.com;mauricio.calderon@ucaldas.edu.co;
odaguirre@ses.com.co

Abstract. Proper management of pain in the course of labor prevents complications during the process that may put the health of the mother-child binomial at risk. In this sense, epidural neuraxial analgesia helps attenuate the stress response and provides local insensibility during and after labor. However, specialists often assess anesthetic effectiveness using conventional approaches such results are highly subjective, exhaustive, and invasive to the patient. To address these issues, we present FeetGUI, a low-cost, easy-to-use, and portable computer vision-based tool that allows users to estimate the effectiveness of epidural anesthesia. FeetGUI comprises a set of functionalities based on computer vision techniques for infrared thermography imaging analysis, resulting in a support tool that improves anesthetic evaluation procedures.

1 Problem Statement

Uncontrolled pain during birth-giving produces significant complications related to the healthy of binomial mother-son [1]. In this sense, epidural anesthesia helps to attenuate the stress response and provides local insensibility during and after labor [2]. Nevertheless, in some cases, this type of anesthesia results ineffective (in up 32%), implying an increase in the repetition of anesthetic doses. In addition, another problem of regional anesthesia is its high subjectivity in performance feedback, leading to high false positive and negative rates [3]. Besides, current procedures evaluates epidural anesthetic performance using [4]: assessment of the sensory thermoalgesic effect, electrophysiological modality through tests, and modalities based on imaging techniques. Nonetheless, none of these methods

© The Author(s), under exclusive license to Springer Nature Switzerland AG 2022
A. C. Bicharra Garcia et al. (Eds.): IBERAMIA 2022, LNAI 13788, pp. 407–409, 2022.
https://doi.org/10.1007/978-3-031-22419-5

possess the requirements of low-cost implementation, minimal invasiveness, and high reliability at the same time.

2 Results and Contributions

With these above factors in mind, we were motivated to create FeetGUI. This Python-based, low-cost, and portable tool employs embedded systems as the technology intending to support the anesthetic protocol by describing temperature changes in feet dermatomes using computer vision algorithms. As a result, we can estimate the effectiveness of epidural anesthesia, which is vital for medical decision-making, like the re-application of pharmacological mixtures in a suitable dose and timing. Feet-GUI features and functionalities are illustrated in Fig. 1. FeetGUI is designed to impact the fields of anesthetic medical research, facilitating behavioral analysis of epidural blockade in this community. With FeetGUI, the capability to evaluate and follow up the anesthesia effectiveness is enhanced and greatly simplified, avoiding the subjectivity issue. In this sense, the researchers can quickly generate new capture sessions in a fraction of the time it takes to connect the thermography camera, turn on the embedded system, and carry out the temperature recordings themselves from FeetGUI. Likewise, the impact on the target population (pregnant women) is significant since a correct characterization of temperature changes in feet dermatomes, generated during epidural anesthetic procedure and minutes after, provides better support and understanding of pain relief during labor.

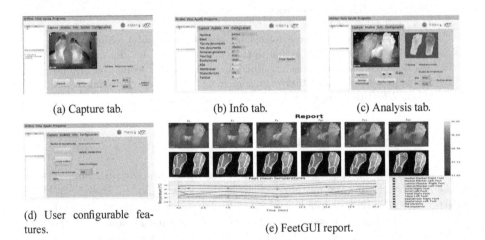

(a) Capture tab.   (b) Info tab.   (c) Analysis tab.

(d) User configurable features.   (e) FeetGUI report.

Fig. 1. FEET-GUI artificial intelligence features overview. Installation instructions can be found in the official repository (https://github.com/UN-GCPDS/FEET-GUI).

References

1. Ullman, R., Smith, L.A., Burns, E., Mori, R., Dowswell, T.: Parenteral opioids for maternal pain relief in labour. Cochrane Database Syst. Rev. **9**, CD007396 (2010)
2. Pedro José Herrera Gómez and Paula Andrea Medina: Los problemas de la analgesia obstétrica. Rev. Colomb. de Anestesiol. **42**(1), 37–39 (2014)
3. Bruins, A.A., Kistemaker, K.R.J., Boom, A., Klaessens, J., Verdaasdonk, R., Boer, C.: Thermographic skin temperature measurement compared with cold sensation in predicting the efficacy and distribution of epidural anesthesia. J. Clin. Monit. Comput. **32**, 335–341 (2018)
4. Nelson, E.W., Woltz, E.M., Wolf, B.J., Gold, M.R.: A survey of current anesthesia trends for electrophysiology procedures. Anesth. Analg. **127**(1), 46–53 (2018)

Proposal of a Software Translator with Interlanguage Translation Resources, Brazilian Sign Language (Libras) – Portuguese

Vagner Luiz Gava[1]([⊠]) ⓘ, Angelina Inácio[1] ⓘ, Felipe A. F. Kleine[1],
Fabio R. L. P. Souza[1] ⓘ, Jorge Bidarra[2] ⓘ, Tânia Martins[2] ⓘ,
Jampierre Rocha[3] ⓘ, and Denis Leite Gomes[3] ⓘ

[1] Instituto de Pesquisas Tecnológicas do Estado de São Paulo – IPT, São Paulo,
Brazil
{vlgava, angelinasi, fkleine, fsouza}@ipt.br
[2] Universidade Estadual Do Oeste Do Paraná (UNIOESTE) Cascavel, Cascavel,
Brazil
{jorge.bidarra, tania.martins}@unioeste.br
[3] Research and Development Lenovo Brasil LTDA Indaiatuba, Indaiatuba,
Brasil
{jrocha2, dgomes2}@lenovo.com

Abstract. The challenge of this work is to propose a solution that is not only capable of helping the deaf in the production of their sentences in Libras but at the same time translating them into Portuguese. It is proposed a translator based solely on morphological and syntactic classifications in the first version. Showing themselves as encouraging, the results are being useful to help UX development in building an interface to support the words typed by deaf users, although on the other side it proved unsuitable for dealing with more elaborate syntactic structures.

Keywords: Accessibility · Brazilian sign language · Libras · Machine translation · Gloss-to-text · Rule- based · Artificial intelligence

1 Problem Statement

The general objective of this article is to develop a tool to help deaf individuals whose primary language is Libras in the insertion of texts in written Portuguese in information and communication technology input devices like notebooks, smartphones, and tablets [1].

2 Results and Contributions

The translation software receives a text as input, and, after a few processing steps, the result is a text close to spoken Portuguese. To do this, the first step is by tokenization and tagging of input sentences (dividing sentences into the smallest feasible unit, such

M. Bicharra Garcia and J. C Ferro (Eds.): IBERAMIA 2022, LNAI 13788, pp. 410–411, 2022.
https://doi.org/10.1007/978-3-031-22419-5

as words, numerals, and symbols), the second step is by an identification of verbal elements and verbal inflection and finally applying logical rules for grammatical sentence construction. A preliminary gloss-to-text translator for Libras was configured in the context of banking situations faced by deaf users to validate the proposed approach [2, 3].

Some examples of the application of these rules are presented in Table 1.

Table 1. Translator Input/Translator Output/Expected Phrases

Translator input (Simple sentences)	Translator output (Simple sentences) **P: Portuguese, E: English**	Expected phrase (Simple sentences)
(P) eu emprestar dinheiro	(P) Eu empresto dinheiro	(P) Eu empresto dinheiro
(E) I to borrow money	(E) I borrow money	(E) I borrow money
Translator input (compound sentences)	Translator output (compound sentences)	Correct sentences (compound sentences)
(P) eu emprestar dinheiro ele devolver amanhã dinheiro	(P) eu emprestarei amanhã dinheiro devolve dinheiro	(P) Eu empresto dinheiro. Ele devolverá dinheiro amanhã
(E) I to borrow money he to return tomorrow money	(E) I will borrow tomorrow money return money	(E) I borrow money. He will return money tomorrow

It can be verified from this table and from the other tests performed that:

- Sentences with more than one paragraph are not correctly translated.
- Additional marking information is needed in sentences entered by the deaf for proper translation (e.g., adverbs of time) and sentence punctuation is not entered.

As contributions to this research, the following stand out:

- The observation of the need to create a grammar as future work to improve the proposed process.
- Need for additional information for marking sentences inserted by the deaf for proper translation. This contribution is especially important when applied to UX development in building an interface to support the words typed by deaf users.

References

1. Bidarra, J.: PORLIBRAS: Fundamentos para a Implementação de Ferramentas Computacionais para suporte ao Desenvolvimento de um Sistema bilíngue de Tradução Automática Português-Libras, Projeto de Pesquisa. Universidade Estadual do Oeste do Paraná (UNIOESTE) (2018). (em curso)
2. Pandey, S., Pandey, S.K.: Applying natural language processing capabilities in computerized textual analysis to measure organizational culture. Organ. Res. Methods **22**(3), 765–797 (2019)
3. Klein, D., Manning, C.D.: Natural language grammar induction with a generative constituent-context model. Pattern Recogn. **38**(9), 1407–1419 (2005)

Heart Disease Diagnoses: A Study Regarding Assigning Credibility to Evidence

L. A. P. A. Ribeiro$^{(\boxtimes)}$, A. C. B. Garcia , and P. S. M. dos Santos

PPGI-Informatics Department, UNIRIO Universidade Federal Do Estado Do Rio de
Janeiro, Rio de Janeiro 22290-240, Brazil
`luiz.ribeiro@uniriotec.br`

Abstract. One of the reasons for fatality due to heart disease is that the
risks are not identified or are only identified at a later stage. Dempster-
Shafer theory approach considers the beliefs attached to each source
to consolidate the hypotheses information to develop a classifier with
higher precision. We propose a novel approach based on Bayesian net and
linear regression to adjust the beliefs. We generate a credibility index to
generate a basic probability assignment. We have used the model in the
heart disease discovery domain using the dataset UCI from Cleaveland
University. The results obtained a precision of 95%.

Keywords: Dempster-shafer theory · Machine learning · Heart
diseases

1 Problem Statement

Some risk factors contributing to cardiovascular disease (CVD) are age, family history, sex and others are controllable as smoking, cholesterol, poor diet,
high blood pressure, obesity, physical inactivity and drinking. Diabetes and high
blood pressure are hereditary risk factors for CVD [4].

Is chest pain a symptom or effect of CVD? The diagnosis of CVD involves
many aspects of uncertainty. Evidence theory is an ally of machine learning
methods to improve the diagnostic task. An accurate and adequate diagnosis of
the risk of CVD in patients becomes necessary to reduce the risk of progression
of CVD [1]. We instantiate the created model in [3] and propose a novel approach based on the Bayesian net and linear regression to generate a credibility
index that more accurately generates basic probability assignment at Dempster-
Shafer's belief.

2 Contributions and Results

The database *UCI heart disease* Cleveland Heart Disease Database (CHDD) is
the most used database for CVD research [2] gathers general patient attributes
and fields containing values from different blood measurements and information

A. C. Bicharra Garcia et al. (Eds.): IBERAMIA 2022, LNAI 13788, pp. 412–413, 2022.
https://doi.org/10.1007/978-3-031-22419-5

about patient risk conditions or diseases. There are two demographic attribute variables "age" and "sex". The variable "num" is the class variable. Of the fourteen fields of dataset, were used 5, divided into three categories:

- Blood factor: Variables used to include blood analysis in the CI (fields "trestbps","chol","thalach")
- Diseases: Variable used to include diabetes factor as a patient's illness (field "fbs")
- Time factor: electrocardiogram(ECG) results are summarized in a linear regression analysis reflecting the trend (from field "restecg").

The contributions of this method provide a causal and a temporal component. The causal component created bayesian net demonstrated that the model allowed the identification of causal relationships for CVD, with heart rate "thal" and "sex" being the variables identified as causal for the diagnosis and detection of CVD "num. The causal structure can be inferred: $"Sex" \implies "tal" \implies "num"$. On the other hand, it is observed that CVDs are the causes of pain or discomfort in the chest field "exang", another inference $"num" \implies "exang"$ is characterized.

The temporal analysis generated results of the ECG are summarized in a linear regression analysis reflecting the trend by the field "restecg". With an expected frequency "tal" $= 3$, there is a greater probability of not suffering from any disease (p $= 38.9\%$). In contrast, with altered frequency ("tal" $= 7$) the most significant risk is concentrated in the worsened presence (p=21.1%). The results of probabilistic inference show the probability of suffering from CVD ("num">0) when there is an average heart rate ("thal"=3), or there are defects in the rate ("thal" $= 6$ or 7).

The relationship $"num" \implies "exang" \implies "cp"$ means that heart conditions cause chest pain, not the other way around. The field "exang" refers to chest pain from exercise and "cp" to chest pain in general, so the former implies the latter. The relationship $"sex" \implies "thal" \implies "num"$ is also natural since sex could determine the predisposition to thalassemia disease and this, in turn, would imply future CVD, The disease's usual symptoms can be confused with CVD. It would be a false positive for CVD. The importance of the "thal" variable to avoid CVD false positives proved relevant to increasing the model's precision.

References

1. Al-Shayea, Q.K.: Artificial neural networks in medical diagnosis. IJCSI **8**(2), 150–154 (2011)
2. Detrano, R.: Cleveland Heart Disease Database. VA Medical Center, Long Beach and Cleveland Clinic Foundation (1989)
3. Ribeiro, L.A.P.A., Garcia, A.C.B., Dos Santos, P.S.M.: Dependency factors in evidence theory: an analysis in an information fusion scenario applied in adverse drug reactions. Sensors **22**(6), 2310 (2022)
4. Sudhakar, K., Manimekalai, D.M.: Study of heart disease prediction using data mining. Int. J. Adv. Res. Comput. Sci. Softw. Eng. **4**(1), 1157–1160 (2014)

Forecasting Key Performance Indicators for Smart Connected Vehicles

David Skiöld[1], Shivani Arora[1], Radu-Casian Mihailescu[1,3(✉)],
and Ramtin Balaghi[2]

[1] Malmö University, Malmö, Sweden
david.skiold.mau@gmail.com, shivani.arora.mau@gmail.com,
radu.c.mihailescu@mau.se
[2] Volvo Cars, Göteborg, Sweden
ramtin.balaghi@volvocars.com
[3] Internet of Things and People Research Centre (IOTAP), Malmö, Sweden

Abstract. As connectivity has been introduced to the car industry, automotive companies have in-use cars which are connected to the internet. A key concern in this context represents the difficulty of knowing how the connection quality changes over time and if there are associated issues. In this work we describe the use of CDR data from connected cars supplied by Volvo to build and study forecasting models that predict how relevant KPIs change over time. Our experiments show promising results for this predictive task, which can lead to improving user experience of connectivity in smart vehicles.

1 Problem Statement

The connected car is a vehicle equipped with wireless communication capabilities, being defined by its devices within, that connect with devices, networks, and services outside the car, including other cars and infrastructure [2]. Apart from providing transportation and communication capabilities, these connected cars also provide additional services in different categories like safety application, efficiency and infotainment. Often, predicting network health is interesting to network service providers to assess the quality of service (QoS), or to car manufacturers, as they are the connection point between customers and the internet service providers (ISP), meaning customers' quality of experience (QoE) is of more interest to them [1].

In this work we investigate the task of predicting connectivity-related KPIs in smart vehicles. These forecasts can then be used for anomaly detection by comparing them to the true values when those are collected in real-time. The dataset used is the CDR (Call Data Record) network data, collected from Volvo's different mobile network operators. The CDR dataset comprises data collected from 103 countries for a period of 2 months, consisting of a total of 509.009 instances. The data is aggregated by country according to a 15 min time-steps and contains the following information: which country the car is in; what operator the car is connected to; unique International Mobile Subscriber Identity (IMSI) number

A. C. Bicharra Garcia et al. (Eds.): IBERAMIA 2022, LNAI 13788, pp. 414–415, 2022.
https://doi.org/10.1007/978-3-031-22419-5

of the SIM embedded in the TCU unit of the car, what type of call is made (which is represented by access point name); how much data is being uploaded and downloaded. The KPIs used in our experiments are as follows: the average upload-data based on country and time; the average download-data based on country and time; the rate of zero-data-sessions (zero data sessions are when a session is established but no data is sent, which indicates some type of fault occurring); the average number of active connections based on time of day, day of the week, and country.

2 Empirical Results

We chose to evaluate the following ML models: LSTM, MLP and Random Forest Regression. All of the abovementioned predictive models are known to perform well in timeseries forecasting and have been described by multiple sources as being useful in predicting similar type of data [3]. The predictors all seem to have worked fairly well for three of our four metrics. On the total count, the average upload, and the average download the RMSE is low for all the models, with similar values. We have observed that the Random Forest regression model performed better than the MLP, but they were both comparable and similar in performance, while the LSTM network was considerably better.

The LSTM performed similarly to the well performing MLP and RF solutions in all regards, but outperformed them on the zero data session rate, where it was one order of magnitude better. The best LSTM model was 1622.06% better than the best MLP model and 1014.61% better than the best Random Forest model. This was not surprising as LSTM networks are usually good at working with time series data. As this was the one KPIs which was considered harder to predict from the start it is not surprising. It is likely that the relation between the zero data session rate and the time features is more complex than it is for the other KPIs. The zero data session rate is the only KPI which seems only somewhat tied to the time of day. It goes up during the hours of 00:00 to 05:00 but it is not uniform. As such the LSTM was positioned better to learn any such relationships.

Acknowledgements. This research was partially funded by Crafoord project grant number 20200953.

References

1. Damaj, I.W., Serhal, D.K., Hamandi, L.A., Zantout, R.N., Mouftah, H.T.: Connected and autonomous electric vehicles: quality of experience survey and taxonomy. Veh. Commun. **28**, 100312 (2021)
2. Kawtar, J., Tomader, M.: Study of connectivity aspect of connected car. In: Proceedings of the 2019 International Conference of Computer Science and Renewable Energies (ICCSRE), pp. 1–7 (2019)
3. Lazaris, A., Prasanna, V.K.: Deep learning models for aggregated network traffic prediction. In: 2019 15th International Conference on Network and Service Management (CNSM), pp. 1–5. IEEE (2019)

Author Index

Printed in the United States
by Baker & Taylor Publisher Services